Lecture Notes in Mathematics

Edited by J.-M. Morel, F. Takens and B. Teissier

Editorial Policy for Multi-Author Publications: Summer Schools / Intensive Courses

1. Lecture Notes aim to report new developments in all areas of mathematics and their applications - quickly, informally and at a high level. Mathematical texts analysing new developments in modelling and numerical simulation are welcome. Manuscripts should be reasonably self-contained and rounded off. Thus they may, and often will, present not only results of the author but also related work by other people. They should provide sufficient motivation, examples and applications. There should also be an introduction making the text comprehensible to a wider audience. This clearly distinguishes Lecture Notes from journal articles or technical reports which normally are very concise. Articles intended for a journal but too long to be accepted by most journals, usually do not have this „lecture notes" character.

2. In general SUMMER SCHOOLS and other similar INTENSIVE COURSES are held to present mathematical topics that are close to the frontiers of recent research to an audience at the beginning or intermediate graduate level, who may want to continue with this area of work, for a thesis or later. This makes demands on the didactic aspects of the presentation. Because the subjects of such schools are advanced, there often exists no textbook, and so ideally, the publication resulting from such a school could be a first approximation to such a textbook.

   Usually several authors are involved in the writing, so it is not always simple to obtain a unified approach to the presentation.

   For prospective publication in LNM, the resulting manuscript should not be just a collection of course notes, each of which has been developed by an individual author with little or no co-ordination with the others, and with little or no common concept. The subject matter should dictate the structure of the book, and the authorship of each part or chapter should take secondary importance. Of course the choice of authors is crucial to the quality of the material at the school and in the book, and the intention here is not to belittle their impact, but simply to say that the book should be planned to be written by these authors jointly, and not just assembled as a result of what these authors happen to submit.

   This represents considerable preparatory work (as it is imperative to ensure that the authors know these criteria before they invest work on a manuscript), and also considerable editing work afterwards, to get the book into final shape. Still it is the form that holds the most promise of a successful book that will be used by its intended audience, rather than yet another volume of proceedings for the library shelf.

3. Manuscripts should be submitted (preferably in duplicate) either to Springer's mathematics editorial in Heidelberg, or to one of the series editors (with a copy to Springer). Volume editors are expected to arrange for the refereeing, to the usual scientific standards, of the individual contributions. If the resulting reports can be forwarded to us (series editors or Springer) this is very helpful. If no reports are forwarded or if other questions remain unclear in respect of homogeneity etc, the series editors may wish to consult external referees for an overall evaluation of the volume. A final decision to publish can be made only on the basis of the complete manuscript; however a preliminary decision can be based on a pre-final or incomplete manuscript. The strict minimum amount of material that will be considered should include a detailed outline describing the planned contents of each chapter.

   Volume editors and authors should be aware that incomplete or insufficiently close to final manuscripts almost always result in longer evaluation times. They should also be aware that parallel submission of their manuscript to another publisher while under consideration for LNM will in general lead to immediate rejection.

Continued on inside back-cover

Lecture Notes in Mathematics

1892

Editors:
J.-M. Morel, Cachan
F. Takens, Groningen
B. Teissier, Paris

Fondazione C.I.M.E. Firenze

C.I.M.E. means Centro Internazionale Matematico Estivo, that is, International Mathematical Summer Center. Conceived in the early fifties, it was born in 1954 and made welcome by the world mathematical community where it remains in good health and spirit. Many mathematicians from all over the world have been involved in a way or another in C.I.M.E.'s activities during the past years.

So they already know what the C.I.M.E. is all about. For the benefit of future potential users and co-operators the main purposes and the functioning of the Centre may be summarized as follows: every year, during the summer, Sessions (three or four as a rule) on different themes from pure and applied mathematics are offered by application to mathematicians from all countries. Each session is generally based on three or four main courses $(24-30$ hours over a period of 6-8 working days) held from specialists of international renown, plus a certain number of seminars.

A C.I.M.E. Session, therefore, is neither a Symposium, nor just a School, but maybe a blend of both. The aim is that of bringing to the attention of younger researchers the origins, later developments, and perspectives of some branch of live mathematics.

The topics of the courses are generally of international resonance and the participation of the courses cover the expertise of different countries and continents. Such combination, gave an excellent opportunity to young participants to be acquainted with the most advance research in the topics of the courses and the possibility of an interchange with the world famous specialists. The full immersion atmosphere of the courses and the daily exchange among participants are a first building brick in the edifice of international collaboration in mathematical research.

C.I.M.E. Director
Pietro ZECCA
Dipartimento di Energetica "S. Stecco"
Università di Firenze
Via S. Marta, 3
50139 Florence
Italy
e-mail: zecca@unifi.it

C.I.M.E. Secretary
Elvira MASCOLO
Dipartimento di Matematica
Università di Firenze
viale G.B. Morgagni 67/A
50134 Florence
Italy
e-mail: mascolo@math.unifi.it

For more information see CIME's homepage: http://www.cime.unifi.it

CIME's activity is supported by:

– Istituto Nazionale di Alta Matematica "F. Severi"
– Ministero dell'Istruzione, dell'Università e della Ricerca
– Ministero degli Affari Esteri, Direzione Generale per la Promozione e la Cooperazione, Ufficio V

A. Baddeley · I. Bárány
R. Schneider · W. Weil

Stochastic Geometry

Lectures given at the
C.I.M.E. Summer School
held in Martina Franca, Italy,
September 13–18, 2004

With additional contributions by
D. Hug, V. Capasso, E. Villa

Editor: W. Weil

 Springer

Fondazione
C.I.M.E.

Authors, Editor and Contributors

Adrian Baddeley
School of Mathematics & Statistics
University of Western Australia
Nedlands WA 6009
Australia
e-mail: adrian@maths.uwa.edu.au

Imre Bárány
Rényi Institute of Mathematics
1364 Budapest Pf. 127
Hungary
e-mail: barany@renyi.hu

and

Mathematics
University College London
Gower Street
London, WC1E 6BT
United Kingdom

Rolf Schneider
Daniel Hug
Mathematisches Institut
Albert-Ludwigs-Universität
Eckerstr. 1
79104 Freiburg i. Br.
Germany
e-mail: rolf.schneider@math.uni-freiburg.de
daniel.hug@math.uni-freiburg.de

Wolfgang Weil
Mathematisches Institut II
Universität Karlsruhe
76128 Karlsruhe
Germany
e-mail: weil@math.uni-karlsruhe.de

Vincenzo Capasso
Elena Villa
Department of Mathematics
University of Milan
via Saldini 50
20133 Milano
Italy
e-mail: vincenzo.capasso@mat.unimi.it
villa@mat.unimi.it

Library of Congress Control Number: 2006931679

Mathematics Subject Classification (2000): Primary 60D05
Secondary 60G55, 62H11, 52A22, 53C65

ISSN print edition: 0075-8434
ISSN electronic edition: 1617-9692
ISBN-10 3-540-38174-0 Springer Berlin Heidelberg New York
ISBN-13 978-3-540-38174-7 Springer Berlin Heidelberg New York

DOI 10.1007/3-540-38174-0

Springer is a part of Springer Science+Business Media
springer.com
© Springer-Verlag Berlin Heidelberg 2007

Typesetting by the authors and SPi using a Springer LaTeX package
Cover design: WMXDesign GmbH, Heidelberg

Printed on acid-free paper   SPIN: 11815334   41/SPi   5 4 3 2 1 0

Preface

The mathematical treatment of random geometric structures can be traced back to the 18th century (the Buffon needle problem). Subsequent considerations led to the two disciplines *Integral Geometry* and *Geometric Probability*, which are connected with the names of Crofton, Herglotz, Blaschke (to mention only a few) and culminated in the book of Santaló (Integral Geometry and Geometric Probability, 1976). Around this time (the early seventies), the necessity grew to have new and more flexible models for the description of random patterns in Biology, Medicine and Image Analysis. A theory of *Random Sets* was developed independently by D.G. Kendall and Matheron. In connection with Integral Geometry and the already existing theory of Point Processes the new field *Stochastic Geometry* was born. Its rapid development was influenced by applications in *Spatial Statistics* and *Stereology*. Whereas at the beginning emphasis was laid on models based on stationary and isotropic Poisson processes, the recent years showed results of increasing generality, for nonisotropic or even inhomogeneous structures and without the strong independence properties of the Poisson distribution. On the one side, these recent developments in Stochastic Geometry went hand-in-hand with a fresh interest in Integral Geometry, namely local formulae for curvature measures (in the spirit of Federer's Geometric Measure Theory). On the other side, new models of point processes (Gibbs processes, Strauss processes, hardcore and cluster processes) and their effective simulation (Markov Chain Monte Carlo, perfect simulation) tightened the close relation between Stochastic Geometry and Spatial Statistics. A further, very interesting direction is the investigation of spatial-temporal processes (tumor growth, communication networks, crystallization processes). The demand for random geometric models is steadily growing in almost all natural sciences or technical fields.

The intention of the Summer School was to present an up-to-date description of important parts of Stochastic Geometry. The course took place in Martina Franca from Monday, September 13, to Friday, September 18, 2004. It was attended by 49 participants (including the lecturers). The main lecturers were Adrian Baddeley (University of Western Australia, Perth), Imre

Bárány (University College, London, and Hungarian Academy of Sciences, Budapest), Rolf Schneider (University of Freiburg, Germany) and Wolfgang Weil (University of Karlsruhe, Germany). Each of them gave four lectures of 90 minutes which we shortly describe, in the following.

Adrian Baddeley spoke on **Spatial Point Processes and their Applications**. He started with an introduction to point processes and marked point processes in $\mathbb{R}^d$ as models for spatial data and described the basic notions (counting measures, intensity, finite-dimensional distributions, capacity functional). He explained the construction of the basic model in spatial statistics, the Poisson process (on general locally compact spaces), and its transformations (thinning and clustering). He then discussed higher order moment measures and related concepts (K function, pair correlation function). In his third lecture, he discussed conditioning of point processes (conditional intensity, Palm distributions) and the important Campbell-Mecke theorem. The Palm distributions lead to G and J functions which are of simple form for Poisson processes. In the last lecture he considered point processes in bounded regions and described methods to fit corresponding models to given data. He illustrated his lectures by computer simulations.

Imre Bárány spoke on **Random Points, Convex Bodies, and Approximation**. He considered the asymptotic behavior of functionals like volume, number of vertices, number of facets, etc. of random convex polytopes arising as convex hulls of n i.i.d. random points in a convex body $K \subset \mathbb{R}^d$. Starting with a short historical introduction (Efron's identity, formulas of Rényi and Sulanke), he emphasized the different limit behavior of expected functionals for smooth bodies K on one side and for polytopes K on the other side. In order to explain this difference, he showed that the asymptotic behavior of the expected missed volume $E(K, n)$ of the random convex hull behaves asymptotically like the volume of a deterministic set, namely the shell between K and the cap body of K (the floating body). This result uses Macbeath regions and the 'economic cap covering theorem' as main tools. The results were extended to the expected number of vertices and of facets. In the third lecture, random approximation (approximation of K by the convex hull of random points) was compared with best approximation (approximation from inside w.r.t. minimal missed volume). It was shown that random approximation is almost as good as best approximation. A further comparison concerned convex hulls of lattice points in K. In the last lecture, for a convex body $K \subset \mathbb{R}^2$, the probability $p(n, K)$ that n random points in K are in convex position was considered and the asymptotic behavior (as $n \to \infty$) was given (extension of the classical Sylvester problem).

The lectures of Rolf Schneider concentrated on **Integral Geometric Tools for Stochastic Geometry**. In the first lecture, the classical results from integral geometry, the principal kinematic formulas and the Crofton formulas were given in their general form, for intrinsic volumes of convex bodies (which were introduced by means of the Steiner formula). Then, Hadwiger's characterization theorem for additive functionals was explained and used to generalize the integral formulas. In the second lecture, local versions of the integral formulas for support measures (curvature measures) and extensions to sets in the convex ring were discussed. This included a local Steiner formula for convex bodies. Extensions to arbitrary closed sets were mentioned. The third lecture presented translative integral formulas, in local and global versions, and their iterations. The occurring mixed measures and functionals were discussed in more detail and connections to support functions and mixed volumes were outlined. The last lecture studied general notions of k-dimensional area and general Crofton formulas. Relations between hyperplane measures and generalized zonoids were given. It was shown how such relations can be used in stochastic geometry, for example, to give estimates for the intersection intensity of a general (non-stationary) Poisson hyperplane process in $\mathbb{R}^d$.

Wolfgang Weil, in his lectures on **Random Sets (in Particular Boolean Models)**, built upon the previous lectures of A. Baddeley and R. Schneider. He first gave an introduction to random closed sets and particle processes (point processes of compact sets, marked point processes) and introduced the basic model in stochastic geometry, the Boolean model (the union set of a Poisson particle process). He described the decomposition of the intensity measure of a stationary particle process and used this to introduce the two quantities which characterize a Boolean model (intensity and grain distribution). He also explained the role of the capacity functional (Choquet's theorem) and its explicit form for Boolean models which shows the relation to Steiner's formula. In the second lecture, mean value formulas for additive functionals were discussed. They lead to the notion of density (quermass density, density of area measure, etc.) which was studied then for general random closed sets and particle processes. The principal kinematic and translative formulas were used to obtain explicit formulas for quermass densities of stationary and isotropic Boolean models as well as for non-isotropic Boolean models (with convex or polyconvex grains) in $\mathbb{R}^d$. Statistical consequences were discussed for $d = 2$ and $d = 3$ and ergodic properties were shortly mentioned. The third lecture was concerned with extensions in various directions: densities for directional data and their relation to associated convex bodies (with an application to the mean visible volume of a Boolean model), interpretation of densities as Radon-Nikodym derivatives of associated random measures, density formulas for non-stationary Boolean models. In the final lecture, random closed sets and Boolean models were investigated from outside by means of contact distributions. Recent extensions of this concept were discussed (generalized directed contact distributions) and it was explained that in some cases they suffice to determine the grain distribution of a Boolean model completely. The role of convexity for explicit formulas of contact distributions was discussed and, as the final result, it was explained that the polynomial behavior of the logarithmic linear contact distribution of a stationary and isotropic Boolean model characterizes convexity of the grains.

Since the four lecture series could only cover some parts of stochastic geometry, two additional lectures of 90 minutes were included in the program, given by D. Hug and V. Capasso. Daniel Hug (University of Freiburg) spoke on **Random Mosaics** as special particle processes. He presented formulas for the different intensities (number and content of faces) for general mosaics and for Voronoi mosaics and then explained a recent solution to Kendall's conjecture concerning the asymptotic shape of large cells in a Poisson Voronoi mosaic. Vincenzo Capasso (University of Milano) spoke on **Crystallization Processes** as spatio-temporal extensions of point processes and Boolean models and emphasized some problems arising from applications.

The participants presented themselves in some short contributions, at one afternoon, as well as in two evening sessions.

The attendance of the lectures was extraordinarily good. Most of the participants had already some background in spatial statistics or stochastic geometry. Nevertheless, the lectures presented during the week provided the audience with a lot of new material for subsequent studies. These lecture notes contain (partially extended) versions of the four main courses (and the two additional lectures) and are also intended as an information of a wider readership about this important field. I thank all the authors for their careful preparation of the manuscripts.

I also take the opportunity, on behalf of all participants, to thank C.I.M.E. for the effective organization of this summer school; in particular, I want to thank Vincenzo Capasso who initiated the idea of a workshop on stochastic geometry. Finally, we were all quite grateful for the kind hospitality of the city of Martina Franca.

Karlsruhe, August 2005   Wolfgang Weil

Contents

Spatial Point Processes and their Applications
Adrian Baddeley 1
1 Point Processes 2
    1.1 Point Processes in 1D and 2D 2
    1.2 Formulation of Point Processes 3
    1.3 Example: Binomial Process 6
    1.4 Foundations 7
    1.5 Poisson Processes 8
    1.6 Distributional Characterisation 12
    1.7 Transforming a Point Process 16
    1.8 Marked Point Processes 19
    1.9 Distances in Point Processes 21
    1.10 Estimation from Data 23
    1.11 Computer Exercises 24
2 Moments and Summary Statistics 26
    2.1 Intensity 26
    2.2 Intensity for Marked Point Processes 30
    2.3 Second Moment Measures 32
    2.4 Second Moments for Stationary Processes 35
    2.5 The K-function 38
    2.6 Estimation from Data 39
    2.7 Exercises 40
3 Conditioning 42
    3.1 Motivation 42
    3.2 Palm Distribution 44
    3.3 Palm Distribution for Stationary Processes 49
    3.4 Nearest Neighbour Function 51
    3.5 Conditional Intensity 52
    3.6 J-function 55
    3.7 Exercises 56
4 Modelling and Statistical Inference 57

4.1 Motivation 57
4.2 Parametric Modelling and Inference 58
4.3 Finite Point Processes 61
4.4 Point Process Densities 62
4.5 Conditional Intensity 64
4.6 Finite Gibbs Models 66
4.7 Parameter Estimation 69
4.8 Estimating Equations 70
4.9 Likelihood Devices 72
References 73

Random Polytopes, Convex Bodies, and Approximation

Imre Bárány 77
1 Introduction 77
2 Computing $\mathbb{E}\phi(K_n)$ 79
3 Minimal Caps and a General Result 80
4 The Volume of the Wet Part 82
5 The Economic Cap Covering Theorem 84
6 Macbeath Regions 84
7 Proofs of the Properties of the M-regions 87
8 Proof of the Cap Covering Theorem 89
9 Auxiliary Lemmas from Probability 92
10 Proof of Theorem 3.1 95
11 Proof of Theorem 4.1 96
12 Proof of (4) 98
13 Expectation of $f_k(K_n)$ 101
14 Proof of Lemma 13.2 102
15 Further Results 104
16 Lattice Polytopes 108
17 Approximation 109
18 How It All Began: Segments on the Surface of K 114
References 115

Integral Geometric Tools for Stochastic Geometry

Rolf Schneider 119
Introduction 119
1 From Hitting Probabilities to Kinematic Formulae 120
   1.1 A Heuristic Question on Hitting Probabilities 120
   1.2 Steiner Formula and Intrinsic Volumes 123
   1.3 Hadwiger's Characterization Theorem for Intrinsic Volumes..... 126
   1.4 Integral Geometric Formulae 129
2 Localizations and Extensions 136
   2.1 The Kinematic Formula for Curvature Measures 136
   2.2 Additive Extension to Polyconvex Sets 143
   2.3 Curvature Measures for More General Sets 148

3 Translative Integral Geometry 151
   3.1 The Principal Translative Formula for Curvature Measures 152
   3.2 Basic Mixed Functionals and Support Functions 157
   3.3 Further Topics of Translative Integral Geometry............. 163
4 Measures on Spaces of Flats 164
   4.1 Minkowski Spaces 165
   4.2 Projective Finsler Spaces 173
   4.3 Nonstationary Hyperplane Processes 177
References 181

Random Sets (in Particular Boolean Models)
Wolfgang Weil.. 185
Introduction... 185
1 Random Sets, Particle Processes and Boolean Models 186
   1.1 Random Closed Sets 187
   1.2 Particle Processes 190
   1.3 Boolean Models 194
2 Mean Values of Additive Functionals............................ 198
   2.1 A General Formula for Boolean Models 199
   2.2 Mean Values for RACS 204
   2.3 Mean Values for Particle Processes 208
   2.4 Quermass Densities of Boolean Models 210
   2.5 Ergodicity 214
3 Directional Data, Local Densities, Nonstationary Boolean Models.... 214
   3.1 Directional Data and Associated Bodies 215
   3.2 Local Densities... 222
   3.3 Nonstationary Boolean Models 224
   3.4 Sections of Boolean Models 230
4 Contact Distributions 231
   4.1 Contact Distribution with Structuring Element 232
   4.2 Generalized Contact Distributions 237
   4.3 Characterization of Convex Grains 241
References 243

Random Mosaics
Daniel Hug... 247
1 General Results 247
   1.1 Basic Notions 247
   1.2 Random Mosaics 248
   1.3 Face-Stars 250
   1.4 Typical Cell and Zero Cell................................ 252
2 Voronoi and Delaunay Mosaics 253
   2.1 Voronoi Mosaics... 253
   2.2 Delaunay Mosaics 255
3 Hyperplane Mosaics 257

4 Kendall's Conjecture 260
   4.1 Large Cells in Poisson Hyperplane Mosaics 261
   4.2 Large Cells in Poisson-Voronoi Mosaics 263
   4.3 Large Cells in Poisson-Delaunay Mosaics 264
References 265

**On the Evolution Equations of Mean Geometric Densities
for a Class of Space and Time Inhomogeneous Stochastic
Birth-and-growth Processes**

Vincenzo Capasso, Elena Villa.................................. 267
1 Introduction 267
2 Birth-and-growth Processes 268
   2.1 The Nucleation Process 268
   2.2 The Growth Process 270
3 Closed Sets as Distributions 271
   3.1 The Deterministic Case 271
   3.2 The Stochastic Case 273
4 The Evolution Equation of Mean Densities for the Stochastic
   Birth-and-growth Process 274
   4.1 Hazard Function and Causal Cone........................ 276
References 280

Spatial Point Processes and their Applications

Adrian Baddeley

School of Mathematics & Statistics, University of Western Australia
Nedlands WA 6009, Australia
e-mail: adrian@maths.uwa.edu.au

A spatial point process is a random pattern of points in d-dimensional space (where usually $d = 2$ or $d = 3$ in applications). Spatial point processes are useful as statistical models in the analysis of observed patterns of points, where the points represent the locations of some object of study (e..g. trees in a forest, bird nests, disease cases, or petty crimes). Point processes play a special role in stochastic geometry, as the building blocks of more complicated random set models (such as the Boolean model), and as instructive simple examples of random sets.

These lectures introduce basic concepts of spatial point processes, with a view toward applications, and with a minimum of technical detail. They cover methods for constructing, manipulating and analysing spatial point processes, and for analysing spatial point pattern data. Each lecture ends with a set of practical computer exercises, which the reader can carry out by downloading a free software package.

Lecture 1 ('Point Processes') gives some motivation, defines point processes, explains how to construct point processes, and gives some important examples. Lecture 2 ('Moments') discusses means and higher moments for point processes, especially the intensity measure and the second moment measure, along with derived quantities such as the K-function and the pair correlation function. It covers the important Campbell formula for expectations. Lecture 3 ('Conditioning') explains how to condition on the event that the point process has a point at a specified location. This leads to the concept of the Palm distribution, and the related Campbell-Mecke formula. A dual concept is the conditional intensity, which provides many new results. Lecture 4 ('Modelling and Statistical Inference') covers the formulation of statistical models for point patterns, model-fitting methods, and statistical inference.

1 Point Processes

In this first lecture, we motivate and define point processes, construct examples (especially the **Poisson process** [28]), and analyse important properties of the Poisson process. There are different ways to mathematically construct and characterise a point process (using finite-dimensional distributions, vacancy probabilities, capacity functional, or generating function). An easier way to construct a point process is by transforming an existing point process (by thinning, superposition, or clustering) [43]. Finally we show how to use existing software to generate simulated realisations of many spatial point processes using these techniques, and analyse them using vacancy probabilities (or 'empty space functions').

1.1 Point Processes in 1D and 2D

A **point process** in one dimension ('time') is a useful model for the sequence of random times when a particular event occurs. For example, the random times when a hospital receives emergency calls may be modelled as a point process. Each emergency call happens at an instant, or point, of time. There will be a random number of such calls in any period of time, and they will occur at random instants of time.

Fig. 1. A point process in time.

A **spatial point process** is a useful model for a random pattern of points in d-dimensional space, where $d \geq 2$. For example, if we make a map of the locations of all the people who called the emergency service during a particular day, this map constitutes a random pattern of points in two dimensions. There will be a random number of such points, and their locations are also random.

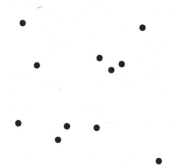

Fig. 2. A point process in two dimensions.

We may also record both the locations and the times of the emergency calls. This may be regarded as a point process in three dimensions (space × time), or alternatively, as a point process in two dimensions where each point (caller location) is labelled or **marked** by a number (the time of the call).

Spatial point processes can be used directly, to model and analyse data which take the form of a point pattern, such as maps of the locations of trees or bird nests ('statistical ecology' [16, 29]); the positions of stars and galaxies ('astrostatistics' [1]); the locations of point-like defects in a silicon crystal wafer (materials science [34]); the locations of neurons in brain tissue; or the home addresses of individuals diagnosed with a rare disease ('spatial epidemiology' [19]). Spatial point processes also serve as a basic model in random set theory [42] and image analysis [41]. For general surveys of applications of spatial point processes, see [16, 42, 43]. For general theory see [15].

1.2 Formulation of Point Processes

There are some differences between the theory of one-dimensional and higher-dimensional point processes, because one-dimensional time has a natural ordering which is absent in higher dimensions.

A one-dimensional point process can be handled mathematically in many different ways. We may study the **arrival times** $T_1 < T_2 < \ldots$ where T_i is the time at which the ith point (emergency call) arrives. Using these random variables is the most direct way to handle the point pattern, but their use is complicated by the fact that they are strongly dependent, since $T_i < T_{i+1}$.

Fig. 3. Arrival times T_i.

Alternatively we may study the **inter-arrival times** $S_i = T_{i+1} - T_i$. These have the advantage that, for some special models (Poisson and renewal processes), the random variables $S_1, S_2, \ldots$ are independent.

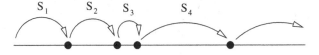

Fig. 4. Inter-arrival times S_i.

Alternatively it is common (especially in connection with martingale theory) to formulate a point process in terms of the cumulative **counting process**

$$N_t = \text{number of points arriving up to time } t$$
$$= \sum_{i=1}^{\infty} \mathbf{1}\{T_i \leq t\},$$

for all $t \geq 0$, where $\mathbf{1}\{\ldots\}$ denotes the indicator function, equal to 1 if the statement "..." is true, and equal to 0 otherwise. This device has the advantage of converting the process to a random function of continuous time t, but has the disadvantage that the values N_t for different t are highly dependent.

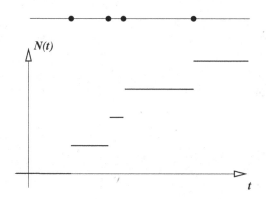

Fig. 5. The counting process N_t associated with a point process.

Alternatively one may use the interval counts

$$N(a, b] = N_b - N_a$$

for $0 \leq a \leq b$ which count the number of points arriving in the interval $(a, b]$. For some special processes (Poisson and independent-increments processes) the interval counts for **disjoint** intervals are stochastically independent.

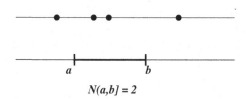

$N(a,b] = 2$

Fig. 6. Interval count $N(a, b]$ for a point process.

In higher dimensions, there is no natural ordering of the points, so that there is no *natural* analogue of the inter-arrival times S_i nor of the counting process N_t. Instead, the most useful way to handle a spatial point process is to generalise the interval counts $N(a, b]$ to the region counts

$$N(B) = \text{number of points falling in } B$$

defined for each bounded closed set $B \subset \mathbb{R}^d$.

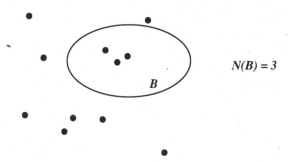

$N(B) = 3$

Fig. 7. Counting variables $N(B)$ for a spatial point process.

Rather surprisingly, it is often sufficient to study a point process using only the **vacancy indicators**

$$V(B) = 1\{N(B) = 0\}$$
$$= 1\{\text{there are no points falling in } B\}.$$

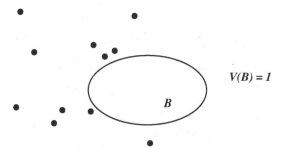

$V(B) = 1$

Fig. 8. Vacancy indicators $V(B)$ for a spatial point process.

The counting variables $N(B)$ are natural for exploring additive properties of a point process. For example, suppose we have two point processes, of 'red' and 'blue' points respectively, and we superimpose them (forming a single point process by discarding the colours). If $N_{\text{red}}(B)$ and $N_{\text{blue}}(B)$ are the counting variables for red and blue points respectively, then the counting variable for the superimposed process is $N(B) = N_{\text{red}}(B) + N_{\text{blue}}(B)$.

The vacancy indicators $V(B)$ are natural for exploring geometric and 'multiplicative' properties of a point process. If $V_{\text{red}}(B)$ and $V_{\text{blue}}(B)$ are the vacancy indicators for two point processes, then the vacancy indicator for the superimposed process is $V(B) = V_{\text{red}}(B) \, V_{\text{blue}}(B)$.

1.3 Example: Binomial Process

To take a very simple example, let us place a fixed number n of points at random locations inside a bounded region $W \subset \mathbb{R}^2$. Let $X_1, \ldots, X_n$ be i.i.d. (independent and identically distributed) random points which are uniformly distributed in W. Hence the probability density of each X_i is

$$f(x) = \begin{cases} 1/\lambda_2(W) & \text{if } x \in W \\ 0 & \text{otherwise} \end{cases}$$

where $\lambda_2(W)$ denotes the area of W. A realisation of this process is shown in Figure 9.

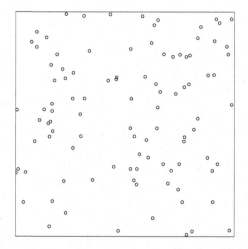

Fig. 9. Realisation of a binomial point process with $n = 100$ in the unit square.

Since each random point X_i is uniformly distributed in W, we have for any bounded set B in $\mathbb{R}^2$

$$\mathbb{P}(X_i \in B) = \int_B f(x) \, \mathrm{d}x$$
$$= \frac{\lambda_2(B \cap W)}{\lambda_2(W)}.$$

The variables $N(B)$ and $V(B)$ may be represented explicitly as

$$N(B) = \sum_{i=1}^{n} \mathbf{1}\{X_i \in B\}$$

$$V(B) = \min_{i=1}^{n} \mathbf{1}\{X_i \notin B\}$$

It follows easily that $N(B)$ has a binomial distribution with parameters n and $p = \lambda_2(B \cap W)/\lambda_2(W)$, hence the process is often called the binomial process.

Note that the counting variables $N(B)$ for different subsets B are not independent. If B_1 and B_2 are disjoint, then

$$N(B_1) + N(B_2) = N(B_1 \cup B_2) \leq n$$

so that $N(B_1)$ and $N(B_2)$ must be dependent. In fact, the joint distribution of $(N(B_1), N(B_2))$ is the multinomial distribution on n trials with success probabilities (p_1, p_2) where $p_i = \lambda_2(B_i \cap W)/\lambda_2(W)$.

1.4 Foundations

Foundations of the theory of point processes in $\mathbb{R}^d$ are expounded in detail in [15]. The following is a very brief and informal introduction.

Random Measure Formalism

The values of the counting variables $N(B)$ for all subsets B give us sufficient information to reconstruct completely the positions of all the points in the process. Indeed the points of the process are those locations x such that $N(\{x\}) > 0$. Hence we may as well *define* a point process as a collection of random variables $N(B)$ indexed by subsets B.

The counting variables $N(B)$ for different sets B satisfy certain relationships, including additivity

$$N(A \cup B) = N(A) + N(B)$$

whenever A and B are disjoint sets $(A \cap B = \emptyset)$ and of course

$$N(\emptyset) = 0$$

where $\emptyset$ denotes the empty set. Furthermore, they are continuous in the sense that, if A_n is a decreasing sequence of closed, bounded sets $(A_n \supseteq A_{n+1})$ with limit $\bigcap_n A_n = A$, then we must have

$$N(A_n) \to N(A).$$

These properties must hold for each realisation of the point process, or at least, with probability 1. They amount to the requirement that N is a *measure* (or at least, that with probability 1, the values $N(B)$ can be extended to a measure). This is the concept of a **random measure** [26, 42].

Formally, then, a point process may be defined as a random measure in which the values $N(B)$ are nonnegative integers [15, 42]. We usually also assume that the point process is **locally finite**:

$$N(B) < \infty \quad \text{with probability 1}$$

for all bounded $B \subset \mathbb{R}^d$. That is, any bounded region contains only a finite number of points, with probability 1. We also assume that the point process is **simple**:

$$N(\{x\}) \leq 1 \quad \text{for all } x \in \mathbb{R}^d$$

with probability 1. That is, with probability 1, no two points of the process are coincident. A simple point process can be regarded as a random set of points.

For example, the binomial process introduced in Section 1.3 is locally finite (since $N(B) \leq n$ for all B) and it is simple because there is zero probability that two independent, uniformly distributed random points coincide:

$$\mathbb{P}(X_1 = X_2) = \mathbb{E}\left[\mathbb{P}\left(X_1 = X_2 \mid X_2\right)\right] = 0.$$

Hence the binomial process is a point process in the sense of this definition.

Random Set Formalism

A *simple* point process can be formulated in a completely different way since it may be regarded as a random set $\mathbf{X}$. Interestingly, the vacancy indicators $V(B)$ contain complete information about the process. If we know the value of $V(B)$ for all sets B, then we can determine the exact location of each point x in the (simple) point process $\mathbf{X}$. To do this, let G be the union of all open sets B such that $V(B) = 1$. The complement of G is a locally finite set of points, and this identifies the random set $\mathbf{X}$.

The vacancy indicators must satisfy

$$V(A \cup B) = \min\{V(A), V(B)\}$$

for any sets A, B, and have other properties analogous to those of the count variables $N(B)$. Thus we could alternatively define a simple point process as a random function V satisfying these properties almost surely. This approach is intimately related to the theory of random closed sets [27, 31, 32].

In the rest of these lectures, we shall often swap between the notation $\mathbf{X}$ (for a point process when it is considered as a random set) and N or $N_{\mathbf{X}}$ (for the counting variables associated with the same point process).

1.5 Poisson Processes

One-dimensional Poisson Process

Readers may be familiar with the concept of a **Poisson point process** in one-dimensional time (e.g. [28, 37]). Suppose we make the following assumptions:

1. The number of points which arrive in a given time interval has expected value proportional to the duration of the interval:

$$\mathbb{E}N(a, b] = \beta(b - a)$$

where $\beta > 0$ is the **rate** or **intensity** of the process;

2. Arrivals in disjoint intervals of time are independent: if $a_1 < b_1 < a_2 < b_2 < \ldots < a_m < b_m$ then the random variables $N(a_1, b_1], \ldots, N(a_m, b_m]$ are independent;
3. The probability of two or more arrivals in a given time interval is asymptotically of uniformly smaller order than the length of the interval:

$$\mathbb{P}(N(a, a + h] \geq 2) = o(h), \qquad h \downarrow 0.$$

For example these would be reasonable assumptions to make about the arrival of cosmic particles at a particle detector, or the occurrence of accidents in a large city.

From these assumptions it *follows* that the number of points arriving in a given time interval must have a Poisson distribution:

$$N(a, b] \sim \mathsf{Poisson}(\beta(b - a))$$

where $\mathsf{Poisson}(\mu)$ denotes the Poisson distribution with mean μ, defined by

$$\mathbb{P}(N = k) = e^{-\mu} \frac{\mu^k}{k!}, \qquad k = 0, 1, 2, \ldots \tag{1}$$

This conclusion follows by splitting the interval $(a, b]$ into a large number n of small intervals. The number of arrivals in each small interval is equal to 0 or 1, except for an event of small probability. Since $N(a, b]$ is the sum of these numbers, it has an approximately binomial distribution. Letting $n \to \infty$ we obtain that $N(a, b]$ must have a Poisson distribution.

Definition 1.1. *The* **one-dimensional Poisson process**, *with uniform intensity* $\beta > 0$, *is a point process in* $\mathbb{R}$ *such that*

[PP1] *for every bounded interval* $(a, b]$, *the count* $N(a, b]$ *has a Poisson distribution with mean* $\beta(b - a)$;
[PP2] *if* $(a_1, b_1], \ldots, (a_m, b_m]$ *are disjoint bounded intervals, then the counts* $N(a_1, b_1], \ldots, N(a_m, b_m]$ *are independent random variables.*

Other properties of the one-dimensional Poisson process include

1. The inter-arrival times S_i have an exponential distribution with rate β:

$$\mathbb{P}(S_i \leq s) = 1 - e^{-\beta s}, \qquad s > 0.$$

2. The inter-arrival times S_i are independent.
3. The ith arrival time T_i has an Erlang or Gamma distribution with parameters $\alpha = i$ and β. The $\mathsf{Gamma}(\alpha, \beta)$ probability density is

$$f(t) = \frac{\beta^\alpha}{\Gamma(\alpha)} t^{\alpha - 1} e^{-\beta t}$$

for $t > 0$, and 0 otherwise.

Fig. 10. Realisation of the one-dimensional Poisson process with uniform intensity 1 in the time interval $[0, 30]$. Tick marks indicate the arrival times.

Properties 1 and 2 above suggest an easy way to generate simulated realisations of the Poisson process on $[0, \infty)$. We simply generate a sequence of independent, exponentially distributed, random variables $S_1, S_2, \ldots$ and take the arrival times to be $T_i = \sum_{1 \le j \le i} S_j$.

We may also study **inhomogeneous Poisson processes** in which the number of arrivals in $(a, b]$ is

$$\mathbb{E}\, N(a, b] = \int_a^b \beta(t)\, \mathrm{d}t$$

where $\beta(t) > 0$ is a function called the (instantaneous) **intensity function**. The probability that there will be a point of this process in an infinitesimal interval $[t, t + \mathrm{d}t]$ is $\beta(t)\, \mathrm{d}t$. Arrivals in disjoint time intervals are independent.

Spatial Poisson Process

The Poisson process can be generalised to two-dimensional space.

Definition 1.2. *The* **spatial Poisson process**, *with uniform intensity* $\beta > 0$, *is a point process in* $\mathbb{R}^2$ *such that*

> **[PP1]** *for every bounded closed set B, the count $N(B)$ has a Poisson distribution with mean $\beta \lambda_2(B)$;*
> **[PP2]** *if $B_1, \ldots, B_m$ are disjoint regions, then $N(B_1), \ldots, N(B_m)$ are independent.*

Here $\lambda_2(B)$ again denotes the area of B.

It turns out that these two properties uniquely characterise the Poisson process. The constant β is the expected number of points per unit area. It has dimensions length^{-2} or "points per unit area".

As in the one-dimensional case, the spatial Poisson process can be derived by starting from a few reasonable assumptions: that $\mathbb{E}N(B) = \beta \lambda_2(B)$; that $\mathbb{P}(N(B) > 1) = o(\lambda_2(B))$ for small $\lambda_2(B)$; and that events in disjoint regions are independent.

An important fact about the Poisson process is the following.

Lemma 1.1 (Conditional Property). *Consider a Poisson point process in $\mathbb{R}^2$ with uniform intensity $\beta > 0$. Let $W \subset \mathbb{R}^2$ be any region with $0 < \lambda_2(W) < \infty$. Given that $N(W) = n$, the conditional distribution of $N(B)$ for $B \subseteq W$ is binomial:*

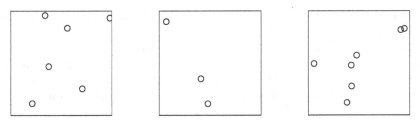

Fig. 11. Three different realisations of the Poisson process with uniform intensity 5 in the unit square.

$$\mathbb{P}\left(N(B) = k \mid N(W) = n\right) = \binom{n}{k} p^k (1-p)^{n-k}$$

where $p = \lambda_2(B)/\lambda_2(W)$. *Furthermore the conditional joint distribution of* $N(B_1), \ldots, N(B_m)$ *for any* $B_1, \ldots, B_m \subseteq W$ *is the same as the joint distribution of these variables in a binomial process.*

In other words, *given that there are n points of the Poisson process in W, these n points are conditionally independent and uniformly distributed in W.*

Proof. Let $0 \le k \le n$. Then

$$\mathbb{P}\left(N(B) = k \mid N(W) = n\right) = \frac{\mathbb{P}(N(B) = k,\ N(W) = n)}{\mathbb{P}(N(W) = n)}$$

$$= \frac{\mathbb{P}(N(B) = k,\ N(W \setminus B) = n - k)}{\mathbb{P}(N(W) = n)}.$$

By the independence property (PP2) the numerator can be rewritten

$$\mathbb{P}(N(B) = k,\ N(W \setminus B) = n - k) = \mathbb{P}(N(B) = k)\,\mathbb{P}(N(W \setminus B) = n - k)$$

We may then evaluate the numerator and denominator using (PP1) to give

$$\mathbb{P}\left(N(B) = k \mid N(W) = n\right) = \frac{e^{-\beta\lambda_2(B)} \frac{(\beta\lambda_2(B))^k}{k!} e^{-\beta\lambda_2(W \setminus B)} \frac{(\beta\lambda_2(W \setminus B))^{n-k}}{(n-k)!}}{e^{-\beta\lambda_2(W)} \frac{(\beta\lambda_2(W))^n}{n!}}$$

$$= \frac{n!}{k!\,(n-k)!} \left(\frac{\lambda_2(B)}{\lambda_2(W)}\right)^k \left(\frac{\lambda_2(W \setminus B)}{\lambda_2(W)}\right)^{n-k}$$

$$= \binom{n}{k} p^k (1-p)^{n-k}$$

where $p = \lambda_2(B)/\lambda_2(W)$. $\square$

Thus, for example, Figure 9 can also be taken as a realisation of a Poisson process in the unit square W, in which it happens that there are exactly 100 points in W. The only distinction between a binomial process and a Poisson process in W is that *different realisations* of the Poisson process will consist of different numbers of points.

The conditional property also gives us a direct way to simulate Poisson processes. To generate a realisation of a Poisson process of intensity β in W, we first generate a random variable M with a Poisson distribution with mean $\beta\lambda_2(W)$. Given $M = m$, we then generate m independent uniform random points in W.

General Poisson Process

To define a uniform Poisson point process in $\mathbb{R}^d$, or an inhomogeneous Poisson process in $\mathbb{R}^d$, or a Poisson point process on some other space S, the following general definition can be used.

Definition 1.3. *Let S be a space, and Λ a measure on S. (We require S to be a locally compact metric space, and Λ a measure which is finite on every compact set and which has no atoms.)*

The **Poisson process** *on S with intensity measure Λ is a point process on S such that*

[PP1] *for every compact set $B \subset S$, the count $N(B)$ has a Poisson distribution with mean $\Lambda(B)$;*
[PP2] *if $B_1, \ldots, B_m$ are disjoint compact sets, then $N(B_1), \ldots, N(B_m)$ are independent.*

Example 1.1 (Poisson process in three dimensions). The uniform Poisson process on $\mathbb{R}^3$ with intensity $\beta > 0$ is defined by taking $S = \mathbb{R}^3$ and $\Lambda(B) = \beta\lambda_3(B)$.

Example 1.2 (Inhomogeneous Poisson process). The inhomogeneous Poisson process on $\mathbb{R}^2$ with intensity function $\beta(u)$, $u \in \mathbb{R}^2$ is defined by taking $S = \mathbb{R}^2$ and $\Lambda(B) = \int_B \beta(u)\, du$. See Figure 12.

Example 1.3 (Poisson process on the sphere). Take S to be the unit sphere (surface of the unit ball in three dimensions) and $\Lambda = \beta\mu$, where $\beta > 0$ and μ is the uniform area measure on S with total mass 4π. This yields the uniform Poisson point process on the unit sphere, with intensity β. This process has a finite number of points, almost surely. Indeed the total number of points $N(S)$ is a Poisson random variable with mean $\Lambda(S) = \beta\mu(S) = 4\pi\beta$. See Figure 13.

1.6 Distributional Characterisation

In Section 1.5 we discussed the fact that a Poisson process in a bounded region W, conditioned on the total number of points in W, is equivalent to a binomial process. This was expressed somewhat vaguely, because we do not yet have the tools needed to determine whether two point processes are 'equivalent' in distribution. We now develop such tools.

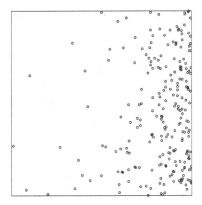

Fig. 12. Realisation of an inhomogeneous Poisson process in the unit square, with intensity function $\beta(x, y) = \exp(2 + 5x)$.

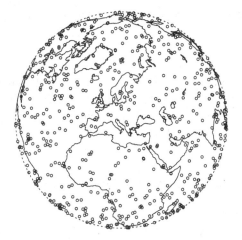

Fig. 13. Uniform Poisson point process on the surface of the Earth. Intensity is $\beta = 100$ points per solid radian; the expected total number of points is $4\pi \times 100 = 1256.6$. Orthogonal projection from a position directly above Martina Franca.

Space of Outcomes

Like any random phenomenon, a point process can be described in statistical terms by defining the space of possible outcomes and then specifying the probabilities of different *events* (an event is a subset of all possible outcomes).

The space of realisations of a point process in $\mathbb{R}^d$ is N, the set of all counting measures on $\mathbb{R}^d$, where a **counting measure** is a nonnegative integer valued measure which has a finite value on every compact set.

A basic event about the point process is the event that there are exactly k points in the region B,

$$E_{B,k} = \{N(B) = k\} = \{N \in \mathsf{N} : \ N(B) = k\}$$

for compact $B \subset \mathbb{R}^d$ and integer $k = 0, 1, 2, \ldots$.

Definition 1.4. *Let* N *be the set of all counting measures on* $\mathbb{R}^d$. *Let* $\mathcal{N}$ *be the* σ-*field of subsets of* N *generated by all events of the form* $E_{B,k}$. *The space* N *equipped with its* σ-*field* $\mathcal{N}$ *is called the* **canonical space** *or* **outcome space** *for a point process in* $\mathbb{R}^d$.

The σ-field $\mathcal{N}$ includes events such as

$$E_{B_1,k_1} \cap \ldots \cap E_{B_m,k_m} = \{N \in \mathsf{N} : \ N(B_1) = k_1, \ldots, N(B_m) = k_m\},$$

i.e. the event that there are exactly k_i points in region B_i for $i = 1, \ldots, m$. It also includes, for example, the event that the point process has no points at all,

$$\{N \equiv 0\} = \{N \in \mathsf{N} : \ N(B) = 0 \text{ for all } B\}$$

since this event can be represented as the intersection of the countable sequence of events $E_{b(0,n),0}$ for $n = 1, 2, \ldots$. Here $b(0, r)$ denotes the ball of radius r and centre 0 in $\mathbb{R}^d$.

A point process $\mathbf{X}$ may now be defined formally, using its counting measure $N = N_{\mathbf{X}}$, as a measurable map $N : \Omega \to \mathsf{N}$ from a probability space $(\Omega, \mathcal{A}, \mathbb{P})$ to the outcome space $(\mathsf{N}, \mathcal{N})$. Thus, each elementary outcome $\omega \in \Omega$ determines an outcome $N^{\omega} \in \mathsf{N}$ for the entire point process. Measurability is the requirement that, for any event $E \in \mathcal{N}$, the event

$$\{N \in E\} = \{\omega \in \Omega : \ N^{\omega} \in E\}$$

belongs to $\mathcal{A}$. This implies that any such event has a well-defined probability $\mathbb{P}(N \in E)$. For example, the probability that the point process is empty, $\mathbb{P}(N \equiv 0)$, is well defined.

The construction of $\mathcal{N}$ guarantees that, if N is a point process on a probability space $(\Omega, \mathcal{A}, \mathbb{P})$, then the variables $N(B)$ for each compact set B are random variables on the same probability space. In fact $\mathcal{N}$ is the minimal σ-field on N which guarantees this.

Definition 1.5. *The* **distribution** *of a point process* $\mathbf{X}$ *is the probability measure* $\mathbf{P}_{\mathbf{X}}$, *on the outcome space* $(\mathsf{N}, \mathcal{N})$, *defined by*

$$\mathbf{P}_{\mathbf{X}}(A) = \mathbb{P}(N_{\mathbf{X}} \in A), \qquad A \in \mathcal{N}.$$

For example, the distribution of a point process specifies the values of joint probabilities

$$\mathbb{P}(N(B) = k \text{ and } N(B') = k')$$

for two sets B, B' and integers k, k'; it also specifies the probability that the entire point process is empty,

$$\mathbb{P}(N \equiv 0) = \mathbb{P}(\mathbf{X} = \emptyset).$$

Characterisations of a Point Process Distribution

The distribution of a point process may be characterised using either the joint distributions of the variables $N(B)$, or the marginal distributions of the variables $V(B)$. First we consider the count variables $N(B)$.

Definition 1.6. *The* **finite-dimensional distributions** *or* **fidis** *of a point process are the joint probability distributions of*

$$(N(B_1), \ldots, N(B_m))$$

for all finite integers $m > 0$ and all compact $B_1, B_2, \ldots$.

Equivalently, the fidis specify the probabilities of all events of the form

$$\{N(B_1) = k_1, \ldots, N(B_m) = k_m\}$$

involving finitely many regions.

Clearly the fidis of a point process convey only a subset of the information conveyed in its distribution. Probabilities of events such as $\{\mathbf{X} = \emptyset\}$ are not specified in the fidis, since they cannot be expressed in terms of a finite number of compact regions. However, it turns out that the fidis are sufficient to characterise the entire distribution.

Theorem 1.1. *Let $\mathbf{X}$ and $\mathbf{Y}$ be two point processes. If the fidis of $\mathbf{X}$ and of $\mathbf{Y}$ coincide, then $\mathbf{X}$ and $\mathbf{Y}$ have the same distribution.*

Corollary 1.1. *If $\mathbf{X}$ is a point process satisfying axioms (PP1) and (PP2) then $\mathbf{X}$ is a Poisson process.*

A *simple* point process (Section 1.4) can be regarded as a random set of points. In this case the vacancy probabilities are useful. The **capacity functional** of a simple point process $\mathbf{X}$ is the functional

$$T(K) = \mathbb{P}(N(K) > 0), \qquad K \text{ compact.}$$

This is a very small subset of the information conveyed by the fidis, since $T(K) = 1 - \mathbb{P}(E_{K,0})$. However, surprisingly, it turns out that the capacity functional is sufficient to determine the entire distribution.

Theorem 1.2. *Suppose $\mathbf{X}$ and $\mathbf{Y}$ are two simple point processes whose capacity functionals are identical. Then their distributions are identical.*

Corollary 1.2. *A simple point process is a uniform Poisson process of intensity β if and only if its capacity functional is*

$$T(K) = 1 - \exp\{-\beta \lambda_d(K)\}$$

for all compact $K \subset \mathbb{R}^d$.

Corollary 1.3. *A simple point process is a binomial process (of n points in W) if and only if its capacity functional is*

$$T(K) = 1 - \left(1 - \frac{\lambda_d(K \cap W)}{\lambda_d(W)}\right)^n$$

for all compact $K \subset \mathbb{R}^d$.

This characterisation of the binomial process now makes it easy to prove the conditional property of the Poisson process described in the last section.

Note that the results above do not provide a simple way to construct a point process *ab initio*. Theorem 1.1 does not say that any given choice of finite dimensional distributions will automatically determine a point process distribution. On the contrary, the fidis must satisfy a suite of conditions (self-consistency, continuity) if they are to correspond to a point process. Hence, the fidis are not a very practical route to the *construction* of point processes. More practical methods of construction are described in Section 1.7.

The concept of a **stationary** point process plays an important role.

Definition 1.7. *A point process $\mathbf{X}$ in $\mathbb{R}^d$ is called **stationary** if, for any fixed vector $v \in \mathbb{R}^d$, the distribution of the shifted point process $\mathbf{X} + v$ (obtained by shifting each point $x \in \mathbf{X}$ to $x + v$) is identical to the distribution of $\mathbf{X}$.*

Lemma 1.2. *A point process is stationary if and only if its capacity functional is invariant under translations, $T(K) = T(K+v)$ for all compact sets $K \subset \mathbb{R}^d$ and all $v \in \mathbb{R}^d$.*

For example, the uniform Poisson process is stationary, since its capacity functional $T(K)$ is clearly invariant under translation.

Similarly, a point process is called **isotropic** if its distribution is invariant under all rotations of $\mathbb{R}^d$. The uniform Poisson process is isotropic.

1.7 Transforming a Point Process

One pragmatic way to construct a new point process is by transforming or changing an existing point process. Convenient transformations include **mapping, thinning, superposition**, and **clustering**.

Mapping

Figure 14 sketches in one dimension the concept of **mapping** a point process $\mathbf{X}$ to another point process by applying a fixed transformation $s : \mathbb{R}^d \to \mathbb{R}^d$ to each individual point of $\mathbf{X}$. The resulting point process is thus $\mathbf{Y} = \bigcup_{x \in \mathbf{X}} s(x)$. For example, the mapping $s(x) = ax$ where $a > 0$ would rescale the entire point process by the constant scale factor a.

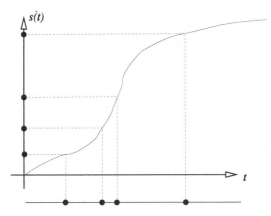

Fig. 14. Application of a transformation s to each individual point in a point process

A vector translation $s(x) = x + v$, where $v \in \mathbb{R}^d$ is fixed, shifts all points of $\mathbf{X}$ by the same vector v. If the original process $\mathbf{X}$ is a uniform Poisson process, then the translated point process $\mathbf{Y}$ is also a uniform Poisson process with the same intensity, as we saw above.

Any mapping s which has a continuous inverse, or at least which satisfies

$$0 < \lambda_d(s^{-1}(B)) < \infty \text{ whenever } B \text{ is compact} \tag{2}$$

transforms a uniform Poisson process into another Poisson process, generally an inhomogeneous one.

An important caution is that, if the transformation s does not satisfy (2), then in general we cannot even be sure that the transformed point process $\mathbf{Y}$ is well defined, since the points of $\mathbf{Y}$ may not be locally finite. For example, consider the projection of the cartesian plane onto the x-axis, $s(x, y) = x$. If $\mathbf{X}$ is a uniform Poisson process in $\mathbb{R}^2$ then the projection onto the x-axis is everywhere dense: there are infinitely many projected points in any open interval (a, b) in the x-axis, almost surely, since $s^{-1}((a, b)) = (a, b) \times \mathbb{R}$. Hence, the projection of $\mathbf{X}$ onto the x-axis is not a well-defined point process.

Thinning

Figure 15 sketches the operation of **thinning** a point process $\mathbf{X}$, by which some of the points of $\mathbf{X}$ are deleted. The remaining, undeleted points form the thinned point process $\mathbf{Y}$. We may formalise the thinning procedure by supposing that each point $x \in \mathbf{X}$ is labelled with an indicator random variable I_x taking the value 1 if the point x is to be retained, and 0 if it is to be deleted. Then the thinned process consists of those points $x \in \mathbf{X}$ with $I_x = 1$.

Independent thinning is the case where the indicators I_x are independent. If a uniform Poisson process is subjected to independent thinning, the resulting thinned process is also Poisson.

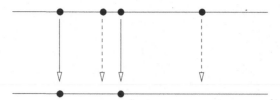

Fig. 15. Thinning a point process. Points of the original process (above) are either retained (solid lines) or deleted (dotted lines) to yield a thinned process (below).

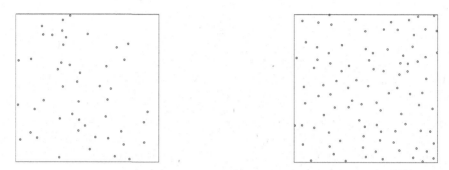

Fig. 16. Dependent thinning: simulated realisations of Matérn's Model I (left) and Model II (right). Both are derived from a Poisson process of intensity 200 in the unit square, and have the same inhibition radius $r = 0.05$.

Examples of **dependent thinning** are the two models of Matérn [30] for spatial inhibition between points. In **Model I**, we start with a uniform Poisson process $\mathbf{X}$ in $\mathbb{R}^2$, and delete any point which has a close neighbour (closer than a distance r, say). Thus $I_x = 1$ if $||x - x'|| \le r$ for any $x' \in \mathbf{X}$. In **Model II**, we start with a uniform Poisson process $\mathbf{X}$ in $\mathbb{R}^2 \times [0, 1]$, interpreting this as a process of two-dimensional points $x \in \mathbb{R}^2$ with 'arrival times' $t \in [0, 1]$. Then we delete any point which has a close neighbour whose arrival time was earlier than the point in question. Thus $I_{(x,t)} = 1$ if $||x - x'|| \le r$ and $t > t'$ for any $(x', t') \in \mathbf{X}$. The arrival times are then discarded to give us a point process in $\mathbb{R}^2$. Simulated realisations of these two models are shown in Figure 16.

Superposition

Figure 17 sketches the **superposition** of two point processes $\mathbf{X}$ and $\mathbf{Y}$ which consists of all points in the union $\mathbf{X} \cup \mathbf{Y}$. If we denote by $N_{\mathbf{X}}(B)$ and $N_{\mathbf{Y}}(B)$ the numbers of points of $\mathbf{X}$ and $\mathbf{Y}$ respectively in a region $B \subset \mathbb{R}^d$, then the superposition has $N_{\mathbf{X} \cup \mathbf{Y}}(B) = N_{\mathbf{X}}(B) + N_{\mathbf{Y}}(B)$ assuming there are no coincident points. Superposition can thus be viewed either as the union of sets or as the sum of measures.

If $\mathbf{X}$ and $\mathbf{Y}$ are *independent*, with capacity functionals $T_{\mathbf{X}}, T_{\mathbf{Y}}$, then the superposition has capacity functional $T_{\mathbf{X} \cup \mathbf{Y}}(K) = 1 - (1 - T_{\mathbf{X}}(K))(1 - T_{\mathbf{Y}}(K))$.

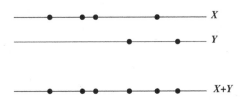

Fig. 17. Superposition of two point processes

The superposition of two *independent* Poisson processes **X** and **Y**, say uniform Poisson processes of intensity μ and ν respectively, is a uniform Poisson process of intensity $\mu + \nu$.

Cluster Formation

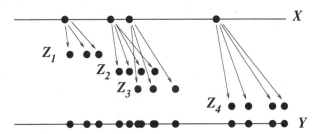

Fig. 18. Schematic concept of the formation of a cluster process.

Finally, in a **cluster process**, we start with a point process **X** and replace each point $x \in \mathbf{X}$ by a random finite set of points Z_x called the cluster associated with x. The superposition of all clusters yields the process $\mathbf{Y} = \bigcup_{x \in \mathbf{X}} Z_x$. See Figure 18.

Usually it is assumed that the clusters Z_x for different parent points x are independent processes. A simple example is the **Matérn cluster process** in which the 'parent' process **X** is a uniform Poisson process in $\mathbb{R}^2$, and each cluster Z_x consists of a random number M_x of points, where $M_x \sim \mathsf{Poisson}(\mu)$, independently and uniformly distributed in the disc $b(x, r)$ of radius r centred on x. Simulated realisations of this process are shown in Figure 19.

1.8 Marked Point Processes

Earlier we mentioned the idea that the points of a point process might be labelled with extra information called **marks**. For example, in a map of the locations of emergency calls, each point might carry a label stating the time of the call and the nature of the emergency.

A marked point can be formalised as a pair (x, m) where x is the point location and m is the mark attached to it.

Fig. 19. Simulated realisations of the Matérn cluster process in the unit square. *Left:* parent intensity $\beta = 5$, mean cluster size $\mu = 20$, cluster radius $r = 0.07$. *Right:* $\beta = 50$, $\mu = 2$, $r = 0.07$. Both processes have an average of 100 points in the square.

Definition 1.8. *A* **marked point process** *on a space S with marks in a space M is a point process $\mathbf{Y}$ on $S \times M$ such that $N_Y(K \times M) < \infty$ a.s. for all compact $K \subset S$. That is, the corresponding projected process (of points without marks) is locally finite.*

Note that the space of marks M can be very general. It may be a finite set, a continuous interval of real numbers, or a more complicated space such as the set of all convex polygons.

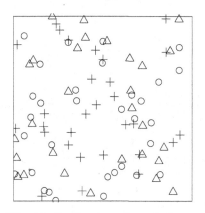 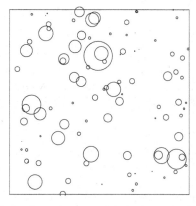

Fig. 20. Realisations of marked point processes in the unit square. *Left:* finite mark space $M = \{a, b, c\}$, marks plotted as symbols $\triangle, O, +$. *Right:* continuous mark space $M = [0, \infty)$, marks plotted as radii of circles.

Example 1.4. Let $\mathbf{Y}$ be a uniform Poisson process in $\mathbb{R}^3 = \mathbb{R}^2 \times \mathbb{R}$. This *cannot* be interpreted as a marked point process in $\mathbb{R}^2$ with marks in $\mathbb{R}$, because the finiteness condition fails. The set of marked points (x, m) which project into a given compact set $K \subset \mathbb{R}^2$ is the solid region $K \times \mathbb{R}$, which has infinite volume, and hence contains infinitely many marked points, almost surely.

Example 1.5. Let $\mathbf{Y}$ be a uniform Poisson process on the three-dimensional slab $R^2 \times [0, a]$ with intensity β. This can be interpreted as a marked point process on $\mathbb{R}^2$ with marks in $M = [0, a]$. The finiteness condition is clearly satisfied. The projected point process (i.e. obtained by ignoring the marks) is a uniform Poisson process in $\mathbb{R}^2$ with intensity βa. By properties of the uniform distribution, the marks attached to different points are independent and uniformly distributed in $[0, a]$.

A marked point process formed by attaching independent random marks to a Poisson process of locations, is equivalent to a Poisson process in the product space.

Theorem 1.3. *Let $\mathbf{Y}$ be a marked point process on S with marks in M. Let $\mathbf{X}$ be the projected process in S (of points without marks). Then the following are equivalent:*

1. $\mathbf{X}$ *is a Poisson process in S with intensity μ, and given $\mathbf{X}$, the marks attached to the points of $\mathbf{X}$ are independent and identically distributed with common distribution Q on M;*
2. $\mathbf{Y}$ *is a Poisson process in $S \times M$ with intensity measure $\mu \otimes Q$.*

See e.g. [28]. This result can be obtained by comparing the capacity functionals of the two processes.

Marked point processes are also used in the formal description of operations like thinning and clustering. For example, thinning a point process $\mathbf{X}$ is formalised by construct a marked point process with marks in $\{0, 1\}$. The mark I_x attached to each point x indicates whether the point is to be retained (1) or deleted (0).

1.9 Distances in Point Processes

One simple way to analyse a point process is in terms of the distances between points. If $\mathbf{X}$ is a point process, let $\mathsf{dist}(u, \mathbf{X})$ for $u \in \mathbb{R}^d$ denote the shortest distance from the given location u to the nearest point of $\mathbf{X}$. This is sometimes called the **contact distance**. Note the key fact that

$$\mathsf{dist}(u, \mathbf{X}) \leq r \text{ if and only if } N(b(u, r)) > 0$$

where $b(u, r)$ is the disc of radius r centred at x. Since $N(b(u, r))$ is a random variable for fixed u and r, the event $\{N(b(u, r)) > 0\}$ is measurable, so the event $\{\mathsf{dist}(u, \mathbf{X}) \leq r\}$ is measurable for all r, which implies that the contact distance $\mathsf{dist}(u, \mathbf{X})$ is a well-defined random variable.

If $\mathbf{X}$ is a uniform Poisson process in $\mathbb{R}^d$ of intensity β, then this insight also gives us the distribution of $\mathsf{dist}(u, \mathbf{X})$:

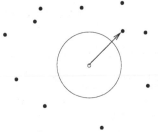

Fig. 21. The contact distance $\mathsf{dist}(u, \mathbf{X})$ from a fixed location ($\circ$) to the nearest random point ($\bullet$) satisfies $\mathsf{dist}(u, \mathbf{X}) > r$ if and only if there are no random points in the disc of radius r centred on the fixed location.

$$\mathbb{P}(\mathsf{dist}(u, \mathbf{X}) \leq r) = \mathbb{P}(N(b(u, r)) > 0)$$
$$= 1 - \exp(-\beta \lambda_d(b(u, r)))$$
$$= 1 - \exp(-\beta \kappa_d r^d)$$

where $\kappa_d = \lambda_d(b(0, 1))$ is the volume of the unit ball in $\mathbb{R}^d$.

One interesting way to rephrase this is that $V = \kappa_d \mathsf{dist}(u, \mathbf{X})^d$ has an exponential distribution with rate β,

$$\mathbb{P}(V \leq v) = 1 - \exp(-\beta v).$$

Notice that V is the volume of the ball of random radius $\mathsf{dist}(u, \mathbf{X})$, or equivalently, the volume of the largest ball centred on u that contains no points of $\mathbf{X}$.

Definition 1.9. *Let* $\mathbf{X}$ *be a stationary point process in* $\mathbb{R}^d$. *The* **contact distribution function** *or* **empty space function** F *is the cumulative distribution function of the distance*

$$R = \mathsf{dist}(u, \mathbf{X})$$

from a fixed point u *to the nearest point of* $\mathbf{X}$. *That is*

$$F(r) = \mathbb{P}(\mathsf{dist}(u, \mathbf{X}) \leq r)$$
$$= \mathbb{P}(N(b(u, r)) > 0).$$

By stationarity this does not depend on u.

Notice that $F(r) = T(b(0, r)) = T(b(u, r))$, where T is the capacity functional of $\mathbf{X}$. Thus the empty space function F gives us the values of the capacity functional $T(K)$ for all discs K. This does not fully determine T, and hence does not fully characterise $\mathbf{X}$. However, F gives us a lot of qualitative information about $\mathbf{X}$. The empty space function is a simple property of the point process that is useful in data analysis.

1.10 Estimation from Data

In applications, spatial point pattern data usually take the form of a finite configuration of points $\mathbf{x} = \{x_1, \dots, x_n\}$ in a region (**window**) W, where $x_i \in W$ and where $n = n(\mathbf{x}) \geq 0$ is not fixed. The data would often be treated as a realisation of a stationary point process $\mathbf{X}$ inside W. It is then important to estimate properties of the process $\mathbf{X}$.

An unbiased estimator of F is

$$\widehat{F}(r) = \frac{1}{\lambda_d(W)} \int_W \mathbf{1}\{\mathrm{dist}(u, \mathbf{X}) \leq r\} \, du. \qquad (3)$$

This is an unbiased estimator of $F(r)$, for each fixed value of r, since

$$
\begin{aligned}
\mathbb{E}\left[\widehat{F}(r)\right] &= \frac{1}{\lambda_d(W)} \mathbb{E}\left[\int_W \mathbf{1}\{\mathrm{dist}(u, \mathbf{X}) \leq r\} \, du\right] \\
&= \frac{1}{\lambda_d(W)} \int_W \mathbb{E}\mathbf{1}\{\mathrm{dist}(u, \mathbf{X}) \leq r\} \, du \\
&= \frac{1}{\lambda_d(W)} \int_W \mathbb{P}(\mathrm{dist}(u, \mathbf{X}) \leq r) \, du \\
&= \frac{1}{\lambda_d(W)} \int_W F(r) \, du \\
&= F(r)
\end{aligned}
$$

where the penultimate line follows by the stationarity of $\mathbf{X}$.

A practical problem is that, if we only observe $\mathbf{X} \cap W$, the integrand in (3) is not observable. When u is a point close to the boundary of the window W, the point of $\mathbf{X}$ nearest to u may lie outside W. More precisely, we have $\mathrm{dist}(u, \mathbf{X}) \leq r$ if and only if $n(\mathbf{X} \cap b(u, r)) > 0$. But our data are a realisation of $\mathbf{X} \cap W$, so we can only evaluate $n(\mathbf{X} \cap W \cap b(u, r))$.

It was once a common mistake to ignore this, and simply to replace $\mathbf{X}$ by $\mathbf{X} \cap W$ in (3). But this results in a negatively biased estimator of F. Call the estimator $\widehat{F}_W(r)$. Since $n(\mathbf{X} \cap W \cap b(u, r)) \leq n(\mathbf{X} \cap b(u, r))$, we have

$$\mathbf{1}\{n(\mathbf{X} \cap W \cap b(u, r)) > 0\} \leq \mathbf{1}\{n(\mathbf{X} \cap b(u, r)) > 0\}$$

so that $\mathbb{E}\widehat{F}_W(r) \leq F(r)$. This is called a **bias due to edge effects**.

One simple strategy for eliminating the edge effect bias is the **border method**. When estimating $F(r)$, we replace W in equation (3) by the erosion

$$W_{-r} = W \ominus b(0, r) = \{x \in W : \mathrm{dist}(x, \partial W) \geq r\}$$

consisting of all points of W that are at least r units away from the boundary ∂W. Clearly, $u \in W_{-r}$ if and only if $b(u, r) \subset W$. Thus, $n(\mathbf{x} \cap b(u, r))$ is observable when $u \in W_{-r}$. Thus we estimate $F(r)$ by

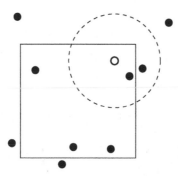

Fig. 22. Edge effect problem for estimation of the empty space function F. If we can only observe the points of $\mathbf{X}$ inside a window W (bold rectangle), then for some reference points u in W (open circle) it cannot be determined whether there is a point of $\mathbf{X}$ within a distance r of u. This problem occurs if u is closer than distance r to the boundary of W.

$$\widehat{F}_b(r) = \frac{1}{\lambda_2(W_{-r})} \int_{W_{-r}} \mathbf{1}\{\mathrm{dist}(u, \mathbf{x}) \le r\} \, du. \tag{4}$$

This is observable, and by the previous argument, it is an unbiased estimator of $F(r)$.

For a survey of corrections for edge effects, see [2].

1.11 Computer Exercises

Software is available for generating simulated realisations of point processes as shown above. The user needs access to the statistical package R, which can be downloaded free from the R website [13] and is very easy to install. Introductions to R are available at [23, 38].

We have written a library `spatstat` in the R language for performing point pattern data analysis and simulation. See [8] for an introduction. The `spatstat` library should also be downloaded from the R website [13], and installed in R.

The following commands in R will then generate and plot simulations of the point processes shown in Figures 9, 11, 12, 16, 19 and 20 above.

```
library(spatstat)
X <- runifpoint(100)
plot(X)
X <- rpoispp(5)
plot(X)
X <- rpoispp(function(x, y) { exp( 2 + 5 * x) })
plot(X)
plot(rMaternI(200, 0.05))
plot(rMaternII(200, 0.05))
```

```
plot(rMatClust(5,  0.07, 20))
plot(rMatClust(50, 0.07, 2))
X <- rpoispp(100)
M <- sample(1:3, X$n, replace=TRUE)
plot(X %mark% M)
M <- rexp(X$n)
plot(X %mark% M)
```

Further information on each *command* can be obtained by typing
help(*command*) in R.

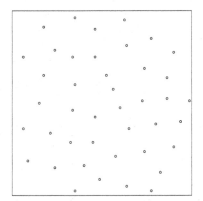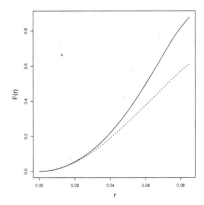

Fig. 23. *Left:* the `cells` point pattern dataset. *Right:* estimated empty space function $F(r)$ plotted against r (solid lines) together with the empty space function of a Poisson process (dotted lines).

The `spatstat` library also contains point pattern datasets and techniques for analysing them. In particular the function `Fest` will estimate the contact distribution function or empty space function F (defined in Section 1.9) from an observed realisation of a stationary point process. The following commands access the `cells` point pattern dataset, plot the data, then compute an estimate of F and plot this function.

```
data(cells)
plot(cells)
Fc <- Fest(cells)
plot(Fc)
```

The resulting plots are shown in Figure 23. There is a striking discrepancy between the estimated function F and the function expected for a Poisson process, indicating that the data cannot be treated as Poisson.

2 Moments and Summary Statistics

In this lecture we describe the analogue, for point processes, of the moments (expected value, variance and higher moments) of a random variable. These quantities are useful in theoretical study of point processes and in statistical inference about point patterns.

The **intensity** or first moment of a point process is the analogue of the expected value of a random variable. **Campbell's formula** is an important result for the intensity. The 'second moment measure' is related to the variance or covariance of random variables. The **K function** and **pair correlation** are derived second-moment properties which have many applications in the statistical analysis of spatial point patterns [16, 43]. The second-moment properties of some point processes will be found here. In the computer exercises we will compute statistical estimates of the K function from spatial point pattern data sets.

2.1 Intensity

Definition 2.1. Let **X** be a point process on $S = \mathbb{R}^d$ (or on any locally compact metric space S). Writing

$$\nu(B) = \mathbb{E}[N_X(B)], \qquad B \subset S,$$

defines a measure ν on S, called the **intensity measure** of **X**, provided $\nu(B) < \infty$ for all compact B.

Example 2.1 (Binomial process). The binomial point process (Section 1.3) of n points in a region $W \subset \mathbb{R}^d$ has $N(B) \sim$ binomial(n, p) where $p = \lambda_d(B \cap W)/\lambda_d(W)$ so

$$\nu(B) = \mathbb{E}N(B) = np = n\frac{\lambda_d(B \cap W)}{\lambda_d(W)}.$$

Thus $\nu(B)$ is proportional to the volume of $B \cap W$.

Example 2.2 (Poisson process). The uniform Poisson process of intensity $\beta > 0$ has $N(B) \sim$ Poisson$(\beta\lambda_d(B))$ so

$$\nu(B) = \beta\lambda_d(B).$$

Thus $\nu(B)$ is proportional to the volume of B.

Example 2.3 (Translated grid). Suppose U_1, U_2 are independent random variables uniformly distributed in $[0, s]$. Let **X** be the point process consisting of all points with coordinates $(U_1 + ms, U_2 + ns)$ for all integers m, n. A realisation of this process is a square grid of points in $\mathbb{R}^2$, with grid spacing s, which has been randomly translated. See Figure 24. It is easy to show that

$$\nu(B) = \mathbb{E}N(B) = \frac{1}{s^2}\lambda_2(B)$$

for any set B in $\mathbb{R}^2$ of finite area. This principle is important in applications to **stereology** [4].

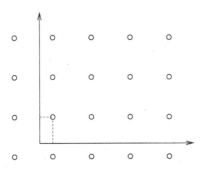

Fig. 24. A randomly translated square grid.

If **X** is a stationary point process in $\mathbb{R}^d$, then

$$\nu(B+v) = \mathbb{E}N(B+v) = \mathbb{E}N(B) = \nu(B)$$

for all $v \in \mathbb{R}^d$. That is, the intensity measure of a stationary point process is invariant under translations. But we know that the only such measures are multiples of Lebesgue measure:

Theorem 2.1. *If ν is a translation-invariant measure on $\mathbb{R}^d$ then $\nu(B) = c\lambda_d(B)$ for some $c \geq 0$.*

Corollary 1 *If **X** is a stationary point process in $\mathbb{R}^d$, then its intensity measure ν is a constant multiple of Lebesgue measure λ_d.*

The constant c in Corollary 1 is often called the **intensity** of **X**.

Definition 2.2. *Suppose the intensity measure ν of a point process **X** in $\mathbb{R}^d$ satisfies*

$$\nu(B) = \int_B \beta(u) \, \mathrm{d}u$$

*for some function β. Then we call β the **intensity function** of **X**.*

If it exists, the intensity function has the interpretation that in a small region $\mathrm{d}x \subset \mathbb{R}^d$

$$\mathbb{P}(N(\mathrm{d}x) > 0) \sim \mathbb{E}N(\mathrm{d}x) \sim \beta(x) \, \mathrm{d}x.$$

For the uniform Poisson process with intensity $\beta > 0$, the intensity function is obviously $\beta(u) \equiv \beta$. The randomly translated square grid (Example 2.3) is a stationary process with intensity measure $\nu(B) = \beta\lambda_2(B)$, so it has an intensity function, $\beta(u) \equiv 1/s^2$.

Theorem 2.2 (Campbell's Formula). *Let* $\mathbf{X}$ *be a point process on* S *and let* $f : S \to \mathbb{R}$ *be a measurable function. Then the random sum*

$$T = \sum_{x \in \mathbf{X}} f(x)$$

is a random variable, with expected value

$$\mathbb{E}\left[\sum_{x \in \mathbf{X}} f(x)\right] = \int_S f(x)\, \nu(\mathrm{d}x). \tag{5}$$

In the special case where $\mathbf{X}$ is a point process on $\mathbb{R}^d$ with an intensity function β, Campbell's Formula becomes

$$\mathbb{E}\left[\sum_{x \in \mathbf{X}} f(x)\right] = \int_{\mathbb{R}^d} f(x)\beta(x)\, \mathrm{d}x.$$

Campbell's Formula applies even to non-simple point processes (i.e. where points may have a multiplicity greater than 1) if the terms in the sum in (5) are counted with their multiplicity.

Proof. The result (5) is true when f is a step function, i.e. a function of the form

$$f = \sum_{i=1}^{m} c_i 1_{B_i}$$

for $B_i \subset S$ compact and $c_i \in \mathbb{R}$, because in that case

$$T = \sum_{x \in \mathbf{X}} f(x) = \sum_{x} \sum_{i} c_i 1_{B_i}(x) = \sum_{i} c_i N_{\mathbf{X}}(B_i)$$

so

$$\mathbb{E}T = \mathbb{E}\left[\sum_{i} c_i N_{\mathbf{X}}(B_i)\right] = \sum_{i} c_i \mathbb{E}N(B_i) = \sum_{i} c_i \nu(B_i) = \int_S f(x)\, \nu(\mathrm{d}x).$$

The result for general f follows by monotone approximation. $\qquad\square$

Example 2.4 (Monte Carlo integration). Suppose we want to compute the integral

$$I = \int_W f(x)\, \mathrm{d}x$$

where $W \subset \mathbb{R}^d$ and f is a nonnegative, integrable, real-valued function. Take any point process $\mathbf{X}$ with intensity

$$\lambda(x) = \begin{cases} c & \text{if } x \in W \\ 0 & \text{if } x \notin W \end{cases}$$

Evaluate the function f at the random points of $\mathbf{X}$, and estimate the integral I by the discrete sum approximation

$$\widehat{I} = \frac{1}{c} \sum_{x \in \mathbf{X}} f(x).$$

Then Campbell's formula (5) gives

$$\mathbb{E}[\widehat{I}] = \frac{1}{c}\mathbb{E}\left[\sum_{x \in \mathbf{X}} f(x)\right] = \frac{1}{c}\int_{\mathbb{R}^d} f(x)\lambda(x)\,\mathrm{d}x = \int_W f(x)\,\mathrm{d}x = I$$

so that $\widehat{I}$ is an *unbiased estimator* of I.

Example 2.5 (Olbers' Paradox). In 1826, the astronomer Heinrich Olbers pointed out a physical paradox in the fact that the sky is dark at night. Suppose we make the following assumptions: (i) the universe exists in 3-dimensional Euclidean space $\mathbb{R}^3$; (ii) the stars currently visible from Earth (with a given absolute magnitude) constitute a stationary point process in $\mathbb{R}^3$; and (iii) the observed brilliance of the light reaching Earth from a star at location $x \in \mathbb{R}^3$ is $a/||x||^2$ where a is constant (the inverse square law). Then the expected total brilliance of the night sky is infinite:

$$\mathbb{E}\left[\sum_{x \in \mathbf{X}\setminus\text{Earth}} \frac{a}{||x||^2}\right] = \lambda \int_{\mathbb{R}^3\setminus\text{Earth}} \frac{a}{||x||^2}\,\mathrm{d}x = \infty.$$

By this argument, 19th century physicists realized that, in a stable, infinite universe with an even distribution of stars, the entire universe should gradually heat up. The paradox led to a review of the theory of thermodynamics.

Example 2.6. Suppose $\mathbf{X}$ consists of a fixed, finite number of random points in $\mathbb{R}^d$, say $\mathbf{X} = \{X_1, \ldots, X_n\}$. Assume X_i has a marginal probability density $f_i(u), u \in \mathbb{R}^d$. Then $\mathbf{X}$ has intensity function $\beta(u) = \sum_{i=1}^n f_i(u)$.

Example 2.7. Consider a Poisson cluster process $\mathbf{Y}$ (Section 1.7). This is formed by taking a uniform Poisson process $\mathbf{X}$ of parent points, with intensity α, and replacing each $x \in \mathbf{X}$ by a random cluster Z_x which is a finite point process.

Suppose Z_x has intensity function $f(u \mid x)$. Then *conditional on* $\mathbf{X}$, the process $\mathbf{Y}$ has intensity function

$$\beta_{\mathbf{Y}|\mathbf{X}}(u) = \sum_{x \in \mathbf{X}} f(u \mid x)$$

It is not hard to show that the (unconditional) intensity function β of $\mathbf{Y}$ is the expectation with respect to $\mathbf{X}$,

$$\beta(u) = \mathbb{E}\left[\beta_{\mathbf{Y}|\mathbf{X}}(u)\right]$$
$$= \mathbb{E}\sum_{x\in\mathbf{X}} f(u \mid x)$$
$$= \alpha \int_{\mathbb{R}^d} f(u \mid x)\, dx$$

by Campbell's formula.

For example, in Matérn's cluster process, a cluster Z_x consists of a Poisson(μ) random number of points, uniformly distributed in the disc $b(x,r)$ of radius r centred on x. This has intensity $f(u \mid x) = \mu/(\pi r^2)$ if $u \in b(x,r)$ and 0 otherwise. Now

$$\int_{\mathbb{R}^d} f(u \mid x)\, dx = \frac{\mu}{\pi r^2}\int_{\mathbb{R}^d} \mathbf{1}\{u \in b(x,r)\}\, dx$$
$$= \frac{\mu}{\pi r^2}\int_{\mathbb{R}^d} \mathbf{1}\{x \in b(u,r)\}\, dx$$
$$= \mu.$$

Hence Matérn's cluster process has intensity $\beta(u) = \alpha\mu$.

2.2 Intensity for Marked Point Processes

Marked point processes were introduced in Section 1.8. Let $\mathbf{Y}$ be a marked point process on the space S with marks in a space M. Viewing $\mathbf{Y}$ as a point process on $S \times M$, we may extend the definition of intensity measure to marked point processes without further work.

The intensity measure of $\mathbf{Y}$ is (by Definition 2.1) a measure ν on $S \times M$ defined by

$$\nu(U) = \mathbb{E}N_{\mathbf{Y}}(U), \qquad U \subset S \times M.$$

It is completely determined by the values

$$\nu(B \times C) = \mathbb{E}N_{\mathbf{Y}}(B \times C)$$
$$= \mathbb{E}\sum_{(x,m)\in\mathbf{Y}} \mathbf{1}\{x \in B\}\mathbf{1}\{m \in C\}$$

for all compact $B \subset S$ and measurable $C \subset M$.

For marked point processes, Campbell's Formula takes the form

$$\mathbb{E}\sum_{(x,m)\in\mathbf{Y}} f(x,m) = \int_{S\times M} f(x,m)\,\nu(dx, dm) \qquad (6)$$

where $f : S \times M \to \mathbb{R}$ is a measurable function.

Differences between marked and unmarked point processes arise with regard to the concept of stationarity.

Definition 2.3. *A marked point process on $\mathbb{R}^d$ with marks in M is* **stationary** *if its distribution is invariant under shifts of $\mathbb{R}^d$ only*

$$(x, m) \mapsto (x + v, m)$$

for all $v \in \mathbb{R}^d$.

Note that the shift operation changes the location of a point but does not alter the mark attached to it.

Theorem 2.3. *Let Y be a stationary marked point process in $\mathbb{R}^d$. Assume the corresponding process of unmarked points has finite intensity (that is $\mathbb{E}N_Y(K \times M) < \infty$ for all compact $K \subset \mathbb{R}^d$).*

Then the intensity measure ν of Y takes the form

$$\nu(A \times B) = \beta \lambda_d(A) Q(B) \tag{7}$$

for all $A \subset \mathbb{R}^d$, $B \subset M$, where $\beta \geq 0$ is the intensity (expected number of points per unit volume), and Q is a probability measure on M called the **distribution of the typical mark**.

As a simple example of (7), consider a point process consisting of points of three colours. This may be formalised as a marked point process in $\mathbb{R}^2$ in which the marks are colours, $M = \{\text{red}, \text{green}, \text{blue}\}$. For a region $A \subset \mathbb{R}^2$, the quantity $\nu(A \times \{\text{red}\})$ is the expected number of red points in A, and by equation (7), this is equal to $\beta \lambda_2(A) Q(\{\text{red}\})$, a constant times the area of A times the probability of the colour red.

Proof. Since Y is stationary, ν is invariant under shifts of $\mathbb{R}^d$,

$$\nu(A \times B) = \nu((A + v) \times B)$$

for all $A \subset \mathbb{R}^d$, $B \subset M$ and all translation vectors $v \in \mathbb{R}^d$. If we fix B and define

$$\mu_B(A) = \nu(A \times B)$$

for all $A \subset \mathbb{R}^d$, then μ_B is a measure on $\mathbb{R}^d$ which is invariant under translations. It follows from Theorem 2.1 that, for fixed B,

$$\nu(A \times B) = c_B \lambda_d(A)$$

for all $A \subset \mathbb{R}^d$, where c_B is a constant depending on B.

On the other hand, if we fix A to be the unit cube, and define $\kappa(B) = \nu(A \times B) = c_B \lambda_d(A) = c_B$, then κ is a measure on M satisfying $\kappa(M) = \nu(A \times M) = \mathbb{E}N_Y(A \times M) < \infty$ by assumption. Letting $\beta = \kappa(M)$ and $Q(B) = \kappa(B)/\beta$ yields the result. $\qquad\square$

The argument we have just seen is often called **factorisation** or **disintegration**. It exploits the property that the intensity measure is invariant with respect to translations on the first factor of the product $\mathbb{R}^d \times M$. We shall have occasion to use the same argument many times.

For a stationary marked point process, Campbell's formula becomes

$$\mathbb{E}\left[\sum_{(x,m)\in\mathbf{Y}} f(x,m)\right] = \beta\mathbb{E}_Q\left[\int_{\mathbb{R}^d} f(x,K)\right] \tag{8}$$

where K denotes a random mark (a random element of M) with distribution Q. As an exercise, the reader may like to use this to prove Olbers' Paradox (Example 2.5) in greater generality, treating the stars in the universe as a stationary marked point process in $\mathbb{R}^3$, with the marks indicating the absolute brightness of each star.

2.3 Second Moment Measures

Let $\mathbf{X}$ be a point process. We are interested in the variance of the count $N(B)$,

$$\operatorname{var} N(B) = \mathbb{E}\left[N(B)^2\right] - [\mathbb{E}N(B)]^2$$

and the covariance of two such counts,

$$\operatorname{cov}[N(B_1),N(B_2)] = \mathbb{E}\left[N(B_1)N(B_2)\right] - [\mathbb{E}N(B_1)]\,[\mathbb{E}N(B_2)].$$

A key observation is that $N(B_1)N(B_2)$ is equal to the number of ordered pairs (x,x') of points in the process $\mathbf{X}$ such that $x \in B_1$ and $x' \in B_2$.

Definition 2.4. *Let $\mathbf{X}$ be a point process on a space S. Then $\mathbf{X}\times\mathbf{X}$ is a point process on $S\times S$ consisting of all ordered pairs (x,x') of points $x,x' \in \mathbf{X}$. The intensity measure ν_2 of $\mathbf{X}\times\mathbf{X}$ is a measure on $S\times S$ satisfying*

$$\nu_2(A\times B) = \mathbb{E}\left[N_{\mathbf{X}}(A)N_{\mathbf{X}}(B)\right].$$

*This measure ν_2 is called the **second moment measure** of $\mathbf{X}$.*

Clearly, the second moment measure contains all information about the variances and covariances of the variables $N_{\mathbf{X}}(A)$. Campbell's formula applied to $\mathbf{X}\times\mathbf{X}$ becomes

$$\mathbb{E}\left[\sum_{x\in\mathbf{X}}\sum_{y\in\mathbf{X}} f(x,y)\right] = \int_S\int_S f(x,y)\,\nu_2(\mathrm{d}x,\mathrm{d}y)$$

for a measurable function $f : S\times S \to \mathbb{R}$.

Example 2.8. For the uniform Poisson point process of intensity $\beta > 0$ in $\mathbb{R}^d$, the second moment measure satisfies

$$\nu_2(A \times B) = \beta^2 \lambda_d(A) \, \lambda_d(B) + \beta \lambda_d(A \cap B).$$

Geometrically this means that the measure ν_2 consists of two components: there is a constant density β^2 on all of $\mathbb{R}^d \times \mathbb{R}^d$, plus a positive mass on the diagonal $\Delta = \{(x, x) : x \in \mathbb{R}^d\}$. The mass on the diagonal arises from the fact that $\mathbf{X} \times \mathbf{X}$ includes pairs (x, x) of identical points. We could write the second moment measure informally as

$$\nu_2(\mathrm{d}x, \, \mathrm{d}y) = \beta^2 \, \mathrm{d}x \, \mathrm{d}y + \beta . \delta(x - y) \, \mathrm{d}x$$

where δ is the delta function. More formally

$$\nu_2 = \beta^2 \lambda_d \otimes \lambda_d + \beta \mathrm{diag}^{-1} \lambda_d$$

where $\mathrm{diag}(x, x) = x$.

To remove the mass on the diagonal, and also to simplify the calculation of certain moments, we introduce the **second factorial moment measure**

$$\nu_{[2]}(A \times B) = \mathbb{E}[N(A)N(B)] - \mathbb{E}[N(A \cap B)].$$

This is the intensity measure of the process $\mathbf{X} * \mathbf{X}$ of all ordered pairs of *distinct* points of $\mathbf{X}$. It satisfies

$$\mathbb{E}\left[\sum_{x \in \mathbf{X}} \sum_{y \in \mathbf{X}, \, y \neq x} f(x, y) \right] = \int_S \int_S f(x, y) \, \nu_{[2]}(\mathrm{d}x, \, \mathrm{d}y).$$

The name 'factorial' is derived from

$$
\begin{aligned}
\nu_{[2]}(A \times A) &= \mathbb{E}\left[N(A)^2 \right] - \mathbb{E}[N(A)] \\
&= \mathbb{E}\left[N(A)[N(A) - 1] \right].
\end{aligned}
$$

For example, for the uniform Poisson process of intensity β, the second factorial moment measure is $\nu_{[2]} = \beta^2 \lambda_d \otimes \lambda_d$.

Definition 2.5. *A point process $\mathbf{X}$ on $\mathbb{R}^d$ is said to have* **second moment density** g_2 *if*

$$\nu_{[2]}(C) = \int_C g_2(x, y) \, \mathrm{d}x \, \mathrm{d}y \tag{9}$$

for any compact $C \subset \mathbb{R}^d \times \mathbb{R}^d$.

Informally, $g_2(x, y)$ gives the joint probability that there will be points of $\mathbf{X}$ at two specified locations x and y:

$$\mathbb{P}(N(\,\mathrm{d}x) > 0, \ N(\,\mathrm{d}y) > 0) \sim g_2(x,y) \,\mathrm{d}x \,\mathrm{d}y.$$

For example, the uniform Poisson process has second moment density $g_2(x,y)$ $= \beta^2$. The binomial process of n points in W has

$$g_2(x,y) = \frac{n(n-1)}{\lambda_d(W)^2}$$

if $x, y \in W$, and zero otherwise.

Definition 2.6. *Suppose* **X** *is a point process on* $\mathbb{R}^d$ *which has an intensity function* $\beta(x)$ *and a second moment density* $g_2(x,y)$. *Then we define the* **pair correlation function** *of* **X** *by*

$$\rho_2(x,y) = \frac{g_2(x,y)}{\beta(x)\beta(y)}.$$

Example 2.9. For a uniform Poisson process of intensity β, we have $\beta(x) \equiv \beta$ and $g_2 \equiv \beta^2$, so that

$$\rho_2(x,y) \equiv 1.$$

Example 2.10. For a binomial process of n points in a region W, we have

$$\rho_2(x,y) \equiv 1 - \frac{1}{n}.$$

Note that the pair correlation function always satisfies $\rho_2(x,y) \geq 0$. It should be regarded as a 'non-centred' analogue of the usual correlation of random variables. The value $\rho_2 = 1$ corresponds to a lack of correlation in the usual statistical sense: if $\rho_2 \equiv 1$ then $\mathrm{cov}[N(B), N(B')] = 0$ for disjoint sets B, B'.

Example 2.11. Continuing Example 2.6, suppose **X** consists of a fixed, finite number of random points in $\mathbb{R}^d$, say $\mathbf{X} = \{X_1, \ldots, X_n\}$. Let $f_i(u), u \in \mathbb{R}^d$ be the marginal probability density of X_i, and $f_{ij}(u,v), u, v \in \mathbb{R}^d$ the marginal joint density of (X_i, X_j). Then **X** has second moment density

$$g_2(x,y) = \sum_{i \neq j} f_{ij}(x,y)$$

and pair correlation function

$$\rho_2(x,y) = \frac{\sum_{i \neq j} f_{ij}(x,y)}{\left(\sum_i f_i(x)\right)\left(\sum_j f_j(y)\right)}.$$

Example 2.12. Continuing Example 2.7, consider a Poisson cluster process **Y**, formed from a Poisson process **X** of parent points with intensity α. The clusters Z_x for different x are independent processes.

Suppose Z_u has intensity function $f(u \mid x)$ and second moment density $h(u, v \mid x)$. It is not hard to show, by first conditioning on X, that the second moment density of Y is

$$g_2(u, v) = \beta(u)\beta(v) + \alpha \int_{\mathbb{R}^d} h(u, v \mid x)\, dx,$$

where $\beta(u) = \int_{\mathbb{R}^d} f(u \mid x)\, dx$ is the intensity of Y. The integral term arises from pairs of points in Y that come from the same cluster Z_x.

For example, in the Matérn cluster process, the second moment density of a cluster Z_x is (by a simple extension of Example 2.11) $h(u, v \mid x) = \mu^2/(\pi^2 r^4)$ if $u, v \in b(x, r)$, and 0 otherwise. We have

$$\int_{\mathbb{R}^d} \mathbf{1}\{u, v \in b(x, r)\}\, dx = \int_{\mathbb{R}^d} \mathbf{1}\{u \in b(x, r)\}\mathbf{1}\{v \in b(x, r)\}\, dx$$

$$= \int_{\mathbb{R}^d} \mathbf{1}\{x \in b(u, r)\}\mathbf{1}\{x \in b(v, r)\}\, dx$$

$$= \lambda_2(b(u, r) \cap b(v, r)).$$

Hence the second moment density of the Matérn cluster process is

$$g_2(u, v) = \alpha^2 \mu^2 + \alpha \frac{\mu^2}{\pi^2 r^4} \lambda_2(b(u, r) \cap b(v, r)).$$

2.4 Second Moments for Stationary Processes

For a *stationary* point process in $\mathbb{R}^d$, there is a 'disintegration' of the second moment measure. Stationarity implies

$$\mathbb{E}\left[N(A + v)N(B + v)\right] = \mathbb{E}\left[N(A)N(B)\right]$$

for all $v \in \mathbb{R}^d$. Thus $\nu_2, \nu_{[2]}$ are invariant under **simultaneous shifts**

$$(x, y) \mapsto (x + v, y + v).$$

See the left panel in Figure 25.

Let us transform the problem by mapping each pair of points (x, y) to the pair $(x, y - x)$. Thus the first element of the image is the first point x, and the second element $y - x$ is the vector from x to y. This transforms $\mathbb{R}^d \times \mathbb{R}^d$ onto itself by $\Psi(x, y) = (x, y - x)$. Under this transformation, the simultaneous shift $(x, y) \mapsto (x + v, y + v)$ becomes a shift of the first coordinate

$$(s, t) \mapsto (s + v, t).$$

See the right panel in Figure 25.

The image of $\nu_{[2]}$ under Ψ is a measure μ on $\mathbb{R}^d$ which is invariant under translations of the first coordinate

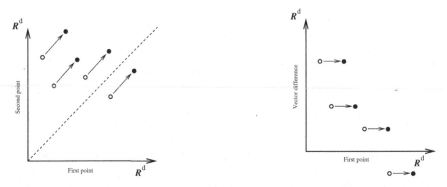

Fig. 25. Disintegration of the second moment measure of a stationary point process. *Left:* The second moment measure is invariant under shifts parallel to the diagonal in $\mathbb{R}^d \times \mathbb{R}^d$. *Right:* If we transform the problem by mapping (x, y) to $(x, y - x)$, the image of the second moment measure is invariant under shifts of the first coordinate. This factorises as a product measure.

$$(s, t) \mapsto (s + v, t)$$

for all $v \in \mathbb{R}^d$. By Theorem 2.1 it follows that

$$\mu = \beta \lambda_d \otimes \mathcal{K}$$

where β is the intensity of the process $\mathbf{X}$, and $\mathcal{K}$ is a measure on $\mathbb{R}^d$ called the **reduced second moment measure** of $\mathbf{X}$.

Retracing our steps and using Campbell's Formula, we find that for an arbitrary integrand f,

$$\mathbb{E}\left[\sum_{x \in \mathbf{X}} \sum_{y \in \mathbf{X}, \, y \neq x} f(x, y)\right] = \int \int f(x, y) \, \nu_{[2]}(\mathrm{d}x, \, \mathrm{d}y)$$

$$= \int \int f(x, x + u) \, \mu(\mathrm{d}x, \, \mathrm{d}u)$$

$$= \beta \int \int f(x, x + u) \, \mathcal{K}(\mathrm{d}u) \, \mathrm{d}x.$$

Theorem 2.4. *Let $\mathbf{X}$ be a stationary point process on $\mathbb{R}^d$ with intensity β. Then there is a measure $\mathcal{K}$ on $\mathbb{R}^d$ such that, for a general integrand f,*

$$\mathbb{E}\left[\sum_{x \in \mathbf{X}} \sum_{y \in \mathbf{X}, \, y \neq x} f(x, y)\right] = \beta \int \int f(x, x + u) \, \mathcal{K}(\mathrm{d}u) \, \mathrm{d}x. \qquad (10)$$

$\mathcal{K}$ *is called the* **reduced second moment measure** *of $\mathbf{X}$.*

To understand the measure $\mathcal{K}$, we notice that for $A, B \subset \mathbb{R}^d$

$$\beta\lambda_d(A)\mathcal{K}(B) = \mu(A \times B)$$

$$= \int\int 1\{s \in A\}1\{t \in B\}\mu(s,t)$$

$$= \int\int 1\{x \in A\}1\{y - x \in B\}\,\nu_{[2]}(dx, dy)$$

$$= \mathbb{E}\left[\sum_{x\in\mathbf{X}}\sum_{y\in\mathbf{X},\,y\neq x} 1\{x \in A\}1\{y - x \in B\}\right]$$

This may also be obtained directly from (10) by taking $f(x,y) = 1\{x \in A\}1\{y - x \in B\}$. Since $\beta\lambda_d(A) = \mathbb{E}N(A)$, we have

$$\mathcal{K}(B) = \frac{\mathbb{E}\sum_{x\in\mathbf{X}\cap A} N((B + x)\setminus x)}{\mathbb{E}N(A)} \tag{11}$$

The right hand side of (11) may be interpreted as the average, over all points x of the process, of the number of other points y of the process such that $y - x \in B$.

Example 2.13. For the uniform Poisson process,

$$\nu_{[2]} = \beta^2\,\lambda_d \otimes \lambda_d$$
$$\mu = \beta^2\,\lambda_d \otimes \lambda_d$$
$$\mathcal{K} = \beta\,\lambda_d$$

Example 2.14. Suppose $\mathbf{X}$ is a stationary process on $\mathbb{R}^d$ which has a second moment density function g_2. Then by comparing (9) with (10) we can see that $g_2(x,y)$ depends only on $y - x$, say

$$g_2(x,y) = g(y - x),$$

for some function g, and we can write

$$\mathcal{K}(B) = \frac{1}{\beta}\int_B g(u)\,du.$$

Example 2.15. The randomly translated square grid was introduced in Example 2.3. This is a stationary process. Following through the derivation above, we find that the reduced second moment measure $\mathcal{K}$ puts mass 1 at each integer point (ns, ms) for all integers n, m, except that there is no atom at $(0,0)$.

Intuitively this reflects the fact that, if we know there is a point of $\mathbf{X}$ at the origin, then this determines the position of the entire grid of points, and we know there will be a point of $\mathbf{X}$ at each location (ns, ms).

This point process does not have a second moment density g_2.

2.5 The K-function

Second moment properties are important in the statistical analysis of spatial point pattern data, just as the sample variance is important in classical statistics.

The reduced second moment measure $\mathcal{K}$ carries important information about the dependence or **interaction** between different points of the process. For practical data analysis, we need some simplification of the measure $\mathcal{K}$. Ripley [39] suggested the function

$$K(t) = \frac{1}{\beta}\mathcal{K}(b(0,t)), \quad t \geq 0. \tag{12}$$

See also Ornstein & Zernike [35].

Using (11) with $B = b(0,t)$, we see that $\beta K(t)$ is the expected number of points y of the process that satisfy $0 < ||y - x|| \leq t$ for a given point x of the process. In other words, $\beta K(t)$ is the expected number of points close to a given point of the process, where 'close' means 'within a distance t'.

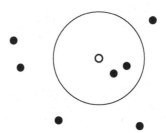

Fig. 26. Concept of the K-function. The value $\beta K(t)$ is the expected number of other points within a circle of radius t centred on a typical point of the process.

Example 2.16. For a uniform Poisson process in $\mathbb{R}^d$,

$$K(t) = \kappa_d t^d, \quad t \geq 0$$

where κ_d is the volume of the unit ball in $\mathbb{R}^d$.

The factor $1/\beta$ in (12) normalises the K-function, making it independent of the intensity β in the Poisson case.

Example 2.17. For a stationary point process in $\mathbb{R}^d$ which has a second moment density, Example 2.14 gives

$$K(t) = \frac{1}{\beta^2}\int_{b(0,t)} g_2(0,x)\,\mathrm{d}x = \int_{b(0,t)} \rho_2(0,x)\,\mathrm{d}x.$$

Lemma 2.1. *Suppose* $\mathbf{X}$ *is a stationary and isotropic point process in* $\mathbb{R}^2$ *which possesses a second moment density* g_2 *and pair correlation function* ρ. *Then* $g_2(x, y)$ *and* $\rho(x, y)$ *depend only on* $||y - x||$, *say*

$$g_2(x, y) = g(||x - y||) \tag{13}$$
$$\rho_2(x, y) = \rho(||x - y||) \tag{14}$$

and the pair correlation can be recovered from the K-function by

$$\rho_2(t) = \frac{\frac{\mathrm{d}}{\mathrm{d}t} K(t)}{2\pi t} \tag{15}$$

Example 2.18. For the uniformly randomly translated grid (Examples 2.3 and 2.15) the K-function is $K(t) = M(t/s) - 1$, where $M(r)$ is the number of points of the integer grid $\mathbb{Z}^2$ inside the disc $b(0, r)$. The function M is studied closely in Prof. Baranyi's lectures in this volume.

Lemma 2.2 (Invariance of K under thinning). *Suppose* $\mathbf{X}$ *is a stationary point process, and* $\mathbf{Y}$ *is obtained from* $\mathbf{X}$ *by random thinning (each point of X is deleted or retained, independently of other points, with retention probability* p). *Then the* K-functions of $\mathbf{X}$ and $\mathbf{Y}$ are identical.

The proof is an exercise.

2.6 Estimation from Data

Assume again that we have observed data in the form of a finite configuration of points $\mathbf{x} = \{x_1, \ldots, x_n\}$ in a window W, where $x_i \in W$ and where $n = n(\mathbf{x}) \geq 0$ is not fixed.

In order to estimate the K-function, consider the identity

$$K(t) = \frac{\mathbb{E} \sum_{x \in \mathbf{X} \cap W} N(b(x, t) \setminus x)}{\beta \, \mathbb{E} N(W)}. \tag{16}$$

Again we have an edge effect problem in applying this identity. If we only observe $\mathbf{X} \cap W$, the random variable in the numerator of (16) is not observable. When x is a point close to the boundary of the window W, the disc $b(x, t)$ may extend outside W. Since the process $\mathbf{X}$ is not observed outside W, the number of points of $\mathbf{X}$ in $b(x, t)$ is not observable.

It is a common mistake to ignore this problem, and estimate the numerator of (16) by

$$\sum_{i=1}^{n} n(\mathbf{x} \cap b(x_i, t) \setminus x_i) = \sum_{i=1}^{n} \sum_{j \neq i} 1\{||x_i - x_j|| \leq t\}. \tag{17}$$

The right hand side of (17) is proportional to the empirical distribution function of the distances $s_{ij} = ||x_i - x_j||$ between all pairs of points. But this is a

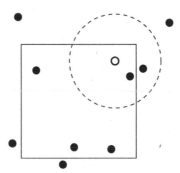

Fig. 27. Edge effect problem for estimation of the K function. If we can only observe the points inside a window W (bold rectangle), then the number of points inside a circle of radius t, centred on a point of the process inside W, is not observable if the circle extends outside W.

biased estimator: the expectation of (17) is less than the numerator of (16), because the observable quantity $n(\mathbf{X} \cap W \cap b(u, t))$ is less than or equal to the desired quantity $n(\mathbf{X} \cap b(u, t))$. This is a bias due to edge effects.

One simple strategy for eliminating the edge effect bias is the border method, introduced in Section 1.10. When estimating $K(t)$, we replace W in equation (16) by the erosion

$$W_{-t} = W \ominus b(0, t) = \{x \in W : \operatorname{dist}(x, \partial W) \ge t\}$$

consisting of all points of W that are at least t units away from the boundary ∂W. Clearly, $u \in W_{-t}$ if and only if $b(u, t) \subset W$. Thus, $n(\mathbf{x} \cap b(x_i, t) \setminus x_i)$ is observable when $x_i \in W_{-t}$. Thus we estimate $K(t)$ by

$$
\begin{aligned}
\widehat{K}(t) &= \frac{\sum_{x \in W_{-t}} N_{\mathbf{X}}(b(x, t) \setminus x)}{\widehat{\beta} n(\mathbf{x} \cap W_{-t})} \\
&= \frac{\sum_{i=1}^{n} \sum_{j \ne i} \mathbf{1}\{\|x_i - x_j\| \le t\}}{\widehat{\beta} n(\mathbf{x} \cap W_{-t})}
\end{aligned}
\tag{18}
$$

where $\widehat{\beta}$ is usually $n(\mathbf{x})/\lambda_2(W)$. This is called the **border method** of edge correction. More sophisticated edge corrections with better performance are discussed in [42, 2].

2.7 Exercises

We again make use of the package `spatstat` described in section 1.11.

The function `Kest` computes estimates of the K function from spatial point pattern data.

```
library(spatstat)
data(cells)
```

```
Kc <- Kest(cells)
plot(Kc)
data(redwood)
plot(Kest(redwood))
```

 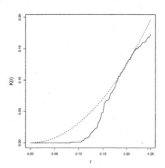

Fig. 28. *Left:* the `cells` point pattern dataset. *Right:* estimated K function plotted against r, together with the theoretical K function for a Poisson process with the same (estimated) intensity.

 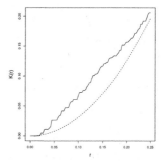

Fig. 29. *Left:* the `redwood` point pattern dataset. *Right:* estimated K plotted against r, together with the empty space function of a Poisson process.

The function `Kmeasure` computes an estimate of (a kernel-smoothed density of) the reduced second moment measure $\mathcal{K}$.

```
KMc <- Kmeasure(cells, sigma=0.03)
plot(KMc)
```

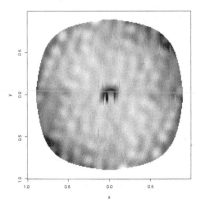

Fig. 30. Kernel-smoothed density estimate of the second moment measure $\mathcal{K}$ of the `cells` dataset. Lighter greys indicate higher estimated densities.

3 Conditioning

In the study of a point process we are often interested in properties relating to a **typical point** of the process. This requires the calculation of conditional probabilities of events given that there is a point of the process at a specified location. It leads to the concept of the **Palm distribution** of the point process, and the related **Campbell-Mecke formula** [42]. These tools allow us to define new characteristics of a point process, such as the nearest neighbour distance distribution function G. A dual concept is the **conditional intensity** which provides many new results about point processes. In the computer exercises we compute statistical estimates of the function G from spatial point pattern data sets.

3.1 Motivation

One simple question about a point process $\mathbf{X}$ is: what is the probability distribution of the distance from a point of $\mathbf{X}$ to its nearest neighbour (the nearest other point of $\mathbf{X}$)?

Note that this is different from the empty space function F introduced in Section 1.9, which is the distribution of the distance $\text{dist}(u, \mathbf{X})$ from a *fixed location* u to the nearest point of $\mathbf{X}$. Here we are asking about the distance from a *point of the process* $\mathbf{X}$ to the nearest other point of the process.

If x is known to be a point of $\mathbf{X}$, then the nearest neighbour distance is $R_x = \text{dist}(x, \mathbf{X} \setminus x)$, and we seek the 'conditional probability'

$$\mathbb{P}\left(R_x \leq r \mid x \in \mathbf{X}\right).$$

The problem is that this is not a conditional probability in the elementary sense, because the event $\{x \in \mathbf{X}\}$ typically has probability zero.

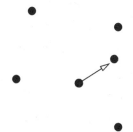

Fig. 31. Concept of nearest neighbour distance.

For some basic examples of point processes, this question can be resolved using classical methods.

Example 3.1. Consider the binomial process (Section 1.3)

$$\mathbf{X} = \{X_1, \ldots, X_n\}$$

where $X_1, \ldots, X_n$ are i.i.d. random points, uniformly distributed in $W \subset \mathbb{R}^2$. For each $i = 1, \ldots, n$ the conditional probability

$$\mathbb{P}\left(R_x \leq r \mid X_i = x\right) = \mathbb{P}\left(\text{dist}(x, \mathbf{X} \setminus X_i) \leq r \mid X_i = x\right)$$

is well-defined (using the classical definition of conditional probability) and equal to

$$\mathbb{P}\left(R_x \leq r \mid X_i = x\right) = 1 - \mathbb{P}\left(R_x > r \mid X_i = x\right)$$
$$= 1 - \mathbb{P}(\mathbf{X}' \cap b(x, r) = \emptyset)$$

where

$$\mathbf{X}' = \mathbf{X} \setminus X_i$$

is a binomial process with $n - 1$ points. Thus

$$\mathbb{P}\left(R_x \leq r \mid X_i = x\right) = 1 - \left[\frac{\lambda_2(b(x, r) \cap W)}{\lambda_2(W)}\right]^{n-1}.$$

The same quantity is obtained for each i, as we might have expected given the exchangeability of $X_1, \ldots, X_n$. Hence it seems reasonable to interpret this to be the value of $\mathbb{P}\left(R_x \leq r \mid x \in \mathbf{X}\right)$.

A similar argument can be used for other point processes which contain a finite number of points, almost surely.

For a stationary point process $\mathbf{X}$, another argument must be used. It is sufficient to consider $x = 0$, that is, to condition on the event that there is a point of $\mathbf{X}$ at the origin 0. One simple way to define and calculate such probabilities would be to condition on the event that there is a point of $\mathbf{X}$ in a small neighbourhood U of the origin 0, and then take the limit as U shrinks down to $\{0\}$.

Example 3.2. Suppose $\mathbf{X}$ is a Poisson process in $\mathbb{R}^2$ with intensity β. For $\epsilon > 0$, let $U = b(0, \epsilon)$ and define $R_{(\epsilon)} = \mathrm{dist}(0, \mathbf{X} \setminus U)$, the distance from 0 to the nearest point of X outside U. Clearly, $R_{(\epsilon)} > r$ iff $\mathbf{X}$ has no points in $b(0, r) \setminus U$. Since U and $b(0, r) \setminus U$ are disjoint,

$$\mathbb{P}\left(R_{(\epsilon)} > r \mid N(U) > 0\right) = \mathbb{P}(R_{(\epsilon)} > r) = \exp\{-\beta\pi(r^2 - \epsilon^2)\}.$$

As $\epsilon \downarrow 0$, this conditional probability converges to the limit $\exp\{-\beta\pi r^2\}$. Note also that

$$\mathbb{P}\left(N(U) = 1 \mid N(U) > 0\right) \to 1 \quad \text{as } \epsilon \downarrow 0$$

so that, for small ϵ, we may effectively assume there is at most one point in U. Additionally, if $\mathbf{X} \cap U = \{x\}$, then

$$\left| \mathrm{dist}(x, \mathbf{X} \setminus x) - \mathrm{dist}(0, \mathbf{X} \setminus U) \right| \le \epsilon$$

so we have some confidence in formally writing

$$\mathbb{P}\left(R_0 \le r \mid 0 \in \mathbf{X}\right) = \exp\{-\beta\pi r^2\}.$$

3.2 Palm Distribution

The **Palm distribution** formalises the concept of conditioning on a point of the process. It was developed by C. Palm (1907-1951) for the study of telephone traffic [36].

Palm Probabilities

The Palm probability $\mathbb{P}^x(A)$ of an event A at a location x is, intuitively speaking, the conditional probability that the event A will occur, given $x \in \mathbf{X}$: that is, given that there is a point of the process X at the specified location x.

An elegant way to define $\mathbb{P}^x(A)$ is the following. Let $(\Omega, \mathcal{A}, P)$ be the underlying probability space. Define the Campbell measure C on $S \times \Omega$ by

$$C(B \times A) = \mathbb{E}\left[N(B)\mathbf{1}_A\right]$$

for all $A \in \mathcal{A}$ and $B \in \mathcal{B}(S)$, then by extension to $\mathcal{A} \otimes \mathcal{B}(S)$. Here $\mathbf{1}_A$ is the indicator random variable of the event A (equal to 1 if the event A occurs and 0 if not) and $\mathcal{B}(S)$ is the Borel σ-field of S. Notice that

$$C(B \times A) \le \mathbb{E}N(B) = \nu(B)$$

where ν is the intensity measure of X (assumed to exist and to be locally finite).

For any fixed A, let $\mu_A(B) = C(B \times A)$ for all B. Then μ_A is a measure, and $\mu_A \le \nu$, so certainly $\mu_A << \nu$. By the Radon-Nikodým Theorem,

$$\mu_A(B) = \int_B f_A(x)\, \nu(\mathrm{d}x)$$

where $f_A : S \to \mathbb{R}_+$ is measurable (and unique up to equality almost everywhere). We shall interpret $f_A(x)$ as the Palm probability $\mathbb{P}^x(A)$.

Under conditions on $(\Omega, \mathcal{A})$, there exist regular conditional probabilities $\mathbb{P}^x(A)$ such that

- for all A, the function $x \mapsto \mathbb{P}^x(A)$ is a version of f_A, i.e.

$$\int_B \mathbb{P}^x(A)\, \nu(\mathrm{d}x) = C(B \times A) = \mathbb{E}\left[N(B)1_A\right]$$

- for almost all x, the map $A \mapsto \mathbb{P}^x(A)$ is a probability measure on $(\Omega, \mathcal{A})$.

Then $\mathbb{P}^x$ is called the **Palm probability measure** associated with the point process $\mathbf{X}$ at the location x. We write $\mathbb{E}^x$ for the expectation with respect to $\mathbb{P}^x$.

Example 3.3 (Poisson process). Let $\mathbf{X}$ be a uniform Poisson process in $\mathbb{R}^d$. Consider the event

$$A = \{N(K) = 0\}$$

where $K \subset \mathbb{R}^d$ is compact. For any closed U disjoint from K we have, by independence properties of the Poisson process,

$$C(U \times A) = \mathbb{E}\left[N(U)1_A\right] = \mathbb{E}[N(U)]\mathbb{P}(A) = \nu(U)\mathbb{P}(A).$$

It follows that $\mathbb{P}^x(A) = \mathbb{P}(A)$ for almost all $x \in \mathbb{R}^d \setminus K$. On the other hand, for $U \subseteq K$ we have $N(U) \le N(K)$ so that

$$C(U \times A) = \mathbb{E}\left[N(U)1_A\right] = 0$$

so that $\mathbb{P}^x(A) = 0$ for almost all $x \in K$.

Now holding x fixed and varying K, and taking the complementary probabilities, we have $\mathbb{P}^x(N(K) > 0) = \mathbb{P}(N(K) > 0)$ if $x \notin K$, and $\mathbb{P}^x(N(K) > 0) = 1$ if $x \in K$. But this is the capacity functional of

$$\mathbf{X} \cup \{x\},$$

the Poisson process $\mathbf{X}$ augmented by a fixed point at the location x.

In other words, under the Palm distribution $\mathbb{P}^x$, the process behaves as if it were a Poisson process superimposed with a fixed point at the location x.

Note that $\mathbb{P}^x$ is a probability measure on the original space $(\Omega, \mathcal{A})$, giving a probability $\mathbb{P}^x(A)$ for any event $A \in \mathcal{A}$, and not just for events defined by the process $\mathbf{X}$.

Example 3.4 (Mixed Poisson process). Suppose Γ is a nonnegative real random variable defined on Ω and that, given $\Gamma = \gamma$, the point process $\mathbf{X}$ is Poisson with intensity γ. The intensity measure of this mixture process is

$$\mathbb{E}N(B) = \mathbb{E}\left[\mathbb{E}[N(B) \mid \Gamma]\right] = \mathbb{E}[\Gamma]\,\lambda_d(B).$$

Let $A = \{\Gamma \leq \gamma\}$ for some fixed $\gamma \geq 0$. Then for $B \subset \mathbb{R}^d$ we have

$$
\begin{aligned}
C(B \times A) &= \mathbb{E}\left[N(B)1_A\right] \\
&= \mathbb{E}\left(\mathbb{E}\left[N(B)1_A \mid \Gamma\right]\right) \\
&= \mathbb{E}\left[\Gamma\lambda_d(B)\mathbf{1}\{\Gamma \leq \gamma\}\right] \\
&= \mathbb{E}\left[\Gamma\mathbf{1}\{\Gamma \leq \gamma\}\right]\lambda_d(B)
\end{aligned}
$$

so that

$$\mathbb{P}^x(\Gamma \leq \gamma) = \frac{\mathbb{E}[\Gamma\mathbf{1}\{\{\Gamma \leq \gamma\}\}]}{\mathbb{E}[\Gamma]}.$$

Thus, the distribution of Γ under $\mathbb{P}^x$ is the Γ-weighted counterpart of its original distribution.

Palm Distribution of Point Process

Many writers consider only the Palm distribution of the point process $\mathbf{X}$ itself, that is, the distribution $\mathbf{P}^x$ on N defined by

$$\mathbf{P}^x(A) = \mathbb{P}^x(\mathbf{X} \in A)$$

for $A \in \mathcal{N}$. Note the distinction between $\mathbf{P}^x$, a point process distribution in the sense of Definition 1.5, and $\mathbb{P}^x$, a probability measure on the original probability space Ω. We sometimes denote the Palm distribution of $\mathbf{X}$ by $\mathbf{P}_\mathbf{X}^x$.

When $\mathbf{X}$ is a homogeneous Poisson process, we have just shown in Example 3.3 that the Palm distribution satisfies

$$\mathbf{P}^x = \mathbf{P} * \Delta_x$$

where $\mathbf{P}$ is the distribution of the original Poisson process, $*$ denotes convolution (superposition of two point processes), and Δ_x is the distribution of the point process consisting of a single point at x.

We sometimes write $\mathbf{X}^x$ for the process governed by the Palm distribution $\mathbf{P}^x$, so that the last equation can be expressed as

$$\mathbf{X}^x \overset{\mathrm{d}}{=} \mathbf{X} \cup \{x\}$$

where $\overset{\mathrm{d}}{=}$ denotes equivalence in distribution. In fact, this property is characteristic of Poisson processes.

Theorem 3.1 (Slivnyak's Theorem). *Let* $\mathbf{X}$ *be a point process with locally finite intensity measure* ν. *Suppose the distribution* $\mathbf{P} = \mathbf{P_X}$ *and the Palm distribution* $\mathbf{P}^x = \mathbf{P_X^x}$ *of* $\mathbf{X}$ *are related by*

$$\mathbf{P}^x = \mathbf{P} * \Delta_x.$$

Then $\mathbf{X}$ *is a Poisson process with intensity measure* ν.

It is often convenient to remove the point x from consideration.

Definition 3.1. *The* **reduced Palm distribution** $\mathbf{P}^{!x}$ *of a point process* $\mathbf{X}$ *is the distribution of* $\mathbf{X} \setminus x$ *under* $\mathbf{P}^x$:

$$\mathbf{P}^{!x}(A) = \mathbf{P}^x(\mathbf{X} \setminus x \in A)$$

for $A \in \mathcal{N}$.

Thus Slivnyak's Theorem states that $\mathbf{X}$ is a Poisson point process if and only if $\mathbf{P_X^{!x}} = \mathbf{P_X}$.

Example 3.5 (Binomial process). Let $\mathbf{Y}_{(n)}$ be the binomial process (Section 1.3) consisting of n independent random points $X_1, \ldots, X_n$ uniformly distributed in a domain W. It is easy to show that the reduced Palm distribution of $\mathbf{Y}^{(n)}$ is identical to the distribution of $\mathbf{Y}^{(n-1)}$.

Example 3.6 (Palm distribution of mixed Poisson process). Let $\mathbf{X}$ be the mixed Poisson process described in Example 3.4. Consider the event

$$A = \{N(K) = 0\}$$

where $K \subset \mathbb{R}^d$ is compact. Following the argument in Example 3.3 we find that if $x \in K$ then $\mathbb{P}^x(A) = 0$, while if $x \notin K$, then

$$\begin{aligned}
\mathbb{P}^x(A) &= \frac{\mathbb{E}[\Gamma \mathbf{1}\{N(K) = 0\}]}{\mathbb{E}[\Gamma]} \\
&= \frac{\mathbb{E}[\Gamma \mathbb{P}(N(K) = 0 \mid \Gamma)]}{\mathbb{E}[\Gamma]} \\
&= \frac{\mathbb{E}[\Gamma \exp(-\Gamma \lambda_d(K))]}{\mathbb{E}[\Gamma]}.
\end{aligned}$$

Hence, the capacity functional of $\mathbf{X}^{!x}$ is the Γ-weighted mean of the capacity functional of a Poisson process with intensity Γ. This is different from the capacity functional of the original process $\mathbf{X}$, which is the unweighted mean $T(K) = \mathbb{E}[\exp(-\Gamma \lambda_d(K))]$.

The distribution $\mathbf{P}$ and reduced Palm distribution $\mathbf{P}^{!x}$ of $\mathbf{X}$ satisfy, for all events $A \in \mathcal{N}$,

$$\mathbf{P}(A) = \mathbb{E}[\pi_\Gamma(A)] \tag{19}$$

$$\mathbf{P}^{!x}(A) = \frac{\mathbb{E}[\Gamma \pi_\Gamma(A)]}{\mathbb{E}[\Gamma]} \tag{20}$$

where π_γ denotes the distribution of the uniform Poisson process with intensity γ.

To put it another way, let $F(t) = \mathbb{P}(\Gamma \le t)$ be the cumulative distribution function of the random intensity of the original process. Define the weighted c.d.f.

$$F_1(t) = \frac{1}{\mathbb{E}[\Gamma]} \int_0^t s \, F(ds).$$

Then $\mathbf{X}^{!x}$ is a mixed Poisson process whose random intensity has the weighted c.d.f. F_1.

An intuitive explanation for the last example is the following. Points of the process are generated with greater intensity when Γ is larger. Hence, by Bayes' Theorem, given that a point was observed to occur, the posterior probability distribution of Γ favours larger values of Γ.

Theorem 3.2 (Campbell-Mecke formula). *For any function $Y : S \times \Omega \mapsto \mathbb{R}_+$ that is integrable with respect to the Campbell measure,*

$$\mathbb{E} \sum_{x \in \mathbf{X}} Y(x) = \int_S \mathbb{E}^x[Y(x)] \, \nu(dx) \tag{21}$$

In particular, if $Y(x) = f(x, \mathbf{X})$, that is, $Y(x, \omega) = f(x, \mathbf{X}(\omega))$, we get

$$\mathbb{E} \left[\sum_{x \in \mathbf{X}} f(x, \mathbf{X}) \right] = \int_S \mathbb{E}^x[f(x, \mathbf{X})] \, \nu(dx). \tag{22}$$

Example 3.7 (dependent thinning). We shall determine the intensity of Matérn's Model I, which was described in Section 1.7. Let $\mathbf{X}$ denote the original Poisson process, of intensity β in $\mathbb{R}^2$, and $\mathbf{Y}$ the thinned process obtained by deleting any point $x \in \mathbf{X}$ such that $\text{dist}(x, \mathbf{X} \setminus x) \le r$, that is, deleting any point which has a neighbour closer than r units. For any $B \subset \mathbb{R}^d$ let

$$f(x, \mathbf{X}) = \mathbf{1}\{x \in B\}\mathbf{1}\{\text{dist}(x, \mathbf{X} \setminus x) \le r\}$$
$$= \mathbf{1}\{x \in B\}\mathbf{1}\{x \in \mathbf{Y}\}.$$

Since $\mathbf{X}$ is a Poisson process we have

$$\mathbb{P}^x(\text{dist}(x, \mathbf{X} \setminus x) \le r) = \mathbb{P}(\text{dist}(x, \mathbf{X}) \le r) = 1 - \exp\{-\beta\pi r^2\}.$$

Hence

$$\mathbb{E}\left[n(\mathbf{Y} \cap B)\right] = \mathbb{E} \sum_{x \in \mathbf{X}} f(x, \mathbf{X})$$

$$= \beta \int_{\mathbb{R}^2} \mathbb{E}^x[f(x, \mathbf{X})] \, dx$$

$$= \beta \lambda_2(B)(1 - \exp\{-\beta \pi r^2\}).$$

It follows that $\mathbf{Y}$ has intensity $\beta(1 - \exp\{-\beta \pi r^2\})$.

Example 3.8 (Boundary length of Boolean model). Consider the union

$$Z = \bigcup_{x \in \mathbf{X}} b(x, r)$$

where $\mathbf{X}$ is a homogeneous Poisson process of intensity β in $\mathbb{R}^2$, and $r > 0$ is fixed. We want to find

$$\mathbb{E}\left[\text{length}(W \cap \partial Z)\right]$$

where ∂ denotes boundary. Write

$$\text{length}(W \cap \partial Z) = \sum_{x \in \mathbf{X}} Y(x)$$

where

$$Y(x) = \text{length}(W \cap \partial b(x, r) \setminus Z_{-x})$$

and

$$Z_{-x} = \bigcup_{y \in \mathbf{X} \setminus \{x\}} b(x, r).$$

Under the Palm probability measure $\mathbb{P}^x$, this random set Z_{-x} is a Boolean model with the same distribution as Z. Hence

$$\mathbb{E}^x[Y(x)] = (1 - p) \, \text{length}(W \cap \partial b(x, r))$$

where $p = 1 - \exp(-\beta \pi r^2)$ is the coverage probability of Z. Hence by Campbell-Mecke

$$\mathbb{E}\left[\text{length}(W \cap \partial Z)\right] = \int_{\mathbb{R}^2} (1 - p) \, \text{length}(W \cap \partial b(x, r))\beta \, dx$$

$$= 2\pi \beta r \exp(-\beta \pi r^2)\lambda_2(W).$$

3.3 Palm Distribution for Stationary Processes

In the case of a stationary point process, the Palm distributions $\mathbf{P}^x$ at different locations x are equivalent under translation.

Lemma 3.1. *If* $\mathbf{X}$ *is a stationary point process in* $\mathbb{R}^d$*, then*

$$\mathbf{X}^x \overset{d}{=} \mathbf{X}^0 + x$$

where $\mathbf{X}^x$ *again denotes a process governed by the Palm probability measure* $\mathbb{P}^x$.

More formally, let T_x denote the effect of translation by a vector $x \in \mathbb{R}^d$ on a counting measure N,

$$T_x N(B) = N(B - x), \qquad B \subset \mathbb{R}^d$$

and correspondingly for events $E \in \mathcal{N}$

$$T_x E = \{N \in \mathsf{N} : T_x N \in E\}$$

and for any point process distribution $\mathbf{Q}$ define $T_x \mathbf{Q}$ by

$$T_x \mathbf{Q}(E) = Q(T_x E), \qquad E \in \mathcal{N}.$$

Then Lemma 3.1 states that

$$\mathbf{P}^x_{\mathbf{X}} = T_x \, \mathbf{P}^0_{\mathbf{X}} \tag{23}$$

for any stationary point process $\mathbf{X}$.

Proof. Apply the Campbell-Mecke formula to functions of the form

$$f(x, \mathbf{X}) = \mathbf{1}\{x \in B\}\mathbf{1}\{\mathbf{X} - x \in A\}$$

where $A \in \mathcal{N}$ is an event, $B \subset \mathbb{R}^d$, and $\mathbf{X} - x = \mathbf{X} + (-x)$ is the result of shifting $\mathbf{X}$ by the vector $-x$. This yields

$$\mathbb{E}\left[\sum_{x \in \mathbf{X} \cap B} \mathbf{1}\{\mathbf{X} - x \in A\}\right] = \beta \int_B \mathbb{P}^x(\mathbf{X} - x \in A) \, \mathrm{d}x.$$

Since $\mathbf{X}$ is stationary, $\mathbf{X}$ has the same distribution as $\mathbf{X} + v$ for any vector v, so

$$\mathbb{E}\left[\sum_{x \in \mathbf{X} \cap B} \mathbf{1}\{\mathbf{X} - x \in A\}\right] = \mathbb{E}\left[\sum_{x \in (\mathbf{X}+v) \cap B} \mathbf{1}\{(\mathbf{X} + v) - x \in A\}\right]$$

$$= \mathbb{E}\left[\sum_{x \in \mathbf{X} \cap T_{-v} B} \mathbf{1}\{\mathbf{X} - x \in A\}\right].$$

Thus

$$\beta \int_B \mathbb{P}^x(\mathbf{X} - x \in A) \, \mathrm{d}x = \beta \int_{T_{-v} B} \mathbb{P}^x(\mathbf{X} - x \in A) \, \mathrm{d}x$$

which implies that for all B

$$\int_B \mathbb{P}^x(\mathbf{X} - x \in A)\, \mathrm{d}x = c\,\lambda_2(B)$$

for some constant c, and hence that $\mathbb{P}^x(\mathbf{X} - x \in A)$ is constant. This proves (23). $\qquad\square$

One way to interpret this result is to construct a marked point process $\mathbf{Y}$ on $\mathbb{R}^d$ with marks in N by attaching to each point $x \in \mathbf{X}$ the mark $\mathbf{X} - x$. That is, the mark attached to the point x is a copy of the entire realisation of the point process, translated so that x is shifted to the origin. The result shows that $\mathbf{Y}$ is a *stationary* marked point process. Hence the intensity measure of $\mathbf{Y}$ factorises,

$$\nu(B \times A) = \beta\,\lambda_d(B)\,Q(A)$$

for $B \subset \mathbb{R}^d$, $A \in \mathcal{N}$ where Q is the **mark distribution**. Clearly Q can be interpreted as the Palm distribution given there is a point at 0. That is, $\mathbf{P}^0 = Q$, and $\mathbf{P}^x = T_{-x}Q$.

This gives us a direct interpretation of the Palm distribution $\mathbf{P}^0$ (but not the Palm probability measure $\mathbb{P}^0$) for a stationary point process in $\mathbb{R}^d$. We have

$$\mathbb{E}\left[\sum_{x \in B} \mathbf{1}\{\mathbf{X} - x \in A\}\right] = \beta\,\lambda_d(B)\,\mathbf{P}^0(A) \qquad (24)$$

for $B \subset \mathbb{R}^d$, $A \in \mathcal{N}$. Thus

$$\mathbf{P}^0(A) = \frac{\mathbb{E}\left[\sum_{x \in B} \mathbf{1}\{\mathbf{X} - x \in A\}\right]}{\mathbb{E}N(B)} \qquad (25)$$

for all $B \subset \mathbb{R}^d$ such that $0 < \lambda_d(B) < \infty$. On the right side of (25), the denominator is the expected number of terms in the numerator, so we can interpret $\mathbf{P}^0(A)$ as the 'average' fraction of points x satisfying $\mathbf{X} - x \in A$.

3.4 Nearest Neighbour Function

In Definition 1.9 we defined the empty space function F of a stationary point process. The function F can be estimated from data, and provides a simple summary of the process. It can be useful in statistical analysis of point patterns.

A related concept is the nearest neighbour distance distribution.

Definition 3.2. *Let $\mathbf{X}$ be a stationary point process in $\mathbb{R}^d$. The **nearest neighbour function** G is the cumulative distribution function of the distance*

$$R' = \mathrm{dist}(x, \mathbf{X} \setminus x)$$

from a typical point $x \in \mathbf{X}$ to the nearest other point of $\mathbf{X}$. That is

$$G(r) = \mathbb{P}^x(\text{dist}(x, \mathbf{X} \setminus x) \le r)$$
$$= \mathbb{P}^x(N(b(x, r) \setminus x) > 0).$$

By stationarity, this does not depend on x.

Example 3.9. For a stationary Poisson process in $\mathbb{R}^d$, since $\mathbf{X}^x \equiv \mathbf{X} \cup \{x\}$, we have

$$G(r) = \mathbb{P}^x(\text{dist}(x, \mathbf{X} \setminus x) \le r)$$
$$= \mathbb{P}(\text{dist}(x, \mathbf{X}) \le r)$$
$$= 1 - \exp(-\beta \kappa_d r^d).$$

In this case $G(r) \equiv F(r)$.

Estimation of the function G from observed point pattern data is hampered by edge effects, similar to those affecting the estimation of F (see Section 1.10). The basic identity for estimation is (25), or more specifically

$$G(r) = \frac{\mathbb{E}\left[\sum_{x \in B} \mathbf{1}\{\text{dist}(x, \mathbf{X} \setminus x) \le r\}\right]}{\mathbb{E}N(B)}. \tag{26}$$

The simplest strategy for avoiding edge effects is an adaptation of the border method. When estimating $G(r)$ we set $B = W_{-r}$ in equation (26) so that the quantities on the right hand side of (26) are observable. This yields an estimator

$$\widehat{G}_b(r) = \frac{\sum_{x \in W_{-r}} \mathbf{1}\{\text{dist}(x, \mathbf{x} \setminus x) \le r\}}{n(\mathbf{x} \cap W_{-r})}. \tag{27}$$

If we write for each $x_i \in \mathbf{x}$

$$d_i = \text{dist}(x_i, \mathbf{x} \setminus x_i)$$
$$b_i = \text{dist}(x_i, \partial W)$$

so that d_i is the observed nearest-neighbour distance and b_i is the distance to the boundary of the observation window, then the estimator can be rewritten

$$\widehat{G}_b(r) = \frac{\sum_i \mathbf{1}\{d_i \le r, \, b_i \ge r\}}{\sum_i \mathbf{1}\{b_i \ge r\}}. \tag{28}$$

Further discussion of edge corrections can be found in [42, 2].

3.5 Conditional Intensity

Return for a moment to the heuristic definition of the Palm probability $\mathbb{P}^x(A)$ as the limit of $\mathbb{P}(A \mid N(U) > 0)$ as $U \downarrow \{x\}$, where U is an open neighbourhood of x in $\mathbb{R}^d$, and A is an event in $\mathcal{A}$. Applying Bayes' Theorem

$$\mathbb{P}\left(N(U) > 0 \mid A\right) = \frac{\mathbb{P}(N(U) > 0)}{\mathbb{P}(A)}\mathbb{P}\left(A \mid N(U) > 0\right)$$

so that, as $U \downarrow \{x\}$,

$$\frac{\mathbb{P}\left(N(U) > 0 \mid A\right)}{\mathbb{P}(N(U) > 0)} \rightarrow \frac{\mathbb{P}^x(A)}{\mathbb{P}(A)}.$$

Suppose $\mathbf{X}$ has an intensity function $\beta(u)$, $u \in \mathbb{R}^d$ which is continuous at x. Then asymptotically

$$\mathbb{P}(N(U) > 0) \sim \mathbb{E}[N(U)] = \int_U \beta(u) \, du \sim \beta(x)\lambda_d(U)$$

so that

$$\frac{\mathbb{P}\left(N(U) > 0 \mid A\right)}{\lambda_d(U)} \rightarrow \beta(x)\frac{\mathbb{P}^x(A)}{\mathbb{P}(A)}. \tag{29}$$

Since we can also write

$$\frac{\mathbb{P}(N(U) > 0)}{\lambda_d(U)} \rightarrow \beta(x),$$

then (29) can be interpreted as a conditional analogue of the intensity $\beta(x)$ given the event A.

This motivates the following definitions.

Definition 3.3. *Let* $\mathbf{X}$ *be a point process on a space* S, *The* **reduced Campbell measure** *of* $\mathbf{X}$ *is the measure* $C^!$ *on* $S \times \mathsf{N}$ *such that*

$$C^![B \times A] = \mathbb{E}\left[\sum_{x \in \mathbf{X}} 1\{x \in B\}1\{\mathbf{X} \setminus x \in A\}\right]$$

for $B \subset \mathbb{R}^d$ *and* $A \in \mathcal{N}$.

Definition 3.4. *Let* $\mathbf{X}$ *be a point process on* $\mathbb{R}^d$, *and suppose its reduced Campbell measure* $C^!$ *is absolutely continuous with respect to* $\lambda_d \otimes \mathbf{P}$ *(where* $\mathbf{P}$ *is the distribution of* $\mathbf{X}$*).*

Then the Radon-Nikodým derivative $\beta^* : \mathbb{R}^d \times \mathcal{N} \rightarrow \mathbb{R}_+$ *of* $C^!$ *with respect to* $\lambda_d \otimes \mathbf{P}$ *is called the* **conditional intensity** *of* $\mathbf{X}$. *It is defined to satisfy*

$$C^![B \times A] = \int_B \mathbb{E}\left[\beta^*(u, \mathbf{X})1\{\mathbf{X} \in A\}\right] \, du \tag{30}$$

for $B \subset \mathbb{R}^d$ *and* $A \in \mathcal{N}$.

If $\mathbf{X}$ has a conditional intensity then, by extension of the last equation, for any integrable $g : \mathbb{R}^d \times \mathsf{N} \rightarrow \mathbb{R}_+$,

$$\mathbb{E}\left[\sum_{x \in \mathbf{X}} g(x, \mathbf{X} \setminus x)\right] = \int_{\mathbb{R}^d} \mathbb{E}\left[\beta^*(x, \mathbf{X})g(x, \mathbf{X})\right] \mathrm{d}x. \tag{31}$$

If $\mathbf{X}$ has an intensity function $\beta(u)$, $u \in \mathbb{R}^d$, then the Campbell-Mecke formula gives

$$\mathbb{E}\left[\sum_{x \in \mathbf{X}} g(x, \mathbf{X} \setminus x)\right] = \int_{\mathbb{R}^d} \mathbb{E}^{!x}[g(x, \mathbf{X})]\beta(x) \, \mathrm{d}x \tag{32}$$

writing $\mathbb{E}^{!x}$ for the expectation with respect to $\mathbf{P}^{!x}$. Comparing the right sides of (31) and (32) shows that, for almost all $x \in \mathbb{R}^d$,

$$\mathbb{E}^{!x}[g(x, \mathbf{X})] = \mathbb{E}\left[\frac{\beta^*(x, \mathbf{X})}{\beta(x)} g(x, \mathbf{X})\right]. \tag{33}$$

Thus, $\beta^*(x, \mathbf{X})/\beta(x)$ is the Radon-Nikodým density of $\mathbf{P}^{!x}$ with respect to $\mathbf{P}$. In particular taking $g \equiv 1$

$$\beta(x) = \mathbb{E}[\beta^*(x, X)]. \tag{34}$$

Example 3.10. If $\mathbf{X}$ is a Poisson process on $\mathbb{R}^d$ with intensity function $\beta(x)$, then we have $\mathbf{P}^{!x} = \mathbf{P}$ for all x, so $\beta^*(x, \mathbf{X})/\beta(x)$ is identically equal to 1, and the conditional intensity is $\beta^*(x, \mathbf{X}) = \beta(x)$.

Our heuristic argument says that

$$\beta^*(x, \mathbf{X}) \, \mathrm{d}x = \mathbb{P}\left(N(\mathrm{d}x) > 0 \mid \mathbf{X} \setminus x\right);$$

that is, roughly speaking, $\beta^*(x, \mathbf{X}) \, \mathrm{d}x$ is the conditional probability that there will be a point of $\mathbf{X}$ in an infinitesimal neighbourhood of x, given the location of all points of $\mathbf{X}$ outside this neighbourhood.

Example 3.11. The binomial process $\mathbf{Y}^{(n)}$ consists of n independent random points, uniformly distributed in a domain W. We saw above that the reduced Palm distribution of $\mathbf{Y}^{(n)}$ is identical to the distribution of $\mathbf{Y}^{(n-1)}$. In this case, $\mathbf{P}^{!x}$ and $\mathbf{P}$ are mutually singular (since, for example, the event that there are exactly n points in the process has probability 1 under $\mathbf{P}$ and has probability 0 under $\mathbf{P}^{!x}$). Hence, this process does not have a conditional intensity.

Example 3.12 (Mixture of binomial processes). Suppose $\mathbf{X}$ consists of a random number N of points, where $\mathbb{P}(N = n) = p(n)$, and that given $N = n$, the points are independent and uniformly distributed in a domain W. The intensity of $\mathbf{X}$ is $\beta(u) = \mathbb{E}[N]/\lambda_d(W)$ for $u \in W$, and $\beta(u) = 0$ for $u \notin W$. If Q_n denotes the distribution of the binomial process with n points, then the distribution of $\mathbf{X}$ is

$$\mathbf{P}(A) = \sum_{n=0}^{\infty} p(n)Q_n(A).$$

It is fairly easy to show that the distribution of $\mathbf{X}^{!x}$ is

$$\mathbf{P}^{!x}(A) = \frac{1}{\mathbb{E}[N]} \sum_{n=0}^{\infty} np(n)Q_{n-1}(A).$$

We saw above that Q_n and Q_m are mutually singular for $n \neq m$. Assume that $p(n) > 0$ implies $p(n-1) > 0$ for any n. Then we must have

$$\frac{\beta^*(x, \mathbf{X})}{\beta(x)} = \frac{d\mathbf{P}^{!x}}{d\mathbf{P}}(\mathbf{X}) = \frac{r(n(\mathbf{X}))}{\mathbb{E}[N]}$$

where

$$r(n) = \frac{(n+1)p(n+1)}{p(n)}.$$

Hence

$$\beta^*(x, \mathbf{X}) = \frac{r(n(\mathbf{X}))}{\lambda_d(W)}.$$

For example, if $\mathbf{X}$ is a uniform Poisson process with intensity α in W, then $N \sim \mathsf{Poisson}(\alpha\lambda_d(W))$, and we get $r(n) = \alpha\lambda_d(W)$, yielding $\beta^*(x, \mathbf{X}) = \alpha$.

Example 3.13. The randomly translated grid in $\mathbb{R}^2$ was studied in Example 2.3. Intuitively, if we know that there is a point of the grid at the location x, this determines the position of the entire grid. The Palm distribution $\mathbf{P}^x$ for this process is completely deterministic: with probability one, $\mathbf{X}^x$ consists of points at the locations $x + (ks, ms)$ for all integers k, m. This can be proved using (25).

It follows that $\mathbf{P}^x$ is not absolutely continuous with respect to $\mathbf{P}$, so this process does not possess a conditional intensity.

3.6 *J*-function

An interesting combination of the empty space function F and the nearest neighbour function G is the following [47].

Definition 3.5. *Let $\mathbf{X}$ be a stationary point process in $\mathbb{R}^d$. The J-function of $\mathbf{X}$ is*

$$J(r) = \frac{1 - G(r)}{1 - F(r)}$$

for all $r \geq 0$ such that $F(r) < 1$.

For a uniform Poisson process, we know that $F(r) \equiv G(r)$ and hence $J(r) \equiv 1$. The J-function of a stationary process can be written explicitly in terms of the conditional intensity:

$$J(r) = \frac{\mathbb{P}^0(\text{dist}(0, \mathbf{X} \setminus 0) > r)}{\mathbb{P}(\text{dist}(0, \mathbf{X}) > r)}$$

$$= \frac{\mathbf{P}^{!0}(\text{dist}(0, \mathbf{X}) > r)}{\mathbb{P}(\text{dist}(0, \mathbf{X}) > r)}$$

$$= \frac{\mathbb{E}\left[\frac{\beta^*(0, \mathbf{X})}{\beta(0)} \mathbf{1}\{\text{dist}(0, \mathbf{X}) > r\}\right]}{\mathbb{P}(\text{dist}(0, \mathbf{X}) > r)}$$

$$= \mathbb{E}\left[\frac{\beta^*(0, \mathbf{X})}{\beta(0)} \mid \text{dist}(0, \mathbf{X}) > r\right].$$

This representation can often be evaluated, while F and G often cannot be evaluated explicitly.

The J-function has good properties with respect to many operations on point processes. For example, suppose $\mathbf{X}$ and $\mathbf{Y}$ are independent stationary point processes, with intensities $\alpha_{\mathbf{X}}, \alpha_{\mathbf{Y}}$ and J-functions $J_{\mathbf{X}}, J_{\mathbf{Y}}$. Then the superposition $\mathbf{X} \cup \mathbf{Y}$ has J-function

$$J_{\mathbf{X} \cup \mathbf{Y}}(r) = \frac{\alpha_{\mathbf{X}}}{\alpha_{\mathbf{X}} + \alpha_{\mathbf{Y}}} J_{\mathbf{X}}(r) + \frac{\alpha_{\mathbf{Y}}}{\alpha_{\mathbf{X}} + \alpha_{\mathbf{Y}}} J_{\mathbf{Y}}(r).$$

3.7 Exercises

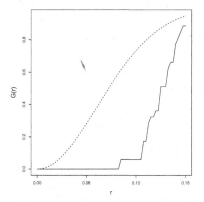 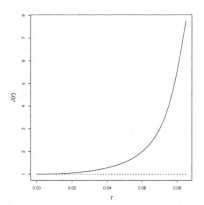

Fig. 32. Estimates of the nearest neighbour function G (Left) and the J-function (Right) for the `cells` dataset. The functions are plotted (as solid lines) against the distance argument r, together with the theoretical expected value for a Poisson process (dotted lines).

Again we use the package `spatstat` described in Sections 1.11 and 2.7. The commands `Gest` and `Jest` compute estimates of the nearest neighbour function G and the J-function, respectively, from spatial point pattern data.

For the `cells` data introduced in Section 1.11, Figure 32 shows an analysis using the G and J functions, produced by the following code.

```
library(spatstat)
data(cells)
G <- Gest(cells)
plot(G)
plot(Jest(cells))
```

Inspection of the plot of $G(r)$ shows that the `cells` data have strong inhibition between points: no point has a nearest neighbour closer than 0.08 units.

4 Modelling and Statistical Inference

This lecture is concerned with statistical models for point pattern data. Point processes in a bounded region of space are considered: in this case it is possible to construct many point process models by writing down their probability densities. We then describe techniques for fitting such models to data [16].

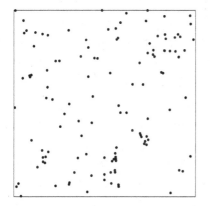 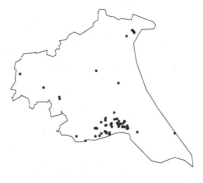

Fig. 33. Examples of point pattern data. *Left:* Locations of 126 pine saplings in a Finnish forest (kindly supplied by S. Kellomaki and A. Penttinen). *Right:* Home address locations of 62 cases of childhood leukaemia and lymphoma in North Humberside, England (from [14]).

4.1 Motivation

Suppose we have observed a point pattern dataset

$$\mathbf{x} = \{x_1, \ldots, x_n\}, \quad n \geq 0, \quad x_i \in W$$

consisting of a finite number of points in a bounded window $W \subset \mathbb{R}^d$ where typically $d = 2, 3$ or 4.

After an initial exploratory phase of data analysis, in which we use tools such as the F, G and K functions to gain qualitative information about the spatial pattern, we may feel confident enough to proceed to a more formal style of statistical analysis. This involves formulating a statistical model, fitting the model to the data, and deciding whether the model is a good fit to the data.

A simple example of a statistical model is the uniform Poisson process with unknown intensity β. This often serves as a reference model for 'complete spatial randomness' [16], since the points of a uniform Poisson process are stochastically independent of each other, and are uniformly distributed over the region of observation. Alternative models may describe various kinds of departures from a completely random pattern.

4.2 Parametric Modelling and Inference

In this lecture we will adopt the approach of *parametric* statistical modelling. Our statistical model will be a spatial point process $\mathbf{X}$. Our model will completely specify the probability distribution of $\mathbf{X}$, except for the unknown value of a parameter θ. (An example would be the uniform Poisson point process with unknown intensity β). The parameter θ ranges over some set Θ of valid parameter values, typically a subset of $\mathbb{R}^k$ for some $k \geq 1$. Write $\mathbb{P}_\theta$ for the probability distribution, and $\mathbb{E}_\theta$ for the expectation operator, under the model. Then we shall estimate θ from the data $\mathbf{x}$ giving an estimate $\widehat{\theta} = \widehat{\theta}(\mathbf{x})$.

Three generic methods for parameter estimation are surveyed below.

Method of Moments

In the Method of Moments, we choose a statistic $T(\mathbf{x})$, and take our estimate $\widehat{\theta}$ to be the solution of

$$\mathbb{E}_\theta[T(\mathbf{X})] = T(\mathbf{x}). \tag{35}$$

That is, we choose θ so that the theoretical expected value $\mathbb{E}_\theta[T(\mathbf{X})]$ matches the observed value $T(\mathbf{x})$.

Example 4.1. Suppose the model is the uniform Poisson process with intensity β. This can be fitted by the method of moments. Let the statistic T be $T(\mathbf{x}) = n(\mathbf{x})$, the number of points in $\mathbf{x}$. Then we have $\mathbb{E}_\beta[T(\mathbf{X})] = \beta \, \lambda_d(W)$. The solution to (35) is

$$\widehat{\beta} = n(\mathbf{x})/\lambda_d(W).$$

The method of moments works well when we have an analytic expression for $\mathbb{E}_\theta[T(\mathbf{X})]$ as a function of θ, when the equation (35) is guaranteed to have a solution, and when the solution is always unique.

A disadvantage of the method of moments in applications to spatial statistics is that the moments of interesting point processes are often difficult to calculate. Usually we have to resort to simulation. That is, for each possible value of θ, we generate a large number of simulated realisations $\mathbf{x}^{(1)}, \ldots, \mathbf{x}^{(N)}$ from the model with parameter θ, and estimate $\mathbb{E}_\theta[T(\mathbf{X})]$ by the sample mean $\overline{T} = \frac{1}{N} \sum_i T(\mathbf{x}^{(i)})$. Since the simulations have to be performed again for each new value of θ under consideration, this computation can be expensive.

As we consider more complex models involving several parameters, the statistic T must also become more complex. If θ is a k-dimensional vector, then the values of $T(\mathbf{x})$ must also be k-dimensional (or higher) in order that the solution of (35) may be unique.

Minimum Contrast Estimation

In the Method of Minimum Contrast, our estimate is

$$\widehat{\theta} = \mathrm{argmin}_\theta \ D(T(\mathbf{x}), \mathbb{E}_\theta[T(\mathbf{X})]). \tag{36}$$

where $T(\mathbf{x})$ is a chosen statistic, as above, and D is a metric. That is, $\widehat{\theta}$ is the parameter value giving an expectation $\mathbb{E}_\theta[T(\mathbf{X})]$ that is closest (as measured by D) to the observed value $T(\mathbf{x})$. This usually avoids difficulties arising when the method of moments estimating equation (35) does not have a solution. It also allows us to make use of a statistic $T(\mathbf{x})$ which takes values in an arbitrary space.

Diggle & Gratton [18] proposed the method of minimum contrast in combination with the K-function. Suppose that the K-function for the model is known analytically as a function of θ, say $K_\theta(r)$. Given point pattern data $\mathbf{x}$, we may first estimate the K-function using the non-parametric estimator $\widehat{K}(t)$ described in Section 2.6. Then we choose θ to minimise

$$\int_a^b |K_\theta(r) - \widehat{K}(r)|^p \, \mathrm{d}r$$

where $0 \le a < b$ and $p > 0$ are chosen values. This is the minimum contrast method based on $T(\mathbf{x}) = (\widehat{K}(r), a \le r \le b)$, a function-valued statistic, and the metric D is an L^p distance.

Example 4.2. The Thomas process is a cluster process (Section 1.7) formed by taking a Poisson process of 'parents' with intensity α, and replacing each parent x by several offspring, where the number of offspring is Poisson with mean μ, and the offspring of a parent x are i.i.d. random points $y_i = x + e_i$ where the displacement vector e_i has coordinates which are independent Normal $N(0, \sigma^2)$ random variables. The K-function for this process is known analytically:

$$K(r) = \pi r^2 + \frac{1}{\alpha} \left(1 - \exp \left(-\frac{r^2}{4\sigma^2} \right) \right).$$

The intensity of the process is $\beta = \alpha \mu$. Diggle & Gratton [18] proposed fitting this model by the method of minimum contrast. See also [16]. Given point pattern data $\mathbf{x}$ in a window W, we first estimate the K-function using the non-parametric estimator $\widehat{K}(t)$ described in Section 2.6. The intensity $\beta = \alpha\mu$ is also estimated by the method of moments, $\widehat{\beta} = n(\mathbf{x})/\lambda_2(W)$. Then we find the values (α, μ, r) which minimise

$$\int_a^b |K_\theta(t) - \widehat{K}(t)|^p \, \mathrm{d}t$$

subject to the constraint $\alpha\mu = n(\mathbf{x})/\lambda_2(W)$.

However, there are very few point processes for which we have analytic expressions for K_θ. Usually one would estimate K_θ by simulation.

It is well known that the K function does not completely characterise a stationary point process. A counterexample is given in [5]. Thus, if we are not careful, our model may be **unidentifiable** from the K-function, in the sense that there are two distinct parameter values θ_1, θ_2 such that $K_{\theta_1} \equiv K_{\theta_2}$.

The method of minimum contrast avoids some of the problems of the method of moments. If D has certain convexity properties, then the minimisation (36) has a unique solution. However the statistical properties of the solution are not well understood in general. Additionally, the numerical behaviour of the algorithm used to find the minimum in (36) may also cause difficulties.

Maximum Likelihood

In the Method of Maximum Likelihood, first articulated by R.A. Fisher, we define the likelihood function

$$L(\theta; \mathbf{x}) = \frac{\mathrm{d}P_\theta}{\mathrm{d}\nu}(\mathbf{x})$$

where ν is some reference measure. After observing data $\mathbf{x}$, we treat L as a function of θ only, and choose $\widehat{\theta}$ to maximise the likelihood:

$$\widehat{\theta} = \mathsf{argmax}_\theta L(\theta; \mathbf{x})$$

is the value of θ for which the likelihood is maximised. There may be difficulties with non-unique maxima.

Maximum likelihood has some good statistical properties, including asymptotic (large-sample) normality and optimality. Hence a major goal in spatial statistics is to apply maximum likelihood methods.

The likelihood is a probability density. Hence we need to *define/construct point process distributions in terms of their probability densities.*

4.3 Finite Point Processes

A point process $\mathbf{X}$ on S with $N(S) < \infty$ a.s. is called a **finite point process**. The binomial process (Section 1.3) is a simple example in which the total number of points is fixed. In general, the total number of points $N(S)$ is a random variable. The distribution of the process can be specified by giving the probability distribution of $N(S)$, and given $N(S) = n$, the conditional joint distribution of the n points.

Example 4.3. Consider a Poisson process with intensity measure μ that is totally finite ($\mu(S) < \infty$). This is equivalent to choosing a random number $K \sim \mathsf{Poisson}(\mu(S))$, then given $K = k$, generating k i.i.d. random points with common distribution $Q(B) = \mu(B)/\mu(S)$.

Realisations of a finite point process $\mathbf{X}$ belong to the space

$$\mathsf{N}^f = \{N \in \mathsf{N} : N(S) < \infty\}$$

of totally finite, simple, counting measures on S. This may be decomposed into subspaces according to the total number of points:

$$\mathsf{N}^f = \mathsf{N}_0 \cup \mathsf{N}_1 \cup \mathsf{N}_2 \cup \ldots$$

where for each $k = 0, 1, 2, \ldots$

$$\mathsf{N}_k = \{N \in \mathsf{N} : N(S) = k\}$$

is the set of all counting measures with total mass k, that is, effectively the set of all configurations of k points. This can be represented more explicitly by introducing the space of ordered k-tuples

$$S^{!k} = \{(x_1, \ldots, x_k) : x_i \in S, \ x_i \neq x_j \text{ for all } i \neq j\}.$$

Define a mapping $I_k : S^{!k} \to \mathsf{N}_k$ by

$$I_k(x_1, \ldots, x_k) = \delta_{x_1} + \ldots + \delta_{x_k}$$

This gives

$$\mathsf{N}_k \equiv S^{!k} / \sim$$

where $\sim$ is the equivalence relation under permutation, i.e.

$$(x_1, \ldots, x_k) \sim (y_1, \ldots, y_k) \quad \Leftrightarrow \quad \{x_1, \ldots, x_k\} = \{y_1, \ldots, y_k\}$$

Using this representation we can give explicit formulae for point process distributions.

Example 4.4 (binomial process). Let $X_1, \ldots, X_n$ be i.i.d. random points uniformly distributed in W. Set $\mathbf{X} = I_n(X_1, \ldots, X_n)$. The distribution of $\mathbf{X}$ is the probability measure $P_{\mathbf{X}}$ on N defined by

$$P_{\mathbf{X}}(A) = \mathbb{P}(I_n(X_1, \ldots, X_n) \in A)$$
$$= \frac{1}{|W|^n} \int_W \cdots \int_W \mathbf{1}\{I_n(x_1, \ldots, x_n) \in A\} \, \mathrm{d}x_1 \ldots \, \mathrm{d}x_n$$

Example 4.5 (finite Poisson process). Let $\mathbf{X}$ be the Poisson process on S with totally finite intensity measure μ. We know that $N(S) \sim \mathsf{Poisson}(\mu(S))$ and that, given $N(S) = n$, the distribution of $\mathbf{X}$ is that of a binomial process of n points i.i.d. with common distribution $Q(B) = \mu(B)/\mu(S)$. Thus

$$P_{\mathbf{X}}(A) = \sum_{n=0}^{n} \mathbb{P}(N(S) = n)\mathbb{P}(I_n(X_1, \ldots, X_n) \in A)$$
$$= \sum_{n=0}^{\infty} e^{-\mu(S)} \frac{\mu(S)^n}{n!} \int_S \cdots \int_S \mathbf{1}\{I_n(x_1, \ldots, x_n) \in A\} \, Q(\mathrm{d}x_1) \ldots Q(\mathrm{d}x_n)$$
$$= e^{-\mu(S)} \sum_{n=0}^{\infty} \frac{1}{n!} \int_S \cdots \int_S \mathbf{1}\{I_n(x_1, \ldots, x_n) \in A\} \, \mu(\mathrm{d}x_1) \ldots \mu(\mathrm{d}x_n). \tag{37}$$

The term for $n = 0$ in the sum should be interpreted as $\mathbf{1}\{\mathbf{0} \in A\}$ (where $\mathbf{0}$ is the zero measure, corresponding to an empty configuration.)

4.4 Point Process Densities

Henceforth we fix a standard measure μ on S. Typically μ is Lebesgue measure on a bounded set W in $\mathbb{R}^d$. Let π_μ denote the distribution of the Poisson process with intensity measure μ.

Definition 4.1. *Let* $f : \mathsf{N}^f \to \mathbb{R}_+$ *be a measurable function satisfying* $\int_{\mathsf{N}} f(\mathbf{x}) \, \pi_\mu(\mathrm{d}\mathbf{x}) = 1$. *Define*

$$\mathbf{P}(A) = \int_A f(\mathbf{x}) \, \pi_\mu(\mathrm{d}\mathbf{x}).$$

for any event $A \in \mathcal{N}$. *Then* $\mathbf{P}$ *is a point process distribution in the sense of Definition 1.5. The function* f *is said to be the* **probability density** *of the point process with distribution* $\mathbf{P}$.

Lemma 4.1. *For a point process* $\mathbf{X}$ *with probability density* f *we have*

$$\mathbb{P}(X \in A) = e^{-\mu(S)} \sum_{n=0}^{\infty} \frac{1}{n!} \int_S \cdots \int_S \mathbf{1}\{I_n(x_1, \ldots, x_n) \in A\} \tag{38}$$
$$f(I_n(x_1, \ldots, x_n)) \, \mu(\mathrm{d}x_1) \ldots \mu(\mathrm{d}x_n)$$

for any event $A \in \mathcal{N}$, and

$$\mathbb{E}[g(\mathbf{X})] = e^{-\mu(S)} \sum_{n=0}^{\infty} \frac{1}{n!} \int_S \cdots \int_S g(I_n(x_1,\ldots,x_n)) \qquad (39)$$
$$f(I_n(x_1,\ldots,x_n))\, \mu(dx_1)\ldots\,\mu(dx_n)$$

for any integrable function $g : \mathsf{N} \to \mathbb{R}_+$.
  We can also rewrite these identities as

$$\mathbb{P}(X \in A) = \mathbb{E}\left[f(\Pi)1_A(\Pi)\right] \qquad (40)$$
$$\mathbb{E}[g(\mathbf{X})] = \mathbb{E}[g(\Pi)f(\Pi)] \qquad (41)$$

where Π is the Poisson process with intensity measure μ.

For some elementary point processes, it is possible to determine the probability density directly.

Example 4.6 (uniform Poisson process). Let $\beta > 0$. Set

$$f(\mathbf{x}) = \alpha\, \beta^{n(\mathbf{x})}$$

where α is a normalising constant and $n(\mathbf{x}) = \mathbf{x}(S) =$number of points in $\mathbf{x}$. Then

$$\mathbf{P}(A) = \alpha e^{-\mu(S)} \sum_{n=0}^{\infty} \frac{1}{n!} \int_S \cdots \int_S \mathbf{1}\{I_n(x_1,\ldots,x_n) \in A\}\beta^n\, \mu(dx_1)\ldots\,\mu(dx_n).$$

But this is the distribution of the Poisson process with intensity β. The normalising constant must be $\alpha = e^{(1-\beta)\mu(S)}$. Thus, the uniform Poisson process with intensity β has probability density

$$f(\mathbf{x}) = \beta^{n(\mathbf{x})}e^{(1-\beta)\mu(S)}. \qquad (42)$$

Example 4.7 (Hard core process). Fix $r > 0$. Let

$$H_n = \left\{(x_1,\ldots,x_n) \in S^{!n} : \|x_i - x_j\| \geq r \text{ for all } i \neq j\right\}$$

and

$$H = \bigcup_{n=0}^{\infty} I_n(H_n).$$

Thus H is the subset of N consisting of all point patterns $\mathbf{x}$ with the property that every pair of distinct points in $\mathbf{x}$ is at least r units apart. Now define the probability density

$$f(\mathbf{x}) = \alpha\mathbf{1}\{\mathbf{x} \in H\}$$

where α is the normalising constant. Then we have, for any event $A \in \mathcal{N}$,

$$\mathbf{P}(A) = \mathbb{E}\left[f(\Pi)\mathbf{1}\{\Pi \in A\}\right]$$
$$= \alpha\mathbb{E}\left[\mathbf{1}\{\Pi \in H\}\mathbf{1}\{\Pi \in A\}\right]$$
$$= \alpha\mathbb{P}(\Pi \in H \cap A).$$

It follows that

$$\alpha = 1/\mathbb{P}(\Pi \in H)$$

and $\mathbf{P}$ is the conditional distribution of the Poisson process Π given that $\Pi \in H$. In other words, the process $\mathbf{X}$ with probability density f is equivalent to a Poisson process conditioned on the event that there are no pairs of points closer than r units apart.

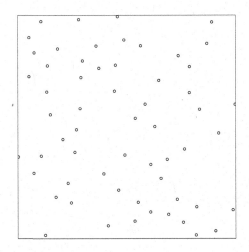

Fig. 34. Realisation of the hard core process with $\beta = 200$ and $r = 0.07$ in the unit square.

Figure 34 shows a realisation of the hard core process with $\beta = 200$ and $r = 0.07$ in the unit square. One simple way to generate such a picture is by the rejection method: we generate a sequence of realisations of the uniform Poisson process with intensity 200, and plot the first realisation which satisfies the constraint H. More efficient simulation methods are described in [33, 46].

4.5 Conditional Intensity

Consider a finite point process $\mathbf{X}$ in a compact set $W \subset \mathbb{R}^d$. Recall that the conditional intensity $\beta^*(u, \mathbf{X})$, if it exists, satisfies

$$\mathbb{E}\sum_{x\in\mathbf{X}} g(x, \mathbf{X} \setminus x) = \int_{\mathbb{R}^d} \mathbb{E}\left[\beta^*(u, \mathbf{X})g(u, \mathbf{X})\right] \, \mathrm{d}u \qquad (43)$$

for any integrable g.

Now suppose $\mathbf{X}$ has probability density $f(\mathbf{x})$ (with respect to the uniform Poisson process Π with intensity 1 on W). Then the expectation of any integrable function $h(\mathbf{X})$ may be written in the form (39) or (41). Applying this to both sides of (43) above, we get

$$\mathbb{E}\left[f(\Pi)\sum_{x\in\Pi}g(x,\Pi\setminus x)\right] = \int_{\mathbb{R}^d}\mathbb{E}\left[\beta^*(u,\Pi)f(\Pi)g(u,\Pi)\right]\,\mathrm{d}u. \qquad (44)$$

If we write

$$h(x,\mathbf{X}) = f(\mathbf{X}\cup\{x\})g(x,\mathbf{X}),$$

then the left side of (44) can be rewritten

$$\mathbb{E}\left[f(\Pi)\sum_{x\in\Pi}g(x,\Pi\setminus x)\right] = \mathbb{E}\left[\sum_{x\in\Pi}h(x,\Pi\setminus x)\right]$$

$$= \int_W \mathbb{E}[h(u,\mathbf{X})]\,\mathrm{d}u,$$

where the last line is obtained by applying equation (43) to the process Π, since the conditional intensity of Π is identically equal to 1 on W. Thus we get

$$\int_{\mathbb{R}^d}\mathbb{E}\left[\beta^*(u,\Pi)f(\Pi)g(u,\Pi)\right]\,\mathrm{d}u = \int_W\mathbb{E}\left[f(\Pi\cup\{u\})g(u,\Pi)\right]\,\mathrm{d}u$$

for all integrable functions g. It follows that

$$\beta^*(u,\Pi)f(\Pi) = f(\Pi\cup u)$$

almost surely, for almost all $u\in W$. Thus we have obtained the following result.

Theorem 4.1. *Let f be the probability density of a finite point process $\mathbf{X}$ in a bounded region W of $\mathbb{R}^d$. Assume that*

$$f(\mathbf{x}) > 0 \quad\Rightarrow\quad f(\mathbf{y}) > 0 \text{ for all } \mathbf{y}\subset\mathbf{x}.$$

Then the conditional intensity of $\mathbf{X}$ exists and equals

$$\beta^*(u,\mathbf{x}) = \frac{f(\mathbf{x}\cup u)}{f(\mathbf{x})} \qquad (45)$$

almost everywhere.

Example 4.8 (Uniform Poisson). The uniform Poisson process on W with intensity β has density

$$f(\mathbf{x}) = \alpha\beta^{n(\mathbf{x})}$$

where α is a certain normalising constant. Applying (45) we get

$$\beta^*(u, \mathbf{x}) = \beta$$

for $u \in W$.

Example 4.9 (Hard core process). The probability density of the hard core process (Example 4.7)

$$f(\mathbf{x}) = \alpha \mathbf{1}\{\mathbf{x} \in H\}$$

yields

$$\lambda(u, \mathbf{x}) = \mathbf{1}\{\mathbf{x} \cup u \in H\}.$$

Lemma 4.2. *Let $\mathbf{X}$ be a finite point process in a bounded region W in $\mathbb{R}^d$. Suppose that $\mathbf{X}$ has a probability density f and a conditional intensity β^*. Then f is completely determined by β^*.*

Proof. We may invert the relationship (45) by starting with the empty configuration $\emptyset$ and adding one point at a time:

$$f(\{x_1, \ldots, x_n\}) = f(\emptyset) \frac{f(\{x_1\})}{f(\emptyset)} \frac{f(\{x_1, x_2\})}{f(\{x_1\})} \cdots \frac{f(\{x_1, \ldots, x_n\})}{f(\{x_1, \ldots, x_{n-1}\})}$$
$$= f(\emptyset)\beta^*(x_1, \emptyset)\beta^*(x_2, \{x_1\}) \ldots \beta^*(x_n, \{x_1, \ldots, x_{n-1}\}).$$

If the values of β^* are known, then this determines f up to a constant $f(\emptyset)$, which is then determined by the normalisation of f. $\qquad\qquad\square$

It is often convenient to formulate a point process model in terms of its conditional intensity $\beta^*(u, \mathbf{x})$, rather than its probability density $f(\mathbf{x})$. The conditional intensity has a natural interpretation (in terms of conditional probability) which may be easier to understand than the density. Using the conditional intensity also eliminates the normalising constant needed for the probability density.

However, we are not free to choose the functional form of $\beta^*(u, \mathbf{x})$ at will. It must satisfy certain consistency relations. The next section describes a large class of models which turns out to characterise the most general functional form of $\beta^*(u, \mathbf{x})$.

4.6 Finite Gibbs Models

Definition 4.2. *A **finite Gibbs process** is a finite point process $\mathbf{X}$ with probability density $f(\mathbf{x})$ of the form*

$$f(\mathbf{x}) = \exp(V_0 + \sum_{x \in \mathbf{x}} V_1(x) + \sum_{\{x,y\} \subset \mathbf{x}} V_2(x, y) + \ldots) \qquad (46)$$

*where $V_k : \mathsf{N}_k \to \mathbb{R} \cup \{-\infty\}$ is called the **potential** of order k.*

Gibbs models arise in statistical physics, where $\log f(\mathbf{x})$ may be interpreted as the **potential energy** of the configuration $\mathbf{x}$. The term $-V_1(u)$ can be interpreted as the energy required to create a single point at a location u. The term $-V_2(u,v)$ can be interpreted as the energy required to overcome a force between the points u and v.

Example 4.10 (Hard core process). Give parameters $\beta, r > 0$, define $V_1(u) = \log \beta$ and

$$V_2(u,v) = \begin{cases} 0 & \text{if } ||u - v|| > r \\ -\infty & \text{if } ||u - v|| \le r \end{cases}$$

and $V_k \equiv 0$ for all $k \ge 3$. Then $\sum_{\{x,y\} \subset \mathbf{x}} V_2(x,y)$ is equal to zero if all pairs of points in $\mathbf{x}$ are at least r units apart, and otherwise this sum is equal to $-\infty$. Taking $\exp(-\infty) = 0$, we find that (46) is

$$f(\mathbf{x}) = \alpha \beta^{n(\mathbf{x})} \mathbf{1}\{\mathbf{x} \in H\}$$

where H is the hard core constraint set defined in Example 4.7, and $\alpha = \exp(V_0)$ is a normalising constant. This is the probability density of the hard core process.

Theorem 4.2. *Let f be the probability density of a finite point process $\mathbf{X}$ in a bounded region W in $\mathbb{R}^d$. Suppose that*

$$f(\mathbf{x}) > 0 \quad \Rightarrow \quad f(\mathbf{y}) > 0 \text{ for all } \mathbf{y} \subset \mathbf{x}. \tag{47}$$

Then f can be expressed in the Gibbs form (46).

Proof. This is a consequence of the Möbius inversion formula (the 'inclusion-exclusion principle'). The functions V_k can be obtained explicitly as

$$V_0 = \log f(\emptyset)$$
$$V_1(u) = \log f(\{u\}) - \log f(\emptyset)$$
$$V_2(u,v) = \log f(\{u,v\}) - \log f(\{u\}) - \log f(\{v\}) + \log f(\emptyset)$$

and in general

$$V_k(\mathbf{x}) = \sum_{\mathbf{y} \subseteq \mathbf{x}} (-1)^{n(\mathbf{x}) - n(\mathbf{y})} \log f(\mathbf{y}).$$

Then equation (46) can be verified by induction on $n(\mathbf{x})$. $\qquad \square$

Any process satisfying (47) also has a conditional intensity, by Theorem 4.1. The corresponding conditional intensity is

$$\beta^*(u, \mathbf{x}) = \exp\left[V_1(u) + \sum_{x \in \mathbf{X}} V_2(u, x) + \sum_{\{x,y\} \subset \mathbf{X}} V_3(u, x, y) + \dots \right] \tag{48}$$

Hence, this is the most general form of a conditional intensity:

Theorem 4.3. *Let* $\mathbf{X}$ *be a finite point process in a bounded region* W *of* $\mathbb{R}^d$. *Assume that* $\mathbf{X}$ *has a probability density* f *satisfying (47).*

A function $\beta^*(u, \mathbf{x})$ *is the conditional intensity of* $\mathbf{X}$ **if and only if** *it can be expressed in the form (48).*

Example 4.11 (Strauss process). For parameters $\beta > 0$, $0 \leq \gamma \leq 1$ and $r > 0$, suppose

$$V_1(u) = \log \beta$$
$$V_2(u, v) = (\log \gamma)\, \mathbf{1}\{\|u - v\| \leq r\}.$$

Then we have

$$\beta^*(u, \mathbf{x}) = \beta \gamma^{t(u, \mathbf{x})}$$

where

$$t(u, \mathbf{x}) = \sum_{x \in \mathbf{x}} \mathbf{1}\{\|u - x\| \leq r\}$$

is the number of points of $\mathbf{x}$ which are close to u. Also

$$f(\mathbf{x}) = \alpha\, \beta^{n(\mathbf{x})} \gamma^{s(\mathbf{x})}$$

where

$$s(\mathbf{x}) = \sum_{x, y \in \mathbf{x}} \mathbf{1}\{\|x - y\| \leq r\}$$

is the number of pairs of close points in $\mathbf{x}$. The normalising constant α is not available in closed form.

When $\gamma = 1$, this reduces to the Poisson process with intensity β. When $\gamma = 0$, we have

$$\beta^*(u, \mathbf{x}) = \mathbf{1}\{\|u - x\| > r \text{ for all } x \in \mathbf{x}\}$$

and

$$f(\mathbf{x}) = \alpha \beta^{n(\mathbf{x})} \mathbf{1}\{\mathbf{x} \in H\}$$

so we get the hard core process.

For $0 < \gamma < 1$ the Strauss process has 'soft inhibition' between neighbouring pairs of points.

Figure 35 shows simulated realisations of the Strauss process with two values of the interaction parameter γ. The case $\gamma = 0$ was already illustrated in Figure 34.

The intensity function of the Strauss process is, applying equation (34),

$$\beta(u) = \mathbb{E}[\beta^*(u, \mathbf{X})]$$
$$= \mathbb{E}[\beta \gamma^{t(u, \mathbf{X})}] \leq \beta$$

It is not easy to evaluate $\beta(u)$ explicitly as a function of β, γ, r.

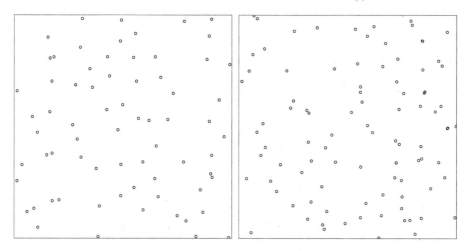

Fig. 35. Realisations of the Strauss process with interaction parameter $\gamma = 0.2$ (Left) and $\gamma = 0.5$ (Right) in the unit square, both having activity $\beta = 200$ and interaction range $r = 0.07$.

4.7 Parameter Estimation

Finally we return to the question of fitting a parametric model to point pattern data. Suppose we have observed a point pattern $\mathbf{x}$ in a bounded window $W \subset \mathbb{R}^2$, and wish to model it as a realisation of a finite point process $\mathbf{X}$ with probability density $f(\mathbf{x}; \theta)$ where θ is the parameter. The likelihood is

$$L(\theta) = f(\mathbf{x}; \theta).$$

Define the maximum likelihood estimator

$$\widehat{\theta} = \mathsf{argmax}_\theta L(\theta).$$

Example 4.12 (Uniform Poisson process). Suppose the model X is a uniform Poisson process in W with intensity β. The probability density of this model was found in Example 4.6 to be

$$f(\mathbf{x}; \beta) = \beta^{n(\mathbf{x})} \exp((1 - \beta)\lambda_2(W)).$$

Thus the log likelihood is

$$L(\beta) = n(\mathbf{x}) \log \beta + (1 - \beta)\lambda_2(W)$$

so the score is

$$U(\beta) = \frac{\mathrm{d}}{\mathrm{d}\beta} \log L(\beta) = \frac{n(\mathbf{x})}{\beta} - \lambda_2(W)$$

so the maximum likelihood estimate of β is

$$\widehat{\beta} = \frac{n(\mathbf{x})}{\lambda_2(W)},$$

the same as the method-of-moments estimator.

Example 4.13 (Hard core process). The probability density for a hard core process was found in Example 4.7 to be

$$f(\mathbf{x}; r) = \frac{\mathbf{1}\{\mathbf{x} \in H_r\}}{\mathbb{P}(\Pi \in H_r)}$$

where H_r is the set of all point patterns $\mathbf{x}$ with the property that every pair of distinct points in $\mathbf{x}$ is at least r units apart. Again Π denotes the uniform Poisson process with intensity 1.

For any configuration $\mathbf{x}$ let $m(\mathbf{x}) = \min\{\|x_i - x_j\| : x_i, x_j \in \mathbf{x}\}$ be the minimum distance between any pair of distinct points in $\mathbf{x}$. Then $H_r = \{\mathbf{x} : m(\mathbf{x}) \geq r\}$. Thus the likelihood can be written

$$L(r) = f(\mathbf{x}; r) = \frac{\mathbf{1}\{m(\mathbf{x}) \geq r\}}{\mathbb{P}(m(\Pi) \geq r)}.$$

The numerator is equal to 1 for all $r \leq m(\mathbf{x})$ and to 0 for $r > m(\mathbf{x})$. The denominator is clearly a decreasing (or at least non-increasing) function of r. Thus, $L(r)$ is a non-decreasing function of r on the range $0 \leq r \leq m(\mathbf{x})$, and is zero for $r > m(\mathbf{x})$. It follows that the maximum likelihood estimate of r is

$$\widehat{r} = m(\mathbf{x}).$$

For more complicated models, we rapidly run into problems in applying the method of maximum likelihood. The likelihood of a cluster process is known analytically [9], but is difficult to maximise. The likelihood of many Gibbs models contains a normalising constant, which is usually an intractable function of θ. Hence, analytic maximisation of L is often difficult [33].

Instead, some form of approximation is often employed as a substitute for exact maximum likelihood estimation. One strategy is to approximate the likelihood itself, for example using a series expansion (such as **virial expansion**), or using simulation (**Monte Carlo maximum likelihood** [21, 33]). Monte Carlo methods are probably the most popular approach, although they have many technical difficulties and are highly computationally-intensive.

Some experts also claim to be able to 'fit by eye' [40].

4.8 Estimating Equations

Another strategy for parameter estimation is to replace the maximum likelihood estimating equations

$$U(\theta) = \frac{\mathrm{d}}{\mathrm{d}\theta} \log L(\theta) = 0 \tag{49}$$

by another system of equations.

Let $\Psi(\theta, \mathbf{X})$ be a function such that, for any θ,

$$\mathbb{E}_\theta[\Psi(\theta, \mathbf{X})] = 0 \tag{50}$$

where $\mathbb{E}_\theta$ denotes the expectation over the distribution of X when θ is the true parameter value. When data $\mathbf{x}$ are observed, suppose we estimate θ by the solution $\widehat{\theta}$ of

$$\Psi(\theta, \mathbf{x}) = 0. \tag{51}$$

Then (51) is called an **unbiased estimating equation** [22].

This concept embraces both maximum likelihood and the method of moments. A key result in the classical theory of maximum likelihood is that (49) is an unbiased estimating equation. The method of moments is characterised by the estimating equation (35), which can be rewritten in the form (51) where $\Psi(\theta, \mathbf{x}) = T(\mathbf{x}) - \mathbb{E}_\theta[T(\mathbf{X})]$. This is clearly an unbiased estimating equation.

The term 'unbiased' should not be misinterpreted as suggesting that $\widehat{\theta}$ is an unbiased estimator of θ. However under reasonable limiting conditions $\widehat{\theta}$ is consistent and asymptotically unbiased.

Another family of estimating equations for point process models was suggested by Tákacs and Fiksel [20, 44, 45]. Consider again the identity (31). In our context the conditional intensity of the point process $\mathbf{X}$ depends on the parameter θ, so we denote the conditional intensity by $\beta_\theta^*(u, \mathbf{x})$, and equation (31) becomes

$$\mathbb{E} \sum_{x \in \mathbf{X}} g(x, \mathbf{X} \setminus x) = \int_{\mathbb{R}^d} \mathbb{E}\left[\beta_\theta^*(x, \mathbf{X}) g(x, \mathbf{X})\right]\, \mathrm{d}x$$

for arbitrary integrable functions g. It follows that if we define

$$\Psi(\theta, \mathbf{x}) = \sum_{x_i \in \mathbf{x} \cap A} g(x_i, \mathbf{x} \setminus \{x_i\}) - \int_A \beta_\theta^*(u, \mathbf{x}) g(u, \mathbf{x})\, \mathrm{d}u \tag{52}$$

for any chosen function g and any $A \subseteq W$, then (50) holds, and we have an unbiased estimating equation.

Example 4.14. To fit the Strauss process (Example 4.11) to a point pattern dataset $\mathbf{x}$, recall that the conditional intensity is $\beta_\theta^*(u, \mathbf{x}) = \beta \gamma^{t(u, \mathbf{x})}$. Assume r is fixed and known. Taking $g \equiv 1$ and $A = W$ in (52) gives

$$\Psi(\theta, \mathbf{x}) = n(\mathbf{x}) - \beta \int_W \gamma^{t(u, \mathbf{x})}\, \mathrm{d}u.$$

Taking $g(u, \mathbf{x}) = t(u, \mathbf{x})$ gives

$$\Psi(\theta, \mathbf{x}) = \sum_{i=1}^{n(\mathbf{x})} t(x_i, \mathbf{x} \setminus \{x_i\}) - \beta \int_W t(u, \mathbf{x}) \gamma^{t(u, \mathbf{x})}\, \mathrm{d}u$$

$$= 2s(\mathbf{x}) - \beta \int_W t(u, \mathbf{x}) \gamma^{t(u, \mathbf{x})}\, \mathrm{d}u$$

where $s(\mathbf{x})$ is the number of unordered pairs of points in $\mathbf{x}$ which are closer than a distance r apart. Equating these two functions Ψ to zero gives us two equations in the two unknowns β and γ which can be solved numerically to give estimates $\widehat{\beta}, \widehat{\gamma}$. To simplify these expressions, notice that $t(u, \mathbf{x})$ has integer values, and let

$$m_k = \int_W 1\{t(u, \mathbf{x}) = k\} \, du$$

for each $k = 0, 1, 2, \ldots$. Then the equations to be solved are

$$n(\mathbf{x}) = \beta \sum_{k=0}^{\infty} \gamma^k m_k$$

$$2s(\mathbf{x}) = \beta \sum_{k=0}^{\infty} k\gamma^k m_k.$$

4.9 Likelihood Devices

Another approach is to replace the likelihood function by another function altogether. An example is the pseudolikelihood proposed by Besag [10, 11].

Definition 4.3. *For a point process* $\mathbf{X}$ *with conditional intensity* $\beta_\theta^*(u, \mathbf{x})$, *where* θ *is the unknown parameter, define the* **pseudolikelihood**

$$\mathsf{PL}(\theta, \mathbf{x}) = \left(\prod_{x_i \in \mathbf{x}} \beta_\theta^*(x_i, \mathbf{x}) \right) \exp \left[- \int_W \beta_\theta^*(u, \mathbf{x}) \, du \right]. \tag{53}$$

The **maximum pseudolikelihood estimator (MPLE)** *of* θ *is the value maximising* $\mathsf{PL}(\theta)$.

A rationale for using this function is offered in [10, 12]. If $\mathbf{x}$ is treated as fixed, and the pseudolikelihood is considered as a function of θ, then the pseudolikelihood has the same functional form as the likelihood of an inhomogeneous Poisson process. The pseudolikelihood for a Poisson process is identical to its likelihood, up to a constant factor. Thus, the pseudolikelihood may be regarded as an approximation to the likelihood which ignores the dependence between points. It may be expected to be a good approximation to the likelihood when interpoint dependence is weak.

Example 4.15. For the Strauss process the log pseudolikelihood is

$$\log \mathsf{PL}(\beta, \gamma, r) = \sum_{i=1}^{n(\mathbf{x})} (\log \beta + t(x_i, \mathbf{x} \setminus \{x_i\}) \log \gamma) - \int_W \beta \gamma^{t(u, \mathbf{x})} \, du$$

$$= n(\mathbf{x}) \log \beta + 2s(\mathbf{x}) \log \gamma - \int_W \beta \gamma^{t(u, \mathbf{x})} \, du.$$

The score components for β, γ are

$$\frac{\partial}{\partial\beta} \log \mathsf{PL} = \frac{n(\mathbf{x})}{\beta} - \int_W \gamma^{t(u,\mathbf{x})} \, du$$

$$\frac{\partial}{\partial\gamma} \log \mathsf{PL} = \frac{2s(\mathbf{x})}{\gamma} - \beta \int_W t(u,\mathbf{x})\gamma^{t(u,\mathbf{x})-1} \, du.$$

The maximum pseudolikelihood estimators of β, γ are the roots of these two functions.

Notice that the maximum pseudolikelihood estimating equations derived in Example 4.15 are equivalent to those derived using the Takacs-Fiksel method in Example 4.14. This is true in greater generality, as shown in [3, 17].

Powerful advantages of the pseudolikelihood are that it avoids the normalising constant present in the likelihood, that it is usually easy to compute and maximise, and that it retains good properties such as consistency and asymptotic normality in a large-sample limit [25, 24]. A disadvantage of the pseudolikelihood is that it is known to be biased and statistically inefficient in small samples.

An algorithm for fitting very general point process models by maximum pseudolikelihood was developed in [6]. It is now possible to perform parametric modelling and inference for spatial point processes using tools similar to those available for other types of data. For further details, see [8, 7].

References

1. Babu, G.J., Feigelson, E.D.: Astrostatistics. Chapman & Hall, London (1996)
2. Baddeley, A.J.: Spatial sampling and censoring. In: Barndorff-Nielsen, O.E., Kendall, W.S., van Lieshout, M.N.M. (eds) Stochastic Geometry. Likelihood and Computation. Monographs on Statistics and Applied Probability, **80**, Chapman & Hall/CRC, Boca Raton, FL (1999)
3. Baddeley, A.: Time-invariance estimating equations. Bernoulli, **6**, 783–808 (2000)
4. Baddeley, A., Jensen, E.B. Vedel: Stereology for Statisticians. Chapman & Hall/CRC, Boca Raton, FL (2005)
5. Baddeley, A.J., Silverman, B.W.: A cautionary example on the use of second-order methods for analyzing point patterns. Biometrics, **40**, 1089–1094 (1984)
6. Baddeley, A., Turner, R.: Practical maximum pseudolikelihood for spatial point patterns (with discussion). Austral. N. Z. J. Stat., **42**(3), 283–322 (2000)
7. Baddeley, A., Turner, R.: Modelling spatial point patterns in R. In: Baddeley, A., Mateu, J., Stoyan, D. (eds) Case Studies in Spatial Point Pattern Modelling. Springer, Berlin Heidelberg New York (to appear)
8. Baddeley, A., Turner, R.: Spatstat: an R package for analyzing spatial point patterns. J. Statist. Software, **12**(6), 1–42 (2005). URL: www.jstatsoft.org, ISSN: 1548-7660
9. Baddeley, A.J., van Lieshout, M.N.M, Møller, J.: Markov properties of cluster processes. Adv. in Appl. Probab., **28**, 346–355 (1996)

10. Besag, J.: Statistical analysis of non-lattice data. The Statistician, **24**, 179–195 (1975)
11. Besag, J.: Some methods of statistical analysis for spatial data. Bull. Internat. Statist. Inst., **44**, 77–92 (1978)
12. Besag, J., Milne, R., Zachary, S.: Point process limits of lattice processes. J. Appl. Probab., **19**, 210–216 (1982)
13. The Comprehensive R Archive Network. URL http://www.cran.r-project.org
14. Cuzick, J., Edwards, R.: Spatial clustering for inhomogeneous populations (with discussion). J. Royal Statist. Soc. Ser. B, **52**, 73–104 (1990)
15. Daley, D.J., Vere-Jones, D.: An Introduction to the Theory of Point Processes. Springer, Berlin Heidelberg New York (1988)
16. Diggle, P.J.: Statistical Analysis of Spatial Point Patterns. Second edition, Arnold, London (2003)
17. Diggle, P.J., Fiksel, T., Grabarnik, P., Ogata, Y., Stoyan, D., Tanemura, M.: On parameter estimation for pairwise interaction processes. Internat. Statist. Rev., **62**, 99–117 (1994)
18. Diggle, P.J., Gratton, R.J.: Monte Carlo methods of inference for implicit statistical models (with discussion). J. Royal Statist. Soc. Ser. B, **46**, 193–227 (1984)
19. Elliott, P., Wakefield, J., Best, N., Briggs, D. (eds): Spatial Epidemiology: Methods and Applications. Oxford University Press, Oxford (2000)
20. Fiksel, T.: Estimation of interaction potentials of Gibbsian point processes. Statistics, **19**, 77–86 (1988)
21. Geyer, C.J.: Likelihood inference for spatial point processes. In: Barndorff-Nielsen, O.E., Kendall, W.S., van Lieshout, M.N.M. (eds) Stochastic Geometry. Likelihood and Computation. Monographs on Statistics and Applied Probability, **80**, Chapman & Hall/CRC, Boca Raton, FL (1999)
22. Godambe, V.P. (ed): Estimating Functions. Oxford Statistical Science Series, **7**, Oxford Science Publications, Oxford (1991)
23. Hornik, K.: The R FAQ: Frequently asked questions on R. URL http://www.ci.tuwien.ac.at/~hornik/R/. ISBN 3-901167-51-X
24. Jensen, J.L., Künsch, H.R.: On asymptotic normality of pseudo likelihood estimates for pairwise interaction processes. Ann. Inst. Statist. Math., **46**, 475–486 (1994)
25. Jensen, J.L., Møller, J.: Pseudolikelihood for exponential family models of spatial point processes. Ann. Appl. Probab., **1**, 445–461 (1991)
26. Kallenberg, O.: Random Measures. Third edition, Akademie Verlag/Academic Press, Berlin/New York (1983)
27. Kendall, D.G.: Foundations of a theory of random sets. In: Harding, E.F., Kendall, D.G. (eds) Stochastic Geometry. John Wiley & Sons, Chichester (1974)
28. Kingman, J.F.C.: Poisson Processes. Oxford University Press, Oxford (1993)
29. Ludwig, J.A., Reynolds, J.F.: Statistical Ecology: a Primer on Methods and Computing. John Wiley & Sons, New York (1988)
30. Matérn, B.: Spatial Variation. Meddelanden från Statens Skogsforskningsinstitut, **49**(5), 1–114 (1960)
31. Matheron, G.: Ensembles fermés aléatoires, ensembles semi-markoviens et polyèdres poissoniens. Adv. in Appl. Probab., **4**, 508–543 (1972)
32. Matheron, G.: Random Sets and Integral Geometry. John Wiley & Sons, New York (1975)

33. Møller, J., Waagepetersen, R.P.: Statistical Inference and Simulation for Spatial Point Processes. Chapman & Hall/CRC, Boca Raton, FL (2003)

34. Ohser, J., Mücklich, F.: Statistical Analysis of Microstructures in Materials Science. John Wiley & Sons, Chichester (2000)

35. Ornstein, L.S., Zernike, F.: Accidental deviations of density and opalesence at the critical point of a single substance. Proceedings, Section of Sciences, Royal Academy of Sciences/ Koninklijke Akademie van Wetenschappen, Amsterdam, **17**, 793–806 (1914)

36. Palm, C.: Intensitätsschwankungen im Fernsprechverkehr. Ericsson Technics, **44**, 1–189 (1943)

37. Parzen, E.: Stochastic Processes. Holden-Day, San Francisco (1962)

38. R Development Core Team. R: A Language and Environment for Statistical Computing. R Foundation for Statistical Computing, Vienna, Austria (2004). ISBN 3-900051-00-3

39. Ripley, B.D.: The second-order analysis of stationary point processes. J. Appl. Probab., **13**, 255–266 (1976)

40. Ripley, B.D.: Spatial Statistics. John Wiley & Sons, New York (1981)

41. Serra, J.: Image Analysis and Mathematical Morphology. Academic Press, London (1982)

42. Stoyan, D., Kendall, W.S., Mecke, J.: Stochastic Geometry and its Applications. Second edition, John Wiley & Sons, Chichester (1995)

43. Stoyan, D., Stoyan, H.: Fractals, Random Shapes and Point Fields. John Wiley & Sons, Chichester (1995)

44. Takacs, R.: Estimator for the pair potential of a Gibbsian point process. Statistics, **17**, 429–433 (1986)

45. Takacs, R., Fiksel, T.: Interaction pair-potentials for a system of ants' nests. Biom. J., **28**, 1007–1013 (1986)

46. van Lieshout, M.N.M.: Markov Point Processes and their Applications. Imperial College Press, London (2000)

47. van Lieshout, M.N.M, Baddeley, A.J.: A nonparametric measure of spatial interaction in point patterns. Statist. Neerlandica, **50**, 344–361 (1996)

Random Polytopes, Convex Bodies, and Approximation

Imre Bárány*

Rényi Institute of Mathematics
H-1364 Budapest Pf. 127 Hungary
e-mail: barany@renyi.hu
and
Mathematics, University College London
Gower Street, London, WC1E 6BT, UK

Assume $K \subset \mathbb{R}^d$ is a convex body and $X_n \subset K$ is a random sample of n uniform, independent points from K. The convex hull of X_n is a convex polytope K_n called random polytope inscribed in K. We are going to investigate various properties of this polytope: for instance how well it approximates K, or how many vertices and facets it has. It turns out that K_n is very close to the so called floating body inscribed in K with parameter $1/n$. To show this we develop and use the technique of cap coverings and Macbeath regions. Its power will be illustrated, besides random polytopes, on several examples: floating bodies, lattice polytopes, and approximation problems.

1 Introduction

We write $\mathcal{K}$ or $\mathcal{K}^d$ for the set of convex bodies in $\mathbb{R}^d$, that is, compact convex sets with nonempty interior in $\mathbb{R}^d$. Assume $K \in \mathcal{K}$ and $x_1, \ldots, x_n$ are random, independent points chosen according to the uniform distribution in K. The convex hull of these points, to be denoted by K_n, is called a **random polytope** inscribed in K. Thus $K_n = [x_1, \ldots, x_n]$ where $[S]$ stands for the convex hull of the set S. The study of random polytopes began with Sylvester's famous "four-point question" [55]. For more information and recent results on the four-point question see [7] and [8].

Starting with the work of Rényi and Sulanke [39] there has been a lot of research to understand the asymptotic behaviour of random polytopes. Most of it has been concentrated on the expectation of various functionals associated with K_n. For instance the number of vertices, $f_0(K_n)$, or more generally, the

* Partially supported by Hungarian National Science Foundation Grants T 037846 and T 046246

number of k-dimensional faces, $f_k(K_n)$, of K_n, or the volume missed by K_n, that is $\mathrm{vol}(K \setminus K_n)$. The latter quantity measures how well K_n approximates K. As usual we will denote the expectation of $f_k(K_n)$ by $\mathbb{E}f_k(K_n)$, and that of $\mathrm{vol}(K \setminus K_n)$ by $\mathbb{E}(K, n)$.

In their 1963 paper [39] Rényi and Sulanke made a surprising discovery. Already in the planar case, the expectation of the number of vertices, $f_0(K_n)$, depends heavily on the boundary structure of K. It is of order $\ln n$ when K is a convex polygon, and is of order $n^{1/3}$ when K is a circle (or any other smooth enough convex body). Similarly, $\mathbb{E}(K, n)$ is of order $n^{-2/3}$ for smooth convex bodies in $\mathbb{R}^2$ and $(\ln n)/n$ for convex polygons. What is the reason for such a different behaviour?

The aim of this survey is to give a thorough introduction to the theory of random polytopes. There are two kinds of results concerning $\mathbb{E}(K, n)$ and $\mathbb{E}f_k(K_n)$: precise asymptotic and order of magnitude. We mainly focus on the second type of results and only mention some precise asymptotics. Along the way we will see why such a different behaviour of $\mathbb{E}(K, n)$ is quite natural. We will introduce the notion of caps, M-regions, and cap coverings of convex bodies. They constitute a method to handle the boundary structure of convex bodies. The technique of M-regions and cap coverings can be used for other problems as well: several applications will be presented, some of them coming from my paper [6]. This survey contains very little new material. I will indicate at the end of some sections where the results come from and mention if they are new.

I am not quite sure I organized the material of the survey in a concise way. There were too many directions to talk about: random polytopes, M-regions, cap coverings, technical preparations and lemmas, probabilistic tools, and further applications. It is difficult to order them linearly (as a survey should be written). Here is the contents, section by section:

1. Introduction
2. Computing $\mathbb{E}\phi(K_n)$
3. Minimal caps and a general result
4. The volume of the wet part
5. The economic cap covering theorem
6. Macbeath regions
7. Proofs of the properties of Macbeath regions
8. Proof of the cap covering theorem
9. Auxiliary lemmas from probability
10. Proof of Theorem 3.1
11. Proof of Theorem 4.1
12. Proof of (4)
13. Expectation of $f_k(K_n)$
14. Proof of Lemma 13.2
15. Further results
16. Lattice polytopes

17. Approximation
18. How it all began: segments on bd K

The next section presents a sketch of the method for computing expectations directly. The main results for random polytopes are contained in Section 3. Notation and terminology, including the wet part and the floating body are introduced there. Important properties of the wet part are given in Section 4. The economic cap covering theorem, together with a corollary, is stated next. Macbeath regions are defined and their properties stated in Section 6. The proofs of these results are given in Sections 7 to 12. Some of them can be skipped on first reading, although the proofs use concepts and methods from proofs from previous sections. We treat separately the expectation of the number of k-dimensional faces of K_n in Section 13 and 14. This proof is new, using the cap covering technique in a slightly different way and avoiding the probabilistic tools. Further results, including some spectacular new theorems, are explained without proof in Section 15. Applications of the cap-covering technique are given in Sections 16 and 17. The final section is devoted to the origins of the method.

2 Computing $\mathbb{E}\phi(K_n)$

The method for computing $\mathbb{E}\phi(K_n)$ goes back to the 1963 paper of Rényi and Sulanke [39] and is the following. Let P be a polytope and write $\mathcal{F}$ for the set of facets of P. Assume the function ϕ is of the form

$$\phi(P) = \sum_{F \in \mathcal{F}} \phi(F).$$

Such functions are f_{d-1}, volume, or surface area. The orientation of the facet is given by the outer normal to P at F. As K_n is a simplicial polytope with probability one, each facet is of the form $[x_{i_1}, \ldots, x_{i_d}]$. We write $\mathbf{1}\{E\}$ for the indicator function of the event E. Then, assuming $\operatorname{vol} K = 1$,

$$
\begin{aligned}
\mathbb{E}\phi(K_n) &= \sum_{1 \le i_1 < \cdots < i_d \le n} \int_K \cdots \int_K \mathbf{1}\{[x_{i_1}, \ldots, x_{i_d}] \in \mathcal{F}]\} \times \\
&\quad \times \phi([x_{i_1}, \ldots, x_{i_d}]) \mathrm{d}x_1 \ldots \mathrm{d}x_n \\
&= \binom{n}{d} \int_K \cdots \int_K \mathbf{1}\{[x_1, \ldots, x_d] \in \mathcal{F}\} \phi([x_1, \ldots, x_d]) \mathrm{d}x_1 \ldots \mathrm{d}x_n.
\end{aligned}
$$

We will denote by $V = V(x_1, \ldots, x_d)$ the volume of the smaller cap cut off from K by $\operatorname{aff}\{x_1, \ldots, x_d\}$ (which is a hyperplane, almost surely). Here, we used aff S for the affine hull of S. Since $F = [x_1, \ldots, x_d]$ is a facet, if and only if $x_{d+1}, \ldots, x_n$ are all on one side of $\operatorname{aff}\{x_1, \ldots, x_d\}$, we have the following theorem.

Theorem 2.1. *Under the above conditions,*

$$\mathbb{E}\phi(K_n) = \binom{n}{d} \int_K \cdots \int_K [(1-V)^{n-d} + V^{n-d}]\phi(F)\mathrm{d}x_1 \ldots \mathrm{d}x_d. \qquad (1)$$

One can give precise estimates for this integral in several special cases. For instance if $\phi = f_{d-1}$, then $\phi(F) = 1$ and the above formula can be directly evaluated for smooth convex sets and for polygons in the plane. In [39], Rényi and Sulanke prove that, for smooth convex sets of area one,

$$\mathbb{E}f_1(K_n) = \left(\frac{2}{3}\right)^{1/3} \Gamma\left(\frac{5}{3}\right) \left(\int_{\mathrm{bd}\,K} \kappa^{1/3}\mathrm{d}s\right) n^{1/3}(1+o(1)),$$

where κ is the curvature and integration is by arc length $\mathrm{d}s$ on the boundary $\mathrm{bd}\,K$ of the convex body. For polygons, direct computation in [39] shows that

$$\mathbb{E}f_1(K_n) = \frac{2}{3}f_0(K)\ln n\,(1+o(1)).$$

Of course, $f_1 = f_0$ in these cases. The computation reveals that, for smooth bodies, the vertices of K_n are distributed evenly near $\mathrm{bd}\,K$, the boundary of K, while for polygons, they are concentrated near the vertices of the original polygon. This is a first level explanation for the different behaviour of $\mathbb{E}f_0(K_n)$.

For smooth convex bodies in higher dimension $\mathbb{E}\phi(K_n)$ can sometimes be evaluated using the Blaschke-Petkantschin [43] integral formula. We will return to this in Section 15.

Remark. Equation (1) was used first by Rényi and Sulanke [39] in the planar case.

3 Minimal Caps and a General Result

Recalling that $\mathcal{K}$ denotes the set of all convex bodies in $\mathbb{R}^d$, we will write $\mathcal{K}_1$ for the set of those $K \in \mathcal{K}$ that have unit volume, $\mathrm{vol}\,K = 1$. This is convenient since then the Lebesgue measure and the uniform probability measure on $K \in \mathcal{K}_1$ coincide.

Assume $a \in \mathbb{R}^d$ is a unit vector and $t \in \mathbb{R}$. Then the halfspace $H = H(a \le t)$ is defined as
$$H(a \le t) = \{x \in \mathbb{R}^d : a \cdot x \le t\},$$

where $a \cdot x$ is the scalar product of a and x.

A **cap** of $K \in \mathcal{K}$ is simply a set of the form $C = K \cap H$ where H is a closed halfspace. The width of the cap, $w(C)$ is the usual width of C in the normal direction of H. We define the function $v : K \to \mathbb{R}$ by

$$v(x) = \min\{\mathrm{vol}(K \cap H) : x \in H \text{ and } H \text{ is a halfspace}\},$$

This function is going to play a central role in what follows. The **minimal cap** belonging to $x \in K$ is a cap $C(x)$ with $x \in C(x)$ and $\operatorname{vol} C(x) = v(x)$. The minimal cap $C(x)$ need not be unique, so our notation is a little ambiguous but this will not cause any trouble.

The level sets of v are defined as

$$K(v \geq t) = \{x \in K : v(x) \geq t\}.$$

The **wet part** of K with parameter $t > 0$ is

$$K(t) = K(v \leq t) = \{x \in K : v(x) \leq t\}.$$

The name comes from the mental picture when K is a three dimensional convex body containing t units of water. We call $K(v \geq t)$ the **floating body** of K with parameter $t > 0$ as, in a similar picture, this is the part of K that floats above water (cf. [10] and [32]). The floating body is the intersection of halfspaces, so it is convex.

The general behaviour of $\mathbb{E}(K, n)$ was described in [10]: $\mathbb{E}(K, n)$ is of the same order of magnitude as the volume of the wet part with $t = 1/n$. This works for general convex bodies $K \in \mathcal{K}$, not only when K is smooth or is a polytope. Precisely, we have the following result.

Theorem 3.1. *For every $d \geq 2$ there are constants $c_0, c_1, c_2 > 0$ such that for every $K \in \mathcal{K}_1$ and $n \geq c_0$*

$$c_1 \operatorname{vol} K(1/n) \leq \mathbb{E}(K, n) \leq c_2 \operatorname{vol} K(1/n).$$

It will be convenient to use the $\ll, \gg$ and $\approx$ notation. For instance, $f(n) \ll g(n)$ means that there is a constant b such that $f(n) \leq b g(n)$ for all values of n. This notation always hides a constant which, as a rule, does not depend on n but may depend on dimension. With this notation, the above theorem can be formulated in the following way.

Theorem 3.2. *For large enough n and for every $K \in \mathcal{K}_1$,*

$$\operatorname{vol} K(1/n) \ll \mathbb{E}(K, n) \ll \operatorname{vol} K(1/n).$$

The content of Theorem 3.2 is that, instead of determining $\mathbb{E}(K, n)$, one can determine the volume of the wet part (which is usually simpler) and obtain the order of magnitude of $\mathbb{E}(K, n)$. The reader will have no difficulty understanding that for the unit ball B^d in $\mathbb{R}^d$ the wet part $B^d(v \leq t)$ is the annulus $B^d \setminus (1 - h)B^d$ where h is of order $t^{2/(d+1)}$. Thus

$$\mathbb{E}(B^d, n) \approx \operatorname{vol} B^d(1/n) \approx n^{-2/(d+1)}.$$

Similarly, for the unit cube Q^d in $\mathbb{R}^d$ the floating body with parameter t (in the subcube $[0, 1/2]^d$) is bounded by the hypersurface $\{x \in \mathbb{R}^d : \prod x_i = d^d t/d!\}$.

From this, the volume of the wet part can be determined easily (see also Section 12),

$$\mathbb{E}(Q^d, n) \approx \text{vol}\, Q^d(1/n) \approx \frac{(\ln n)^{d-1}}{n}.$$

This is the second level of explanation for the very different behaviour of $\mathbb{E}(K, n)$: the volume of the wet part varies heavily depending on the boundary structure of K.

4 The Volume of the Wet Part

By the theorems of the previous section, the order of magnitude of $\mathbb{E}(K, n)$ is determined by that of $\text{vol}\, K(1/n)$. In this section we state several results on the function $t \mapsto \text{vol}\, K(t)$. In particular, we are interested in the cases when this function is maximal and minimal.

The wet part $K(t) = K(v \leq t)$ is a kind of inner parallel body to the boundary of K. We note first that the function $v : K \to \mathbb{R}$ is invariant (or rather equivariant) under non-degenerate linear transformations $A : \mathbb{R}^d \to \mathbb{R}^d$. Precisely, recalling the notation $v(x) = v_K(x)$, we have

$$v_{AK}(Ax) = |\det A| v_K(x)$$

since $C_{AK}(Ax) = A(C_K(x))$. This also shows that the quantity

$$\frac{\text{vol}\, K(v \leq t \,\text{vol}\, K)}{\text{vol}\, K} \tag{2}$$

is invariant under non-degenerate linear transformations.

Theorem 4.1. *Assume* $K \in \mathcal{K}_1$ *and* $t \geq 0$. *Then*

$$\text{vol}\, K(t) \gg t \left(\ln \frac{1}{t}\right)^{d-1}. \tag{3}$$

This theorem is best possible (apart from the implied constant) as shown by polytopes. We need the following definition. A **tower** of a polytope P is a chain of faces $F_0 \subset F_1 \subset \ldots, \subset F_{d-1}$ where F_i is i-dimensional. Write $T(P)$ for the number of towers of P.

Theorem 4.2. *Assume* $P \in \mathcal{K}_1$ *and* $t \geq 0$. *Then*

$$\text{vol}\, P(t) = \frac{T(P)}{d^{d-1}d!} t \left(\ln \frac{1}{t}\right)^{d-1} (1 + o(1)).$$

The result is due to Schütt [50], and independently to Bárány and Buchta [9]. It is used in the proof of the following theorem.

Theorem 4.3. *Assume $P \in \mathcal{K}_1$ and $t \geq 0$. Then*

$$\mathbb{E}(P, n) = \frac{T(P)}{(d+1)^{d-1}(d-1)!} \frac{(\ln n)^{d-1}}{n}(1 + o(1)).$$

This is a difficult theorem whose proof is based on work of Affentranger and Wieacker [1] and Bárány and Buchta [9]. Here we will only prove Theorem 4.2 in the simpler form saying that

$$\operatorname{vol} P(t) \ll T(P)t \left(\ln \frac{1}{t}\right)^{d-1}, \tag{4}$$

where the implied constant depends on dimension only.

Concerning the upper bound on the volume of the wet part, or on $\mathbb{E}(K, n)$, the following result of Groemer [22] gives a complete answer.

Theorem 4.4. *Among all convex bodies in $\mathcal{K}_1$, $\mathbb{E}(K, n)$ is maximal for ellipsoids, and only for ellipsoids.*

The affine isoperimetric inequality (cf. Blaschke [14] and Schütt [52]) expresses a similar extremal property of ellipsoids.

Theorem 4.5. *For all convex bodies in $\mathcal{K}_1$,*

$$\limsup_{t \to 0} t^{-\frac{2}{d+1}} \operatorname{vol} K(t)$$

is maximal for ellipsoids, and only for ellipsoids.

In case of smooth convex bodies in $\mathbb{R}^d$ more precise information is available.

Theorem 4.6. *For a convex body $K \in \mathcal{K}_1$ with $\mathcal{C}^2$ boundary and positive curvature κ at each point of $\operatorname{bd} K$,*

$$\mathbb{E}(K, n) = c(d) \left(\int_{\operatorname{bd} K} \kappa^{\frac{1}{d+1}} \mathrm{d}S\right) n^{-\frac{2}{d+1}}(1 + o(1)),$$

where $\mathrm{d}S$ denotes integration over $\operatorname{bd} K$.

The above results show that one can determine $\mathbb{E}(K, n)$ and $\operatorname{vol} K(t)$ for smooth convex bodies and for polytopes. What happens between these two extreme classes of convex bodies is not a mystery: it is the usual unpredictable behaviour. Using the above results and a general theorem of Gruber [24] one can show the following.

Theorem 4.7. *Assume $\omega(n) \to 0$ and $\Omega(n) \to \infty$. Then for most (in the Baire category sense) convex bodies in $\mathcal{K}_1$ one has, for infinitely many n,*

$$\mathbb{E}(K, n) \geq \omega(n)n^{-\frac{2}{d+1}},$$

and also, for infinitely many n,

$$\mathbb{E}(K, n) \leq \Omega(n)\frac{(\ln n)^{d-1}}{n}.$$

There is, of course, an analogous theorem for $K(t)$ with $\omega(t)$ and $\Omega(t)$ whose formulation and proof are left to the interested reader.

We will only prove Theorem 4.1 and inequality (4).

5 The Economic Cap Covering Theorem

Everything interesting that can happen to a convex body happens near its boundary. The technique of cap coverings and M-regions is a powerful method to deal with the boundary structure of convex bodies. The proof of the economic cap covering theorem (see [10] and [4]) is based on this technique. It says the following.

Theorem 5.1. *Assume* $K \in \mathcal{K}_1$ *and* $0 < \varepsilon < \varepsilon_0 = (2d)^{-2d}$. *Then there are caps* $C_1, \ldots, C_m$ *and pairwise disjoint convex sets* $C_1', \ldots, C_m'$ *such that* $C_i' \subset C_i$, *for each* i, *and*

(i) $\bigcup_1^m C_i' \subset K(\varepsilon) \subset \bigcup_1^m C_i$,
(ii) $\operatorname{vol} C_i' \gg \varepsilon$ *and* $\operatorname{vol} C_i \ll \varepsilon$ *for each* i,
(iii) *for each cap* C *with* $C \cap K(v > \varepsilon) = \emptyset$ *there is a* C_i *containing* C.

The meaning is that the caps C_i cover the wet part, but do not "over cover" it. In particular,

$$m\varepsilon \ll \operatorname{vol} K(\varepsilon) \ll m\varepsilon. \tag{5}$$

The next corollary expresses a certain concavity property of the function $\varepsilon \mapsto \operatorname{vol} K(\varepsilon)$. It says that, apart from the constant implied by the $\gg$ notation, the dth root of $\operatorname{vol} K(\varepsilon)$ is a concave function. This will be sufficient for our purposes, that is, for the proof of Theorem 3.2.

Corollary 5.1. *If* $K \in \mathcal{K}_1$, $\varepsilon \leq \varepsilon_0$, *and* $\lambda \geq 1$, *then*

$$\operatorname{vol} K(\varepsilon) \gg \lambda^{-d} \operatorname{vol} K(\lambda\varepsilon). \tag{6}$$

The proof of the above results relies heavily on the Macbeath regions and their properties. They are defined, with their properties explained, in the next section.

6 Macbeath Regions

Macbeath regions, or M-regions, for short, were introduced in 1952 by A. M. Macbeath [34]: given a convex body $K \in \mathcal{K}^d$, and a point $x \in K$, the corresponding M-region is, by definition,

$$M(x) = M_K(x) = K \cap (2x - K).$$

So $M(x)$ is, again, a convex set. It is centrally symmetric with centre x. We define the blown-up version of the M-region as follows

$$M(x, \lambda) = M_K(x, \lambda) = x + \lambda \left[(K - x) \cap (x - K) \right].$$

This is just a blown-up copy of $M(x)$ from its centre x with scalar $\lambda > 0$.

We define the function $u : K \to R$ by

$$u(x) = \operatorname{vol} M(x).$$

The level sets of u are defined the same way as those of v,

$$K(u \le t) = \{x \in K : u(x) \le t\}, \ K(u \ge t) = \{x \in K : u(x) \ge t\}.$$

We note that the function $u : K \to \mathbb{R}$, just like v, is invariant (or rather equivariant) under non-degenerate linear transformations $A : \mathbb{R}^d \to \mathbb{R}^d$. That is,

$$u_{AK}(Ax) = |\det A| u_K(x)$$

since $M_{AK}(Ax) = A(M_K(x))$. This also shows that the quantity

$$\frac{\operatorname{vol} K(u \le t \operatorname{vol} K)}{\operatorname{vol} K} \tag{7}$$

is invariant under non-degenerate linear transformations, cf. (2).

M-regions have an important property that can often be used with induction on dimension. Namely, assume H is a hyperplane and $x \in K \cap H$. Then, as it is very easy to see,

$$M_{K \cap H}(x) = M_K(x) \cap H. \tag{8}$$

The convexity of $K(u \ge t)$ is not as simple as that of $K(v \ge t)$ and we state it as a separate lemma.

Lemma 6.1. *The set $K(u \ge t)$ is convex.*

Proof (cf. Macbeath [34]). We check first that $\frac{1}{2}(M(x) + M(y)) \subset M(\frac{1}{2}(x+y))$. So assume $a \in M(x)$, that is $a \in K$ and $a \in 2x - K$, or $a = 2x - k_1$ for some $k_1 \in K$. Similarly $b \in M(y)$ implies $b \in K$ and $b = 2y - k_2$ for some $k_2 \in K$. Then, by the convexity of K, $(a + b)/2 \in K$ and

$$\frac{a + b}{2} = x + y - \frac{k_1 + k_2}{2} \in 2\frac{x + y}{2} - K,$$

implying the claim. Now the Brunn-Minkowski inequality [46] together with the containment $\frac{1}{2}(M(x) + M(y)) \subset M(\frac{1}{2}(x + y))$ implies that the function $u^{1/d}$ is concave. Thus, in particular, the level sets $K(u \ge t)$ are convex. □

The computation of $u(x)$ is simpler than that of $v(x)$ since one does not have to minimize. It turns out that $v(x) \approx u(x)$, when x is close to the boundary of K. A word of warning is in place here: closeness to the boundary is to be expressed equivariantly, that is, in terms of how small $v(x)$ or $u(x)$ is, as both u and v are affinely equivariant.

We now list several properties of these functions and their interrelations. The proofs are technical and will be given in the next section which can be skipped on first reading. In each one of these lemmas we assume that K is a convex body in $\mathcal{K}_1$ and $\varepsilon_0 = d^{-1}3^{-d}$.

Lemma 6.2. *If $x, y \in K$ and $M(x, 1/2) \cap M(y, 1/2) \neq \emptyset$, then*

$$M(y, 1) \subset M(x, 5).$$

Lemma 6.3. *We have $u(x) \leq 2v(x)$, for all $x \in K$.*

Lemma 6.4. *If $x \in K$ and $v(x) \leq \varepsilon_0$, then*

$$C(x) \subset M(x, 2d).$$

Lemma 6.5. *If $x \in K$ and $v(x) \leq \varepsilon_0$, then $v(x) < (2d)^d u(x)$.*

Lemma 6.6. *If $x \in K$ and $u(x) \leq (3d)^{-d}\varepsilon_0$, then $v(x) < (2d)^d u(x)$.*

Lemma 6.7. *$K(v \geq \varepsilon)$ contains no line segment on its boundary.*

Lemma 6.8. *Assume C is a cap of K and $C \cap K(v \geq \varepsilon) = \{x\}$, a single point. If $\varepsilon < \varepsilon_0$, then $\operatorname{vol} C \leq d\varepsilon$ and*

$$C \subset M(x, 2d).$$

Lemma 6.9. *Every $y \in K(\varepsilon)$ is contained in a minimal cap $C(x)$ with $\operatorname{vol} C(x) = \varepsilon$ and $x \in \operatorname{bd} K(v \geq \varepsilon)$.*

Lemma 6.10. *If $\varepsilon \leq \varepsilon_0$, then $K(v \leq \varepsilon) \subset K(u \leq 2\varepsilon)$. If $\varepsilon \leq (2d)^{-d}\varepsilon_0$, then $K(u \leq \varepsilon) \subset K(v \leq (2d)^d\varepsilon)$.*

The importance of these lemmas lies in the fact that they show $u \approx v$ near the boundary of K in a strong sense. Namely, under the conditions of Lemma 6.4 the minimal cap is contained in a blown-up copy of the Macbeath region. On the other hand, "half" of the Macbeath region is contained in the minimal cap. Precisely, if $C = K \cap H(a \leq t)$ is a minimal cap, then

$$M(x) \cap H(a \leq t) \subset C(x). \tag{9}$$

This shows that there is a two-way street between $C(x)$ and $M(x)$: $C(x)$ can be replaced by $M(x)$ and $M(x)$ by $C(x)$ whenever it is more convenient to work with the other one.

Remark. Lemma 6.4 was proved first by Ewald, Larman, and Rogers in [21], they show $C \subset M(x, 3d)$. The slightly better constant here is new and so is the proof (given in the next section) and it comes from an effort to use affine-invariant methods when the statement is affinely invariant.

7 Proofs of the Properties of the M-regions

Proof of Lemma 6.2 (from the ground breaking paper by Ewald, Larman, Rogers [21]). Assume a is the common point of $M(x, 1/2)$ and $M(y, 1/2)$. Then

$$a = x + \frac{1}{2}(x - k_1) = y + \frac{1}{2}(k_2 - y)$$

for some $k_1, k_2 \in K$ implying $y = 3x - k_1 - k_2$. Suppose now that $b \in M(y, 1)$. Then $b \in K \subset x + 5(K - x)$ clearly, and $b = y + (y - k_3)$ with some $k_3 \in K$. Consequently

$$b = 2y - k_3 = 6x - 2k_1 - 2k_2 - k_3$$
$$= x + 5\left(x - \left[\frac{2}{5}k_1 + \frac{2}{5}k_2 + \frac{1}{5}k_3\right]\right) \in x + 5(x - K).$$

□

Lemma 6.3 follows from (9). Lemma 6.4 is also from [21], the proof below is a slight improvement on the constant.

Proof of Lemma 6.4. The basic observation is that if $C(x) = K \cap H(a \le t)$ is a minimal cap, then x is the centre of gravity of the section $K \cap H(a = t)$. This can be checked by a routine variational argument. We first prove the following.

Claim 7.1 *Assume $C(x)$ has width w, and K contains a point k in the hyperplane $H(a = t + 2w)$. Then $C(x) \subset M(x, 2d)$.*

Proof. Assume that, on the contrary, there is a point $z \in C(x)$ which is not in $M(x, 2d)$. Then $z \notin x + 2d(x - K)$ implying

$$z^* = x - \frac{1}{2d}(z - x) \notin K.$$

Let L be the two-dimensional plane containing x, k and z, then $z^* \in L$ as well, and our problem has become a simple planar computation. Fix a coordinate system to L with x lying at the origin and the hyperplane $H(a = t)$ intersecting L in the y axis. In this setting $z^* = -\frac{1}{2d}z$. The line aff$\{k, z\}$, resp. aff$\{k, z^*\}$ intersects the y axis at the points $u \in K$ (since $k, z \in K$) and $u^* \notin K$ (since $k \in K$ and $z^* \notin K$). As x is the centre of gravity of the $(d - 1)$-dimensional section, $(d - 1)\|u^*\| > \|u\|$ must hold. Write $k = (k_1, k_2)$ and $z = (z_1, z_2)$; the conditions imply that $k_1 = 2w$ and $z_1 \in [-w, 0]$. It is not hard to check that

$$\|u\| = \frac{|k_1 z_2 - z_1 k_2|}{k_1 - z_1} \quad \text{and} \quad \|u^*\| = \frac{|k_1 z_2 - z_1 y_2|}{2dk_1 + z_1}.$$

Then $(d-1)\|u^*\| \ge \|u\|$ implies $(d-1)(k_1 - z_1) \ge 2dk_1 + z_1$ or $-dz_1 \ge (d+1)k_1$ contradicting $k_1 = 2w$ and $z_1 \in [-w, 0]$.

□

The rest of the proof is what I like to call trivial volume estimates. We show that if $C(x) = K \cap H(a \leq t)$ is a minimal cap of width w and $v(x) \leq \varepsilon_0$, then the width of K in direction a is at least $3w$. Assume the contrary and let

$$A = \max\{\text{vol}\,_{d-1}(K \cap H(a = \tau))\}.$$

Then $1 = \text{vol}\,K \leq 3wA$ and $v(x) \geq wA/(d3^d)$ and so $v(x) \geq 1/(d3^d)$ contradicting $v(x) \leq \varepsilon_0$. □

Lemma 6.5 follows immediately.

Proof of Lemma 6.6. Assume that $u(x) \leq (2d)^{-d}\varepsilon_0$. Let $C(x) = K \cap H(a \leq t)$ be the minimal cap at x. Suppose its width is w. We show that the width of K in direction a is at least $3w$. This implies the lemma via Lemma 6.4. Assume the contrary. With the same setting as in the previous proof one sees that vol $_{d-1}(K \cap H(a = t)) \geq A/3^{d-1}$. As x is the centre of gravity of this section, the Löwner-John theorem implies that

$$\text{vol}\,_{d-1}(M(x) \cap H(a = t)) \geq \frac{A}{(3(d-1))^{d-1}}$$

and $u(x) \geq 2/(3d)^d$ follows. □

Proof of Lemma 6.7. Let $x, y \in \text{bd}\,K(v \geq \varepsilon)$ and assume $z = \frac{1}{2}(x + y)$ is also in bd $K(v \geq \varepsilon)$. Then there is a minimal cap $C(z)$ of volume ε. $C(z)$ cannot contain x (or y) in its interior as otherwise a smaller "parallel" cap would contain x (or y). Then $C(z)$ must contain both x and y in its bounding hyperplane. Then it is a minimal cap for both x and y. But both x and y cannot be the centre of gravity of the section $K \cap H(a = t)$ at the same time unless $x = y$. □

Proof of Lemma 6.8. Denote the set of outer normals to $K(v \geq \varepsilon)$ at $z \in$ bd $K(v \geq \varepsilon)$ by $N(z)$. It is well known (see [41]) that as $K(v \geq \varepsilon)$ is a convex body, $N(z)$ coincides with the cone hull of its extreme rays.

For $b \in S^{d-1}$ define C^b as the unique cap $C^b = K \cap H(b \leq t)$ such that $C^b \cap K(v \geq \varepsilon) \neq \emptyset$ but $C^b \cap \text{int}\,K(v \geq \varepsilon) = \emptyset$.

We show first that if b is the direction of an extreme ray of $N(z)$, then vol $C^b = \varepsilon$. To prove this we use a classical result of Alexandrov (see [46]) stating that at almost every point z on the boundary of a convex body the supporting hyperplane is unique. This shows that if $z \in$ bd $K(v \geq \varepsilon)$ is such a point then $N(z) \cap S^{d-1}$ is a unique vector, to be denoted by $b(z)$. In this case, of course, vol $C^{b(z)} = \varepsilon$.

Notice next that $N(z)$ is the polar of the minimal cone whose apex is z and which contains $K(v \geq \varepsilon)$ (see [41] again). So there is a vector $w \in S^{d-1}$ such that $w \cdot b = 0$ and $w \cdot x < 0$ for all $x \in N(z)$, $x \neq \lambda b$ ($\lambda > 0$) and such that there are points $z(t) \in$ bd $K(v \geq \varepsilon)$ for all small enough $t > 0$ with

$$\|(z(t) - z) - tw\| = o(t),$$

as $t \to 0$. Choose now a subsequence $z_k \in \operatorname{bd} K(v \geq \varepsilon)$ very close to $z(1/k)$ with unique tangent hyperplane to $K(v \geq \varepsilon)$ (using Alexandrov's theorem). We may assume that $\lim b(z_k)$ exists and equals $b_0 \in S^{d-1}$. It is easily seen that $b_0 \in N(z)$. Assume $b_0 \neq b$. Then, since $b(z_k) \in N(z_k)$,

$$0 \geq b(z_k) \cdot (y - z_k),$$

for every $y \in K(v \geq \varepsilon)$. In particular, for $y = z$ we get

$$0 \geq b(z_k) \cdot (y - z_k) = -\frac{1}{k} b(z_k) \cdot u - o(1/k) > -\frac{1}{2k} b(z_k) \cdot u - o(1/k) > 0,$$

for large enough k, a contradiction proving $b_0 = b$. The continuity of the map $b \mapsto \operatorname{vol} C^b$ implies $\operatorname{vol} C^b = \varepsilon$.

Now let $C = K \cap H(a \leq t)$ be the cap in the statement of the lemma. Then $-a \in N(x)$ and thus $-a$ is in the cone hull of extreme rays of $N(x)$. Thus, by Carathéodory's theorem, $-a$ is in the cone hull of $b_1, \ldots, b_d \in S^{d-1}$, where each b_i represents an extreme ray of $N(z)$. Then C is contained in $\cup C^{b_i}$. This implies that $\operatorname{vol} C \leq d \operatorname{vol} C^{b_i} = d\varepsilon$. Also, each C^{b_i} is a minimal cap, so by Lemma 6.4 it is contained in $M(x, 2d)$. Consequently,

$$C \subset \bigcup_{1}^{d} C^{b_i} \subset M(x, 2d).$$

$\square$

Proof of Lemma 6.9. The minimal cap $C(y) = K \cap H(a \leq t)$ is internally disjoint from the floating body $K(v \geq \varepsilon)$. Let τ be the maximal number with $H(a \leq \tau)$ internally disjoint from $K(v \geq \varepsilon)$. By Lemma 6.7 the cap $C = K \cap H(a \leq \tau)$ contains a unique point $x \in \operatorname{bd} K(v \geq \varepsilon)$. The proof of Lemma 6.8 gives that

$$y \in C(y) \subset C \subset \bigcup_{1}^{d} C^{b_i}$$

where each C^{b_i} is a minimal cap. $\square$

Proof of Lemma 6.10. We prove the first inclusion by showing that $K(u > 2\varepsilon) \subset K(v > \varepsilon)$. As both sets are convex, it suffices to see that $x \in \operatorname{bd} K(u > 2\varepsilon)$ implies $v(x) \geq \varepsilon$. The condition says that $u(x) = 2\varepsilon$ and Lemma 6.3 gives $2\varepsilon = u(x) \leq 2v(x)$.

The proof of the second inclusion is similar, just Lemma 6.5 is needed. $\square$

Remark. Most of the results here come from [21], [10] and [4]. But as I mentioned at the end of the previous section, some of the proofs here are new.

8 Proof of the Cap Covering Theorem

We start with a definition. If a cap $C = K \cap H(a \leq t)$ has width w, then $H(a = t - w)$ is a supporting hyperplane to K. The **centre** of the cap is the centre of gravity of the set $K \cap H(a = t - w)$. The blown-up copy of C from its centre by a factor $\lambda > 0$ is denoted by C^λ. It is clear that C^λ lies between hyperplanes $H(a = t - w)$ and $H(a = t - w + \lambda w)$, and convexity implies that

$$K \cap H(a \le t - w + \lambda w) \subset C^\lambda, \tag{10}$$

and so $\operatorname{vol} K \cap H(a \le t - w + \lambda w) \le \lambda^d \operatorname{vol} C$.

Choose a system of points $x_1, \dots, x_m$ on the boundary of the floating body $K(v \ge \varepsilon)$ which is maximal with respect to the property

$$M(x_i, 1/2) \cap M(x_j, 1/2) = \emptyset,$$

for each i, j distinct. Such a maximal system is finite since the M-regions are pairwise disjoint, all of them are contained in K and $\operatorname{vol} M(x_i, 1/2) = 2^{-d} u(x_i) \ge (6d)^{-d} v(x) = (6d)^{-d} \varepsilon$.

Claim 8.1 *For each $y \in K(\varepsilon)$ there is an $x \in \operatorname{bd} K(v \ge \varepsilon)$ with $y \in M(x)$.*

Proof. Assume this is false for some $y \in K(\varepsilon)$. So for each $x \in \operatorname{bd} K(v \ge \varepsilon)$, we have $y \notin M(x) = K \cap (2x - K)$ implying that $2x - y \notin K$, for each $x \in K(v \ge \varepsilon)$. In other words, a homothetic copy of $K[\varepsilon]$ blown up from y by a factor of 2 is disjoint from K. Let $H(a \le t)$ be the halfspace containing K and disjoint from the homothetic copy, and let $H(a \le \tau)$ be the parallel halfspace disjoint from $K[\varepsilon]$ with its bounding hyperplane tangent to $K[\varepsilon]$. The cap $K \cap H(a \le \tau)$ has volume at most $d\varepsilon$ by Lemma 6.8. The width of $K \cap H(a \le t)$ is at most twice the width of $K \cap H(a \le \tau)$. Thus we have, using (10),

$$1 = \operatorname{vol} K = \operatorname{vol}(K \cap H(a \le t)) \le 2^d \operatorname{vol}(K \cap H(a \le \tau)) \le 2^d d\varepsilon,$$

contradicting $\varepsilon \le \varepsilon_0$.   □

It will be easy to see now that

$$K(\varepsilon) \subset \bigcup_1^m M(x_i, 5).$$

Indeed, for each $y \in K(\varepsilon)$ there is an $x \in \operatorname{bd} K[\varepsilon]$ with $y \in M(x)$. By the maximality of the system $x_1, \dots, x_m$, there is an x_i with $M(x, 1/2) \cap M(x_i, 1/2) \ne \emptyset$. Lemma 6.2 shows then that $y \in M(x_i, 5)$.

We have now a covering of $K(\varepsilon)$ with M-regions. We are going to turn it into a covering with caps. The minimal cap at x_i is given by $C(x_i) = K \cap H(a_i \le t_i)$, let w_i be its width. Define

$$C_i' = M(x_i, 1/2) \cap H(a_i \le t_i) \text{ and } C_i = K \cap H(a_i \le t_i + 5w_i).$$

It is evident that the C_i' are pairwise disjoint convex sets, each contained in C_i and $\operatorname{vol} C_i' \ge \frac{1}{2} \operatorname{vol} M(x_i, \frac{1}{2}) \ge \frac{1}{2}(2d)^{-d}\varepsilon$ by Lemma 6.5. On the other hand,

$M(x_i, 5)$ lies between hyperplanes $H(a_i = t_i - w_i)$ and $H(a_i = t_i + 5w_i)$ and so it is contained in C_i. Finally, (10) shows that $\text{vol}\,C_i \le 6^d \,\text{vol}\,C(x_i) = 6^d \varepsilon$.

So far this is the proof of (i) and (ii) of the theorem. We now show how one can enlarge C_i to satisfy (iii).

This is quite simple. With the previous notation, take $C_i = K \cap H(a_i \le t_i + (10d - 1)w_i)$. The new C_i satisfy (i) and (ii) and $\text{vol}\,C_i \ll \varepsilon$. Moreover, $M(x_i, 10d) \subset C_i$.

Consider now a cap C, disjoint from $K(v > \varepsilon)$. We may assume that our C is maximal in the sense that $C \cap K(v \ge \varepsilon)$ is nonempty. Then, by Lemma 6.7, the intersection $C \cap K(v \ge \varepsilon)$ is a single point, say x, and by Lemma 6.8

$$C \subset M(x, 2d).$$

By the maximality of the system $x_1, \ldots, x_m$, there is an x_i with $M(x, 1) \subset M(x_i, 5)$. We claim that $M(x, 2d) \subset M(x_i, 10d)$. This will prove what we need.

The claim follows from a more general statement.

Fact. *Assume A and B are centrally symmetric convex sets with centre a and b respectively. If $B \subset A$ and $\lambda \ge 1$, then*

$$b + \lambda(B - b) \subset a + \lambda(A - a). \tag{11}$$

Proof. We may assume $a = 0$. Let $c \in B$, we have to prove that $b + \lambda(c - b) \in \lambda A$. B is symmetric, so $2b - c \in B \subset A$, and A is symmetric, so $c - 2b \in A$. Also, A is convex and $c \in B \subset A$, thus $(1/2)(c + (c - 2b)) = c - b \in A$. Then $c \in A$ and $c - b \in A$ imply $\lambda c \in \lambda A$ and $\lambda(c - b) \in \lambda A$. But $b + \lambda(c - b)$ lies on the segment connecting λc and $\lambda(c - b)$,

$$b + \lambda(b - c) = \frac{1}{\lambda}(\lambda c) + \left(1 - \frac{1}{\lambda}\right)\lambda(c - b) \in A,$$

proving the fact. □

Proof of Corollary 5.1. Let $C_1, \ldots, C_m$ be the economic cap covering from Theorem 5.1. We will show that

$$K(\lambda \varepsilon) \subset \bigcup_1^m C_i^{d\lambda}.$$

This will prove what we want.

Consider $x \in K(\lambda\varepsilon)$, we may assume $x \notin \bigcup C_i$. The minimal cap $C(x) = K \cap H(a \le t)$ has centre z and width w. The segment $[x, z]$ intersects bd $K(v \ge \varepsilon)$ at the point y, and let $y \in H(a = t - w')$ and set $t' = w - w'$. Now

$$\varepsilon = v(y) \le \operatorname{vol}(K \cap H(a \le t - w')) = \int_{t-w}^{t-w'} \operatorname{vol}_{d-1}(K \cap H(a = \tau))\mathrm{d}\tau$$
$$\le t' \max\{\operatorname{vol}_{d-1}(K \cap (H(a = \tau)) : t - w \le \tau \le t - w'\}$$
$$\le t' \max\{\operatorname{vol}_{d-1}(K \cap (H(a = \tau)) : t - w \le \tau \le t\}.$$

On the other hand

$$\lambda\varepsilon \ge v(x) = \operatorname{vol}(K \cap H(a \le t))$$
$$\ge \frac{1}{d} t' \max\{\operatorname{vol}_{d-1}(K \cap (H(a = \tau)) : t - w \le \tau \le t\},$$

where the last inequality holds since the double cone whose base is the maximal section $K \cap H(a = \tau)$ is contained in $C(x)$. Now $t/t' = \|z - x\|/\|z - y\|$ and we get

$$\|z - x\| \le d\lambda\|z - y\|.$$

Consider now the cap $C_i = K \cap H(a_i \le t_i)$ that contains y. Let z_i be the centre of C_i and write y_i for the intersection of $[z_i, x] \cap H(a_i = t_i)$. The line aff$\{z, x\}$ intersects the hyperplanes $H(a_i = t_i)$, $H(a_i = t_i - w_i)$ respectively at y' and z'. It is easy to check that the points z', z, y, y', x come in this order on aff$\{z, x\}$. Consequently,

$$\frac{\|x - z_i\|}{\|y_i - z_i\|} = \frac{\|x - z'\|}{\|y' - z'\|} \le \frac{\|x - z\| + \|z - z'\|}{\|y - z\| + \|z - z'\|} \le \frac{\|x - z\|}{\|y - z\|} \le d\lambda.$$

So indeed $x \in \bigcup_1^m C_i^{d\lambda}$. □

9 Auxiliary Lemmas from Probability

We will need an upper and lower bound for the quantity $\mathbb{P}\{x \notin K_n\}$ where x is a fixed point of K and the random polytope K_n varies. The lower bound is simple: if $C(x)$ is the minimal cap of x, then clearly

$$\mathbb{P}\{x \notin K_n\} \ge \mathbb{P}\{X_n \cap C(x) = \emptyset\} = (1 - v(x))^n, \tag{12}$$

where X_n is the random sample of n points from K generating K_n.

We mention at once that this implies the lower bound in Theorem 3.2, or, what is the same, in Theorem 3.1.

Proof (of the lower bound in Theorem 3.1). Using the above inequality we get, for all $t > 0$ that

$$\mathbb{E}(K, n) = \int_K \mathbb{P}\{x \notin K_n\} dx \geq \int_K (1 - v(x))^n dx$$

$$\geq \int_{K(t)} (1 - v(x))^n dx \geq \int_{K(t)} (1 - t)^n dx \geq (1 - t)^n \operatorname{vol} K(t).$$

Choosing here $t = 1/n$ gives the lower bound with $c_1 = 1/4$, for instance. Note that c_1 is universal, it does not depend on the dimension.   □

We need an upper bound on $\mathbb{P}\{x \notin K_n\}$,

$$\mathbb{P}\{x \notin K_n\} \leq 2 \sum_{i=0}^{d-1} \binom{n}{i} \left(\frac{u(x)}{2}\right)^i \left(1 - \frac{u(x)}{2}\right)^{n-i}. \tag{13}$$

Proof. We are going to use the following equality which is due to Wendel [58]. Assume M is an 0-symmetric d-dimensional convex body, and let X_n be a random sample of uniform, independent points from M. Then

$$\mathbb{P}\{0 \notin \operatorname{conv} X_n\} = 2^{-n+1} \sum_{i=0}^{d-1} \binom{n-1}{i}. \tag{14}$$

(I will give a proof of this result at the end of the section.)

Let $x \in K$ be fixed and define $N(x) = X_n \cap M(x)$. Setting $n(x) = |N(x)|$ we have

$$\mathbb{P}\{x \notin K_n\} = \sum_{m=0}^{n} \mathbb{P}\{x \notin K_n | n(x) = m\} \mathbb{P}\{n(x) = m\}$$

$$\leq \sum_{m=0}^{n} \mathbb{P}\{x \notin \operatorname{conv} N(x) | n(x) = m\} \mathbb{P}\{n(x) = m\}$$

$$= 2 \sum_{m=0}^{n} 2^{-m} \sum_{i=0}^{d-1} \binom{m-1}{i} \mathbb{P}\{n(x) = m\}.$$

We used Wendel's equality. $\mathbb{P}\{n(x) = m\}$ is a binomial distribution with parameter $u = u(x)$. Thus

$$\mathbb{P}\{x \notin K_n\} \le 2 \sum_{m=0}^{n} 2^{-m} \sum_{i=0}^{d-1} \binom{m-1}{i} \binom{n}{m} u^m (1-u)^{n-m}$$

$$= 2 \sum_{i=0}^{d-1} \sum_{m=0}^{n} \binom{m-1}{i} \binom{n}{m} \left(\frac{u}{2}\right)^m (1-u)^{n-m}$$

$$\le 2 \sum_{i=0}^{d-1} \sum_{m=i+1}^{n} \binom{m}{i} \binom{n}{m} \left(\frac{u}{2}\right)^m (1-u)^{n-m}$$

$$= 2 \sum_{i=0}^{d-1} \binom{n}{i} \sum_{m=i}^{n} \binom{n-i}{m-i} \left(\frac{u}{2}\right)^m (1-u)^{n-m}$$

$$= 2 \sum_{i=0}^{d-1} \binom{n}{i} \sum_{k=0}^{n-i} \binom{n-i}{k} \left(\frac{u}{2}\right)^{k+i} (1-u)^{n-i-k}$$

$$= 2 \sum_{i=0}^{d-1} \binom{n}{i} \left(\frac{u}{2}\right)^i \left(1-\frac{u}{2}\right)^{n-i}.$$

$\square$

Proof of Wendel's equality. We start with the following simple fact. Assume $H_1, \ldots, H_n$ are hyperplanes in $\mathbb{R}^d$ in general position that is, every d of them has exactly one point in common and no $d+1$ of them intersect. The set $\mathbb{R}^d \setminus \cup_1^n H_i$ is the disjoint union of pairwise disjoint open sets, to be called cells. Each cell is a convex polyhedron.

Claim 9.1 *The number of cells is exactly $\sum_{i=0}^{d} \binom{n}{i}$.*

Proof. We prove this by induction on d. Everything is clear when $d = 1$. Assume $d > 1$ and the statement is true in $\mathbb{R}^{d-1}$. Let $a \in \mathbb{R}^d$ be a unit vector in general position and let C be one of the cells. If $\min\{a \cdot x : x \in C\}$ is finite, then it is reached at a unique vertex of C which is the intersection of some d hyperplanes $H_{i_1}, \ldots, H_{i_d}$. There are $\binom{n}{d}$ such minima and each one comes from a different cell. So exactly $\binom{n}{d}$ cells have a finite minimum in direction a. Let K be a number smaller than each of these $\binom{n}{d}$ minima. The rest of the cells are unbounded in this direction, so they all intersect the hyperplane H with equation $a \cdot x = K$. The induction hypothesis can be used in H (which is a copy of $\mathbb{R}^{d-1}$) to show that the number cells, unbounded in direction a is $\sum_{i=0}^{d-1} \binom{n}{i}$. This finishes the proof of the claim. $\square$

Now for the proof of Wendel's equality. The basic observation is that choosing the points $x_1, \ldots, x_n$ and choosing the points $\varepsilon_1 x_1, \ldots, \varepsilon_n x_n$ (where each $\varepsilon_i = \pm 1$) is equally likely. So we want to see that, out of the 2^n such choices, how many will not have the origin in their convex hull. If $0 \notin [\varepsilon_1 x_1, \ldots, \varepsilon_n x_n]$, then all the $\varepsilon_i x_i$ are contained in the open halfspace $\{x \in \mathbb{R}^d : a \cdot x > 0\}$ for some unit vector $a \in \mathbb{R}^d$. The conditions $a \cdot (\varepsilon_i x_i) > 0$ show that all halfspaces containing each $\varepsilon_i x_i$ $(i = 1, \ldots, n$, the ε_i are fixed) have their normal a in the cone

$$\bigcap_1^n \{y \in \mathbb{R}^d : y \cdot (\varepsilon_i x_i) > 0\}.$$

So the question is how many such cones there are. Or, to put it differently, when you delete the hyperplanes $H_i = \{y \in \mathbb{R}^d : y \cdot x_i = 0\}$ $i = 1, \ldots, n$ from $\mathbb{R}^d$ you get pairwise disjoint open cones C_α; how many such cones are there? Surprisingly, this number is independent of the position of the x_i (if they are in general position and, in the given case, they are). We claim that this number is equal to

$$2 \sum_{i=0}^{d-1} \binom{n-1}{i}.$$

This will, of course, prove Wendel's equality (14).

Consider now the hyperplane $H^* = \{y \in \mathbb{R}^d : y \cdot x_n = 1\}$. The cones C_α come in pairs, C_α together with $-C_\alpha$ and only one of them intersects H^*. So the question is this. If you delete the hyperplanes H_i, $i = 1, \ldots, n-1$ from H^*, how many connected components are left? This is answered by Claim 9.1, there are exactly

$$\sum_{i=0}^{d-1} \binom{n-1}{i}$$

such cells.   □

10 Proof of Theorem 3.1

We only have to prove the upper bound. We start with the integral representation of $\mathbb{E}(K, n)$ and use the upper bound from (13),

$$\mathbb{E}(K, n) = \int_K \mathbb{P}\{x \notin K\} dx$$

$$\leq \int_K 2 \sum_{i=0}^{d-1} \binom{n}{i} \left(\frac{u(x)}{2}\right)^i \left(1 - \frac{u(x)}{2}\right)^{n-i} dx$$

$$\leq 2 \sum_{i=0}^{d-1} \binom{n}{i} \int_K \left(\frac{u(x)}{2}\right)^i \left(1 - \frac{u(x)}{2}\right)^{n-i} dx.$$

K is the disjoint union of the sets K_λ, for $\lambda = 1, 2, \ldots, n$, where

$$K_\lambda = K((\lambda - 1)/n \leq u < \lambda/n).$$

We integrate separately on each K_λ using that, on K_λ, $u(x) < \lambda/(2n)$ and $1 - u(x)/2 \leq \exp\{-(\lambda - 1)/(2n)\}$. Thus

$$\int_{K_\lambda} \left(\frac{u(x)}{2}\right)^i \left(1 - \frac{u(x)}{2}\right)^{n-i} dx \ll \left(\frac{\lambda}{2n}\right)^i \exp\{-(\lambda-1)/4\} \operatorname{vol} K(u \le \lambda/n).$$

We continue the inequality for $\mathbb{E}(K,n)$,

$$\mathbb{E}(\overset{.}{K},n) \ll 2\sum_{i=0}^{d-1} \binom{n}{i} \sum_{\lambda=1}^{n} \left(\frac{\lambda}{2n}\right)^i \exp\{-(\lambda-1)/4\} \operatorname{vol} K(u \le \lambda/n)$$

$$\ll \sum_{\lambda=1}^{n} \sum_{i=0}^{d-1} \binom{n}{i} \left(\frac{\lambda}{2n}\right)^i \exp\{-(\lambda-1)/4\} \operatorname{vol} K(u \le \lambda/n)$$

$$= \sum_{\lambda=1}^{\Lambda} .. + \sum_{\lambda=\Lambda+1}^{n} .. ,$$

where $\Lambda = (2d)^{-d}\varepsilon_0 n = d^{-1}(6d)^{-d}n$. Note that

$$\binom{n}{i}\left(\frac{\lambda}{2n}\right)^i \ll \lambda^i.$$

So we have, using Lemma 6.10 and Corollary 5.1,

$$\sum_{\lambda=1}^{\Lambda} .. \ll \sum_{\lambda=1}^{\Lambda} d\lambda^{d-1} \exp\{-(\lambda-1)/4\} \operatorname{vol} K(v \le (2d)^d\lambda/n)$$

$$\ll \sum_{\lambda=1}^{\Lambda} \lambda^{d-1} \exp\{-(\lambda-1)/4\}\lambda^d \operatorname{vol} K(v \le 1/n)$$

$$\ll \operatorname{vol} K(v \le 1/n).$$

Estimating the second sum is simpler, since one can use the trivial inequality $\operatorname{vol} K(u \le \lambda/n) \le 1$ to get

$$\sum_{\Lambda+1}^{n} .. \ll \sum_{\Lambda+1}^{n} \lambda^{d-1} \exp\{-(\lambda-1)/4\} \operatorname{vol} K(v \le (2d)^d\lambda/n)$$

$$\ll \sum_{\Lambda+1}^{n} \lambda^{d-1} \exp\{-(\lambda-1)/4\}$$

$$\ll \operatorname{vol} K(v \le 1/n).$$

Thus we have $\mathbb{E}(K,n) \ll \operatorname{vol} K(1/n)$. $\qquad\square$

Remark. This proof comes from the paper [10].

11 Proof of Theorem 4.1

We start with introducing further notation. Fix $a \in S^{d-1}$ and let $H(a = t_0)$ be the hyperplane whose intersection with K has maximal $(d-1)$-dimensional volume among all hyperplanes $H(a = t)$. Assume the width of K in direction a is at most $2t_0$; if this were not the case we would take $-a$ instead of a. As a will be fixed during this proof we simply write $H(t) = H(a = t)$. Assume further that $H(0)$ is the tangent hyperplane to K. Define

$$Q(t) = H(t) \cap K \text{ and } q(t) = \text{vol}_{d-1} Q(t).$$

The choice of t_0 ensures that, for $t \in [0, t_0]$,

$$q(t) \geq \left(\frac{t}{t_0}\right)^{d-1} q(t_0) \text{ and } 2t_0 q(t_0) \geq \text{vol } K = 1. \tag{15}$$

Claim 11.1 *For $\varepsilon > 0$ and for $t \in [0, t_0]$,*

$$Q(t)\left(u_{Q(t)} \leq \varepsilon/(2t)\right) \subset K(u_K \leq \varepsilon) \cap H(t).$$

Proof. We are going to show that $x \in H(t) \cap K$ implies $u_K(x) \leq 2t u_{Q(t)}(x)$. This of course proves the lemma.

Notice first that $M(x)$ lies between hyperplanes $H(0)$ and $H(2t)$. Thus

$$u(x) = \int_0^{2t} \text{vol}_{d-1}(M(x) \cap H(\tau))d\tau \leq 2t \text{ vol}_{d-1}(M(x) \cap H(t)),$$

since $M(x)$ is centrally symmetric, so its largest section is the middle one. Observe next that

$$M(x) \cap H(t) = M_{Q(t)}(x),$$

which follows from (8). Consequently $u(x) \leq 2t \text{ vol}_{d-1} M_{Q(t)}(x) = 2t u_{Q(t)}(x)$. □

We show next that, for $\varepsilon \in [0, 1]$,

$$\text{vol } K(u \leq \varepsilon) \gg \varepsilon \left(\ln \frac{1}{\varepsilon}\right)^{d-1}. \tag{16}$$

Then Lemma 6.10 implies that, for $\varepsilon \leq (2d)^{-2d}\varepsilon_0$,

$$\text{vol } K(v \leq \varepsilon) \geq \text{vol } K(u \leq (2d)^{-d}\varepsilon) \gg \varepsilon \left(\ln \frac{1}{\varepsilon}\right)^{d-1}.$$

When $\varepsilon \geq (2d)^{-2d}$ the statement of the theorem follows from the fact that $\varepsilon \mapsto \text{vol } K(v \leq \varepsilon)$ is an increasing function of ε.

We prove (16) by induction on d. The case $d = 1$ trivial. We will need the induction hypothesis in its invariant form (7): for $Q \in \mathcal{K}^{d-1}$ and for $\eta > 0$

$$\frac{\text{vol } Q(u_Q \leq \eta \text{ vol } Q)}{\text{vol } Q} \geq c_{d-1}\eta \left(\ln \frac{1}{\eta}\right)^{d-2}.$$

We have

$$\operatorname{vol} K(u \le \varepsilon) \ge \operatorname{vol}\left(K(u \le \varepsilon) \cap H(a \le t)\right)$$

$$= \int_0^{t_0} \operatorname{vol}_{d-1}\left(K(u \le \varepsilon) \cap H(t)\right) dt$$

$$\ge \int_0^{t_0} \operatorname{vol}_{d-1} Q(t)\left(u_{Q(t)} \le \varepsilon/(2t)\right) dt$$

according to Claim 11.1. Define $\eta = \eta(t) = \varepsilon/(2tq(t))$ and let t_1 be the unique solution to $\eta(t) = 1$ between 0 and t_0. The induction hypothesis implies that for $t \in [t_1, t_0]$,

$$\operatorname{vol}_{d-1} Q(t)(u_{Q(t)} \le \eta q(t)) \ge c_{d-1} q(t) \eta \left(\ln \frac{1}{\eta}\right)^{d-2}$$

$$= c_{d-1} \frac{\varepsilon}{2t}\left(\ln \frac{2tq(t)}{\varepsilon}\right)^{d-2}$$

$$\ge c_{d-1}\frac{\varepsilon}{2t}\left(\ln\left(\frac{2t}{\varepsilon}\left(\frac{t}{t_0}\right)^{d-1}\right)\right)^{d-2},$$

where the last inequality comes from (15). We continue with $\operatorname{vol} K(u \le \varepsilon)$,

$$\operatorname{vol} K(u \le \varepsilon) \ge \int_{t_1}^{t_0} c_{d-1}\frac{\varepsilon}{2t}\left(\ln\left(\frac{2t^d q(t_0)}{\varepsilon t_0^{d-1}}\right)\right)^{d-2} dt.$$

Define α by $\alpha^d = 2q(t_0)/(\varepsilon t_0^{d-1})$ and set $t_2 = 1/\alpha$. In view of (15) again, $t_1 \le t_2 \le t_0$. Substitute now $\tau = \alpha t$ with $\tau_i = \alpha t_i$, $i = 0, 2$. We finally have

$$\operatorname{vol} K(u \le \varepsilon) \ge \int_{t_1}^{t_0} c_{d-1}\frac{\varepsilon}{2\tau}(\ln \tau)^{d-2} d\tau$$

$$= \frac{\varepsilon c_{d-1}}{2(d-1)}\left(\ln \frac{t_0(2q(t_0))^{1/d}}{(\varepsilon t_0^{d-1})^{1/d}}\right)^{d-1} \ge \frac{\varepsilon c_{d-1}}{2(d-1)}\left(\frac{1}{d}\ln \frac{1}{\varepsilon}\right)^{d-1},$$

where the last inequality follows from (15).

Remark. This is the only proof known for Theorem 4.1 and it comes from [10]. The best possible constant in the inequality probably goes with the simplex. Note that in the proof we made full use of the two-way street between minimal caps and M-regions.

12 Proof of (4)

This is a repetition of the previous computation, just the inequalities go the other direction. We need to know $\operatorname{vol} \triangle(v \le t)$, where $\triangle$ is the d-dimensional simplex.

Lemma 12.1. *For all* $t \leq e^{-d+1}$,

$$\frac{\operatorname{vol}\triangle(v \leq t\operatorname{vol}\triangle)}{\operatorname{vol}\triangle} \ll t\left(\ln\frac{1}{t}\right)^{d-1}.$$

We remark that the function on the left hand side of this inequality increases with t while the one on the right hand side increases on $[0, e^{-d+1}]$ and decreases afterwards. That is the reason for the condition $t \leq e^{-d+1}$.

Proof. We use induction on d; $d = 1$ is simple, $d = 2$ needs a bit of special care and is left to the reader. We may assume that $\triangle$ is the regular simplex of volume 1, as our inequality is in equivariant form. Let $w_0, \ldots, w_d$ be the vertices and a_i the unit outer normals to the facet opposite to w_i, $i = 0, \ldots, d$. Then $a_i \cdot w_i = h_i$ and $a_i \cdot w_j = h_i^*$ with $h_i^* > h_i$, and for every $x \in \triangle$ there is an i with

$$h_i \leq a_i \cdot x \leq h_i + \frac{d}{d+1}(h_i^* - h_i).$$

This is quite easy to check. Consequently,

$$\operatorname{vol}\triangle(v \leq t) \leq \sum_0^d \operatorname{vol}_{d-1}\{x \in \triangle : v(x) \leq t,\ a_i \cdot x \leq h_i + \frac{d}{d+1}(h_i^* - h_i)\}.$$

Each term in the last sum is the same, so we work with $i = 0$ only. Assume $w_0 = 0$, then $h_0 = 0$ as well and we set $h^* = h_0^*$ and drop the subscript 0. Define

$$Q_h = \triangle \cap H(a = h),$$

which is a regular and $(d-1)$-dimensional simplex. Note that

$$\operatorname{vol}\triangle \cap H(a \leq h) = \frac{h}{d}\operatorname{vol}_{d-1}Q_h = \left(\frac{h}{h^*}\right)^d$$

follows easily. If $x \in Q_h$ and $v(x) \leq t$ with minimal cap $C(x)$, then

$$t \geq v(x) \geq \frac{1}{d}\min(h, h^* - h)\operatorname{vol}_{d-1}Q_h \cap C(x) \geq \frac{1}{d}\min(h, h^* - h)v_{Q_h}(x).$$

For $0 \leq h \leq h^*/2$ the minimum is h, and for $h \geq h^*/2$ it is at least $h^*/(d+1)$. Using this we can estimate now

$$\operatorname{vol}\triangle(v \leq t) \leq (d+1)\int_0^{\frac{d}{d+1}h^*} \operatorname{vol}_{d-1}\{x \in Q_h : v(x) = h,\ a \cdot x = h\}dh$$

$$\ll \int_0^{h^*/2} \operatorname{vol}_{d-1}Q_h(v_{Q_h} \leq td/h)dh +$$

$$+ \int_{h^*/2}^{\frac{d}{d+1}h^*} \operatorname{vol}_{d-1}Q_h(v_{Q_h} \leq td(d+1)/h^*)dh.$$

The induction hypothesis gives us directly that the second integral is bounded by $\ll t(\ln 1/t)^{d-2}$. The first integral is to be split at h^0 which is defined by

$$e^{-d+1} = \frac{td}{h^0 \operatorname{vol}_{d-1} Q_{h^0}} = \left(\frac{h^0}{h^*}\right)^d.$$

The integral below h^0 is smaller than $\operatorname{vol} \triangle \cap H(a \le h^0) = e^{d-1} t$. On the remaining interval the induction hypothesis gives

$$\int_{h^0}^{h^*/2} [..] \le \int_{h_0}^{h^*/2} \frac{td}{h} \left(\ln \frac{h \operatorname{vol}_{d-1} Q_h}{td}\right)^{d-2} dh$$

$$= \frac{td}{d-1} \left[\left(\ln \frac{h^* \operatorname{vol}_{d-1} Q_{h^*/2}}{2td}\right)^{d-1} - \left(\ln \frac{h^0 \operatorname{vol}_{d-1} Q_{h^0}}{td}\right)^{d-1}\right]$$

$$\ll t \left(\ln \frac{1}{t}\right)^{d-1}.$$

$\square$

Remark. The last step of the proof can be used to show that most of $\triangle(v \le t)$ is concentrated near the vertices of the simplex in the following sense;

$$\operatorname{vol}\left[\triangle(v \le t) \cap \{x \in \triangle : \ln \frac{1}{t} \le ax \le \frac{d}{d+1} h^*\}\right] \le t \left(\ln \frac{1}{t}\right)^{d-2} \ln \ln \frac{1}{t}.$$

The proof is straightforward.

Now we turn to the proof of inequality (4).

We triangulate first the polytope P by simplices $\triangle_i$ using vertices of P only. Clearly, if $v(x) \le t$, then $v_{\triangle_i}(x) \le t$ for the simplex containing x. Consequently

$$\operatorname{vol} P(v \le t) \le \sum \operatorname{vol} \triangle_i (v_{\triangle_i} \le t) \le \sum t \left(\ln \frac{\operatorname{vol} \triangle_i}{t}\right)^{d-1}$$

$$\ll t \left(\ln \frac{1}{t}\right)^{d-1}.$$

The implied constant turns out to be proportional to the number of simplices needed for the triangulation. The argument for the slightly better constant in (4) goes as follows. The last remark gives that $P(v \le t)$ is concentrated near the vertices of P. Assume 0 is a vertex of P and $H(a \le 0)$ is a halfspace (with outer unit normal a) intersecting P only at 0. One shows first that, with $\tau = \ln \frac{1}{t}$,

$$P(v \le t) \cap H(a \le \tau) \ll t \left(\ln \frac{1}{t}\right)^{d-1},$$

where the implied constant is proportional to the number of towers of the section $P \cap H(a = h)$, h very small but positive. This can be proved by induction on d using the same integration on the sections $P \cap H(a = h)$ as in the proof of Lemma 12.1. The number of towers of this section is the number of towers of P incident to the vertex 0. Summing over all vertices gives the required constant.

Remark. These proofs are from [10] and [4].

13 Expectation of $f_k(K_n)$

The following simple identity is due to Efron [20]; for $K \in \mathcal{K}_1$,

$$\mathbb{E} f_0(K_n) = n \mathbb{E}(K, n - 1). \tag{17}$$

The proof is straightforward,

$$
\begin{aligned}
\mathbb{E} f_0(K_n) &= \sum_{i=1}^{n} \mathbb{P}\{x_i \text{ is a vertex of } K_n\} \\
&= n \mathbb{P}\{x_1 \text{ is a vertex of } K_n\} = n \mathbb{P}\{x_1 \notin [x_2, \dots, x_n]\} \\
&= n \mathbb{P}\{x \notin K_{n-1}\} = n \mathbb{E}(K, n - 1),
\end{aligned}
$$

where the last probability is taken with both K_{n-1} and x varying.

Theorem 3.2 determines then the order of magnitude of $\mathbb{E} f_0(K_n)$ as well. The expectation of $f_k(K_n)$, for $k = 1, \dots, d - 1$, must be close to that of $\mathbb{E} f_0(K_n)$ since, as n goes to infinity, K_n looks locally like a "random" triangulation of $\mathbb{R}^{d-1}$ where you don't expect vertices of high degree. We have the following theorem from [4].

Theorem 13.1. *For large enough n, for all $K \in \mathcal{K}_1$ and for all $k = 0, 1, \dots, d - 1$,*

$$n \operatorname{vol} K(1/n) \ll \mathbb{E} f_k(K_n) \ll n \operatorname{vol} K(1/n).$$

The lower bound in case $k = 0$ follows from Efron's identity and the lower bound in Theorem 3.1. The following fact will simplify the proof of Theorem 13.1.

Lemma 13.1. *For all $0 \le i < j \le d - 1$,*

$$f_i(K_n) \le \binom{j+1}{i+1} f_j(K_n).$$

Proof. Almost surely K_n is a simplicial polytope. Double counting the pairs (F_i, F_j) where F_i and F_j are faces of dimension i and j of K_n with $F_i \subset F_j$ we have

$$f_i(K_n) = \sum_{F_i} 1 \le \sum_{(F_i, F_j)} 1 \le \binom{j+1}{i+1} f_j(K_n).$$

$\square$

So we see that for the upper bound in Theorem 13.1 it suffices to show the following.

Lemma 13.2. *For large enough n and for all $K \in \mathcal{K}_1$,*

$$\mathbb{E}f_{d-1}(K_n) \ll n \operatorname{vol} K(1/n).$$

For the proof of this lemma we need a corollary to the economic cap covering theorem. To state it, some preparation is necessary.

Assume $x_1, \ldots, x_k \in K$, set $L = \operatorname{aff}\{x_1, \ldots, x_k\}$ and define

$$v(L) = \max\{v(x) : x \in L\}.$$

We write K^k for the set of ordered k-tuples $(x_1, \ldots, x_k)$ with $x_i \in K$ for each i.

Corollary 13.1. *If $K \in \mathcal{K}_1$, $k = 1, 2, \ldots, d$ and $\varepsilon \le \varepsilon_0$, then*

$$\{(x_1, \ldots, x_k) \in K^k : v(L) \le \varepsilon\} \subset \bigcup_1^m (C_i, \ldots, C_i),$$

where $C_1, \ldots, C_m$ is the set of caps from Theorem 5.1.

Proof. This is where we use part (iii) of Theorem 5.1. If $v(L) \le \varepsilon$, then L and $K(v > \varepsilon)$ are disjoint. By separation, there is a halfspace H, containing L which is disjoint from $K(v > \varepsilon)$. Then the cap $C = K \cap H$ is also disjoint from $K(v > \varepsilon)$. Clearly, C contains $x_1, \ldots, x_k$. Consider now C_i from the cap covering with $C \subset C_i$. It is evident that

$$(x_1, \ldots, x_k) \in (C, \ldots, C) \subset (C_i, \ldots, C_i).$$

$\square$

14 Proof of Lemma 13.2

We are going to use (1) when $\phi(F)$ is equal to one if $F = [x_1, \ldots, x_d]$ is a facet of K_n and 0 otherwise. Recall that $V(x_1, \ldots, x_d)$ is the volume of the smaller cap cut off from K by $\operatorname{aff}\{x_1, \ldots, x_d\}$ which is a hyperplane with probability one. Now Theorem 2.1 says that

$$\mathbb{E}f_{d-1}(K_n) = \binom{n}{d} \int_K \cdots \int_K [(1 - V)^{n-d} + V^{n-d}] \mathrm{d}x_1 \ldots \mathrm{d}x_d. \qquad (18)$$

We split the domain of integration into two parts: K_1 is the subset of K^d where the function V is smaller than $(c \ln n)/n$, and K_2 is where $V \geq (c \ln n)/n$. The constant c will be specified soon. Clearly $V \leq 1/2$. The integrand over K_2 is estimated as follows:

$$(1 - V)^{n-d} + V^{n-d} \leq \exp\{-(n-d)V\} + 2^{-(n-d)}$$
$$\leq 2 \exp\{-(n-d)(c \ln n)/n\}$$
$$= 2n^{-c(n-d)/n},$$

which is smaller than $n^{-(d+1)}$ if c is chosen large enough (depending only on d). Then the contribution of the integral on K_2 to $\mathbb{E}f_{d-1}(K_n)$ is at most $1/n$, so it is very small since, trivially, $\mathbb{E}f_{d-1}(K_n)$ is at least one.

Now let h be an integer with $2^{-h} \leq (c \ln n)/n$. For each such h let $\mathcal{M}_h$ be the collection of caps $\{C_1, \ldots, C_{m(h)}\}$ forming the economic cap covering from Theorem 5.1 with $\varepsilon = 2^{-h}$.

Assume now that $(x_1, \ldots, x_d) \in K_1$. We will denote by $C(x_1, \ldots, x_d)$ the cap cut off from K by the hyperplane aff$\{x_1, \ldots, x_d\}$, clearly vol $C(x_1, \ldots, x_d)$ $= V(x_1, \ldots, x_d)$. We associate with $(x_1, \ldots, x_d)$ the maximal h such that, for some $C_i \in \mathcal{M}_h$, $C(x_1, \ldots, x_d) \subset C_i$. It follows that

$$V(x_1, \ldots, x_d) \leq \text{vol } C_i \ll 2^{-h} \tag{19}$$

and, by the maximality of h,

$$V(x_1, \ldots, x_d) \geq 2^{-h-1}, \tag{20}$$

since otherwise $C(x_1, \ldots, x_d)$ would be contained in a cap from $\mathcal{M}_{h+1}$.

For such an $(x_1, \ldots, x_d)$ we have

$$(1-V)^{n-d} + V^{n-d} \leq 2(1-V)^{n-d} \leq 2(1-2^{-h-1})^{n-d} \leq 2\exp\{-(n-d)2^{-h-1}\}.$$

Now we integrate over K_1 by integrating each $(x_1, \ldots, x_d)$ on its associated $C_i \in \mathcal{M}_h$. In the expression (18) the integral on $C_i \in \mathcal{M}_h$ is bounded by

$$2\exp\{-(n-d)2^{-h-1}\}(\text{vol } C_i)^d \ll \exp\{-(n-d)2^{-h-1}\}(2^{-h})^d,$$

as all the x_i come from C_i. Summing this for all $C_i \in \mathcal{M}_h$ and all $h \geq h_0$ where $h_0 = \lfloor (c \ln n)/n \rfloor$ we get that

$$\mathbb{E}f_{d-1}(K_n) \ll \binom{n}{d} \sum_{h_0}^{\infty} \sum_{C_i \in \mathcal{M}_h} \exp\{-(n-d)2^{-h-1}\}2^{-hd}$$

$$\ll \binom{n}{d} \sum_{h_0}^{\infty} \exp\{-(n-d)2^{-h+1}\}2^{-hd}|\mathcal{M}_h|$$

$$\ll \binom{n}{d} \sum_{h_0}^{\infty} \exp\{-(n-d)2^{-h+1}\}2^{-h(d-1)} \text{ vol } K(2^{-h}),$$

where the last inequality follows from (5).

The rest of the proof is a direct computation using Corollary 5.1. We sum first for $h \geq h_1$ where h_1 is defined by $2^{-h_1} \leq 1/n < 2^{-h_1+1}$. The sum from h_1 to infinity is estimated via

$$\sum_{h_1}^{\infty} \cdot \cdot \leq \sum_{h_1}^{\infty} \exp\{-(n-d)2^{-h+1}\}2^{-h(d-1)} \operatorname{vol} K(1/n)$$

$$\leq \operatorname{vol} K(1/n) \sum_{h_1}^{\infty} 2^{-h(d-1)} \leq n^{-(d-1)} \operatorname{vol} K(1/n).$$

When $h_0 \leq h < h_1$, we set $h = h_1 - k$, so k runs from 1 to $k_1 = \ln\ln n + \ln c$. Then we use Corollary 5.1 to show that

$$\operatorname{vol} K(2^{-h}) \leq \operatorname{vol} K(2^k/n) \ll 2^{kd} \operatorname{vol} K(1/n).$$

Thus

$$\sum_{h_0}^{h_1-1} \cdot \cdot \ll \sum_{k=1}^{k_1} \exp\{-(n-d)2^{-h_1+k-1}\}2^{(-h_1+k)(d-1)}2^{kd} \operatorname{vol} K(1/n)$$

$$\ll n^{-(d-1)} \operatorname{vol} K(1/n) \sum_{k=1}^{k_1} \exp\{-(n-d)2^k/n\}2^{k(d-1)}2^{kd}$$

$$\ll n^{-(d-1)} \operatorname{vol} K(1/n) \sum_{k=1}^{\infty} \exp\{-2^{k-1} + 2dk\ln 2\}$$

$$\ll n^{-(d-1)} \operatorname{vol} K(1/n).$$

Remark. This proof shows that $\mathbb{E} f_{d-1}(K_n) \ll \operatorname{vol} K(1/n)$. Then $\mathbb{E} f_0(K_n) \ll \operatorname{vol} K(1/n)$ follows from Lemma 13.1. Efron's identity implies that $\mathbb{E} f_0(K_n) \approx \mathbb{E}(K,n)$. Thus the proof of Lemma 13.2 is a new proof of the upper bound in Theorem 3.2. We mention further that the proof of $\mathbb{E} f_{d-1}(K_n) \ll \operatorname{vol} K(1/n)$ presented here is new and uses the cap covering theorem in a different and apparently more effective way than the previous proof from [4].

15 Further Results

There is a huge number of papers devoted to random polytopes and this survey is too short to explain or even mention most of them. We only consider the random variables $f_i(K_n)$ and $\operatorname{vol} K_n$, or rather $\operatorname{vol}(K \setminus K_n)$, although several other functionals of K_n have been investigated, like surface area, mean width, and other intrinsic volumes, and the Hausdorff distance of K and K_n. The interested reader should consult the survey papers [60], [49].

Also, other models of random polytopes have been thoroughly studied. The random sample may come from the normal distribution, or from the boundary of K. Here a recent result of Schütt and Werner [54] should be mentioned. In a long and intricate proof they show the precise asymptotic behaviour of $\mathrm{vol}(K \setminus K_n)$ when K is a smooth convex body and the random points are chosen from the boundary of K according to some probability distribution.

Many papers have been devoted to deriving precise asymptotic formulae for $\mathbb{E}(K, n)$ and $\mathbb{E}f_i(K_n)$ for special classes of convex bodies: for smooth convex bodies and polytopes. The starting point is usually the formula in Theorem 2.1: assuming $V \leq \frac{1}{2}$ the term V^{n-d} is exponentially small and we obtain

$$\mathbb{E}\phi(K_n) = \binom{n}{d} \int_K \cdots \int_K (1 - V)^{n-d}\phi(F)\mathrm{d}x_1 \ldots \mathrm{d}x_d + O(2^{-n}).$$

Here one can apply an integral transformation using the Blaschke-Petkantschin identity [43] and the integral becomes an integral over all hyperplanes E meeting K;

$$\mathbb{E}\phi(K_n) = C_d \binom{n}{d} \int (1 - V)^{n-d}g_K(E)\mu(\mathrm{d}E) + O(2^{-n}),$$

where $\mu(\mathrm{d}E)$ represents integration over the Grassmannian of hyperplanes and

$$g_K(E) = (d-1)! \int_{K \cap E} \cdots \int_{K \cap E} \phi([x_1, \ldots, x_d]) \times$$
$$\times \mathrm{vol}_{d-1}([x_1, \ldots, x_d])\mathrm{d}x_1 \ldots \mathrm{d}x_d,$$

where the $\mathrm{d}x_i$ now denote integration in E. The main contribution in the integral above arises when V is close to $1/n$ and so it depends on the local boundary properties of K. This works when K is smooth and $\phi = f_{d-1}$ and gives (see Raynaud [35] and Wieacker [59])

$$\mathbb{E}f_{d-1}(K_n) = b_d \int_{\mathrm{bd}\,K} \kappa^{1/(d+1)}\mathrm{d}Sn^{\frac{d-1}{d+1}}(1 + o(1)),$$

where $\kappa > 0$ is the Gauss curvature and b_d is a constant. The method does not quite work for $f_0(K_n)$ for general smooth convex bodies, but it does for the Euclidean ball [59], which can be used to establish the result

$$\mathbb{E}f_0(K_n) = b'_d \int_{\mathrm{bd}\,K} \kappa^{1/(d+1)}\mathrm{d}Sn^{\frac{d-1}{d+1}}(1 + o(1))$$

for smooth enough convex bodies, see Bárány [5] and Schütt [53]. This implies, via Efron's identity (17), a similar asymptotic formula for $\mathbb{E}(K, n)$.

The precise asymptotic formula for $f_0(K_n)$ and $\mathbb{E}(K, n)$ when K is a polytope is given by Theorem 4.2 plus Efron's identity.

Besides many results on the expectation of various functionals of K_n, very little has been known about the distribution of these functionals up to quite recently. A notable exception is Groeneboom's result [23] establishing a central limit theorem in the following from. For a polygon P is the plane, the distribution of $f_0(P_n)$ is close to the normal. Precisely, if P has r vertices, then

$$\frac{f_0(P_n) - \frac{2}{3}r\ln n}{\sqrt{\frac{10}{27}r\ln n}} \to \mathcal{N}(0,1)$$

in distribution, where $\mathcal{N}(0,1)$ is the standard normal distribution. Further, Cabo and Groeneboom [18] proved, in a version suggested by Buchta [17], that

$$\frac{\mathbb{E}(P,n) - \frac{2}{3}r\frac{\ln n}{n}}{\sqrt{\frac{28}{27}r\frac{\ln n}{n^2}}} \to \mathcal{N}(0,1),$$

again in distribution. Groeneboom showed the central limit theorem for the case of the unit disk and f_0, with the variance evaluated numerically. Hsing [26] proved that

$$\frac{\mathbb{E}(B^2,n) - c_1 n^{-2/3}}{\sqrt{c_2 n^{-5/3}}} \to \mathcal{N}(0,1)$$

in distribution, again. The explicit constants c_1 and c_2 have been determined by Buchta [17]. The asymptotic distribution of the Hausdorff distance between a planar convex body K and K_n has been determined with high precision by Bräker, Hsing, and Bingham [16].

In a series of remarkable papers Reitzner [36], [37] has established an upper bound on the variance of the missed volume and $f_i(K_n)$ in the case of smooth convex bodies K:

$$\text{var}\,\text{vol}(K \setminus K_n) \leq c(K)n^{-(d+3)/(d+1)}$$
$$\text{var}\,f_i(K_n) \leq c(K)n^{(d-1)/(d+1)},$$

where the constants $c(K)$ depend on K and dimension only. These estimates imply a strong law of large numbers for the corresponding functionals. In a recent and very interesting paper [38] Reitzner has given a lower bound for the variance (smooth convex bodies) which are of the same order of magnitude as the upper bounds above. Using this he has been able to show that both the missed volume and the number of i-dimensional faces of K_n satisfy the central limit theorem, a real breakthrough result in the theory of random polytopes. His argument is based on several ingredients: (1) a general central limit theorem of Rinott [40] where only partial independence of the random variables in question is required, (2) the right estimate for the variance, and (3) the precise comparison of random polytopes and polytopes obtained from a Poisson process $X(n)$ of intensity n intersecting the smooth convex body K.

These results have been extended to the case when K is a polytope in $\mathbb{R}^d$ by Reitzner and myself [12]. Using geometric properties of polytopes and their u and v functions, combined with the cap-covering technique we could show that

$$\operatorname{var} \operatorname{vol}(K \setminus K_n) \approx n^{-2}(\ln n)^{d-1},$$
$$\operatorname{var} f_i(K_n) \approx (\ln n)^{d-1}.$$

The implied constants depend on the polytope P in question. The central limit theorem, both for missed volume and f_i, follows the same way as in Reitzner's paper [37]. It is more difficult to measure the partial dependence of the underlying graph. These new results open the possibility for the central limit theorem for the missed volume of general convex bodies, not only smooth ones or for polytopes. I conjecture, for instance, that

$$\operatorname{var} \operatorname{vol}(K \setminus K_n) \approx \frac{\operatorname{vol} K(1/n)}{n}$$

for all convex bodies with the implied constants depending only on dimension.

Almost at the same time and from a completely different and unexpected direction, strong concentration results for random polytopes have been proved by Van Vu [57]. He uses a probabilistic and combinatorial technique which has been very powerful in other cases as well (cf. [56], [28]) for proving tail estimates. I state his result only for the random variable $Y_n = \operatorname{vol}(K \setminus K_n)$. The setting is this. Given a $K \in \mathcal{K}_1$, $\varepsilon > 0$ small, and $x \in K$ with $v(x) < \varepsilon$, define $S_{x,\varepsilon}$ as the set of points $y \in K$ such that the segment $[x, y]$ is disjoint from $K(v \geq \varepsilon)$. So $S_{x,\varepsilon}$ is the union of all ε-caps containing x. Define $g(\varepsilon) = \sup\{\operatorname{vol} S_{x,\varepsilon} : x \in K(\varepsilon)\}$ and let $A = 3g(\varepsilon)$ and $B = 36ng(\varepsilon)^2 \operatorname{vol} K(v \leq \varepsilon)$. With this notation the following holds.

Theorem 15.1. *There are positive constants α, c and ε_0 such that for every $K \in \mathcal{K}_1$ and for every n and λ satisfying $(\alpha \ln n)/n < \varepsilon < \varepsilon_0$ and $0 < \lambda \leq B/(4A^2) = n \operatorname{vol} K(v \leq \varepsilon)$ we have*

$$\mathbb{P}\{|Y_n - \mathbb{E}Y_n| \geq \sqrt{\lambda B}\} \leq 2\exp\{-\lambda/4\} + \exp\{-c\varepsilon n\}.$$

This is a very strong result implying, for instance, large deviation inequalities for Y and good bounds on the centred moments of Y, or on how close Y_n is to $\mathbb{E}(Y_n)$. In the particular case of smooth convex bodies, Vu strengthens the above inequality showing that B can be chosen of the same order of magnitude as the variance of Y_n. The interested reader can learn a lot from Vu's excellent paper [57].

It should be mentioned that Calka and Schreiber [19] have recently proved a large deviation inequality for $f_0(K_n)$ in the case when K is the unit ball. The exponent in their estimate is $n^{(d-1)/(3d+5)}$, the same that Vu proves for smooth convex bodies.

16 Lattice Polytopes

Originally, Macbeath [34] introduced M-regions in order to study integer points in convex bodies. He observed the following interesting fact. The **integer convex hull** of K is defined as $[K \cap \mathbb{Z}^d]$, that is, the convex hull of the lattice points in K. If $z \in \mathbb{Z}^d$ is a vertex of the integer convex hull of K, then $u_K(z) < 2^d$. This follows from Minkowski's classical theorem: $M_K(z)$ is a centrally symmetric convex body with centre $z \in \mathbb{Z}^d$. If it has volume at least 2^d, then it contains another lattice point, say y, and by central symmetry, it also contains $2z - y \in \mathbb{Z}^d$. But then z is not a vertex of the integer convex hull because it is contained in the segment $[y, 2z - y]$.

Macbeath mentions that $u_K(x)$ "is the most interesting function that one can associate with a convex set".

In 1963, G. E. Andrews [2] proved a remarkable theorem saying that a lattice polytope $P \subset \mathbb{R}^d$ of volume $V > 0$ cannot have more than

$$\text{const } V^{\frac{d-1}{d+1}}$$

vertices or facets. Alternative proofs were later found by Arnol'd [3], Konyagin and Sevastyanov [31], Schmidt [44], Bárány and Vershik [13]. Here is yet another proof, from [11], based on the technique of M-regions. What we will present here is a sketch of the proof of

$$f_{d-1}(P) \ll V^{\frac{d-1}{d+1}}$$

since this is another application of the technique of M-regions and cap coverings.

We start the proof by fixing $\varepsilon = (2(10d)^d(d+1)!V)^{-1}$. Note that in this way $\varepsilon V < \varepsilon_0$. Let F be a facet of P and let x_F be the point on bd $P(v \geq \varepsilon V)$ where the tangent hyperplane to $P(v \geq \varepsilon V)$ is parallel to F. According to Lemma 6.7, x_F is unique. Let C_F stand for the cap cut off from K by the hyperplane parallel to F and passing through x_F.

Lemma 16.1. *For distinct facets F and G of P*

$$M(x_F, 1/2) \cap M(x_G, 1/2) = \emptyset.$$

To see this assume this intersection is nonempty. Then, by Lemma 6.2, $M(x_G, 1) \subset M(x_F, 5)$. Further, Lemma 6.8 combined with (11) shows that

$$G \subset C_G \subset M(x_G, 2d) \subset M(x_F, 10d),$$

where the last containment is a simple consequence of $M(x_G, 1) \subset M(x_F, 5)$. Even simpler is

$$F \subset C_F \subset M(x_F, 2d) \subset M(x_F, 10d).$$

Now $M(x_F, 10d)$ contains both facets F and G so it contains $d + 1$ affinely independent lattice points. Thus its volume is at least $1/d!$. Then, using again Lemma 6.3,

$$\frac{1}{d!} \leq \text{vol}\, M(x_F, 10d) = (10d)^d u(x_F) \leq (10d)^d 2v(x_F)$$

$$\leq 2(10d)^d \varepsilon V = \frac{1}{(d+1)!}.$$

This is a contradiction (due to the choice of ε), finishing the proof. □

So the half M-regions $M(x_F, 1/2)$, for all facets F, are pairwise disjoint. Their "half" $M(x_F, 1/2) \cap C(F)$ lies completely in $P(\varepsilon)$. Then, by Theorem 4.5 combined with (7),

$$\sum_F \frac{1}{2} \text{vol}\, M(x_F, 1/2) \leq \text{vol}\, P(v \leq \varepsilon V) \ll \varepsilon^{\frac{2}{d+1}} V \ll V^{\frac{d-1}{d+1}},$$

where the summation is taken over all facets F of P. Now, again by Lemma 6.6,

$$\text{vol}\, M(x_F, 1/2) = 2^{-d} u(x_F) \geq 2^{-d}(2d)^{-d} v(x_f) \gg \varepsilon V \gg 1.$$

The last two formulae show that the number of facets of P is $\ll V^{\frac{d-1}{d+1}}$.

We mention that this implies, via a trick of Andrews, the following slightly stronger theorem whose proof can be found in [11].

Theorem 16.1. *For a lattice polytope $P \in \mathbb{R}^d$ with volume $V > 0$*

$$T(P) \ll V^{\frac{d-1}{d+1}}.$$

Remark. The result and the proof is originally from [10], and the presentation here is close to the one in [6]. The same applies to the next section.

17 Approximation

There are two types of problems in the theory of approximation of a $K \in \mathcal{K}$ by polytopes belonging to a certain class $\mathcal{P}$ of polytopes. The first type is asking for a lower bound, that is, a statement of the form: no polytope $P \in \mathcal{P}$ approximates K better than some function of K and $\mathcal{P}$. The second is asking for the existence of a polytope $P \in \mathcal{P}$ which approximates K well, hopefully as well as the previous function. To be less vague, we consider inscribed polytopes only (that is $P \subset K$) and we measure approximation by the relative missed volume, that is, by

$$\text{appr}(K, P) = \frac{\text{vol}(K \setminus P)}{\text{vol}\, K}.$$

In this section we show how the cap-covering technique can be used to attack both type of approximation problems. As expected, the results do not give precise constants but tell the right order of magnitude.

We start with the problem of the second type. C. Schütt [52] proved two very neat and general results.

Theorem 17.1. *Given $K \in \mathcal{K}_1$ and $t \in (0, t_0]$ (where t_0 depends only on the dimension), there is a polytope P with $K(v \geq t) \subset P \subset K$ for which*

$$f_0(P) \ll \frac{\operatorname{vol} K(t)}{t}.$$

Theorem 17.2. *Given $K \in \mathcal{K}_1$ and $t \in (0, t_0]$ (where t_0 depends only on the dimension), there is a polytope P with $K(v \geq t) \subset P \subset K$ for which*

$$f_{d-1}(P) \ll \frac{\operatorname{vol} K(t)}{t}.$$

This means that $\operatorname{appr}(K, P) \leq \operatorname{vol} K(t)$ and the lost volume is "t per vertex", and "t per facet", respectively. We will see below that, for smooth bodies, this is the best possible order of magnitude. Schütt's proof of these theorems is direct and technical. Here I present a simple argument showing the power and efficacy of the cap covering method. Nevertheless, this argument gives weaker constants than Schütt's original theorem and does not extend to approximation by circumscribed polytopes (cf. [52]).

As the theorems are affinely invariant, we may assume that K is in **standard position**, i.e., it is sandwiched between two balls, both centred at the origin with the radius of the larger at most d times that of the smaller. For both theorems, start with setting $\tau = \lambda t$ (where λ is a constant depending on d only) and choose a system of points $\{x_1, \ldots, x_m\}$ from $\operatorname{bd} K(v \geq \tau)$ maximal with respect to the property that the $M(x_i, 1/2)$ are pairwise disjoint. The economic cap covering argument shows that

$$m \ll \frac{\operatorname{vol} K(\tau)}{\tau} \ll \frac{\operatorname{vol} K(t)}{t}.$$

We start with the case of the facets which is simpler. Fix $\lambda = 6^{-d}$. Let $C(x_i)$ be a minimal cap, and define

$$P = K \setminus \cup_1^m C(x_i)^6.$$

We will show that (1) no $z \in \operatorname{bd} K$ belongs to P, and (2) $K(v \geq t) \subset P$. This clearly suffices for the facet case.

To see (1), assume $z \in \operatorname{bd} K$, and let $z^* \in \operatorname{bd} K(v \geq \tau)$ be the point on the segment connecting z and the origin (which is inside $K(v \geq \tau)$ if t is small enough). Maximality implies the existence of i with $M(z^*, 1/2) \cap M(x_i, 1/2) \neq \emptyset$, and by Lemma 6.2, $M(z^*, 1) \subset M(x_i, 5)$. It is easy to see, using the standard position of K, that $z \in M(z^*, 1)$, and so

$$z \in M(z^*, 1) \subset M(x_i, 5) \subset C(x_i)^6.$$

To check that (2) also holds, we write

$$\operatorname{vol} C(x_i)^6 \leq 6^d \operatorname{vol} C(x_i) = 6^d \tau = t,$$

so $u(x) \leq t$ for every point x cut off from K by one of the caps $C(x_i)^6$.

For the proof of the vertex case set $\lambda = d^{-1}6^{-d}$. Let y_i be the intersection, with bd K, of the halfline through x_i starting from the origin, and define

$$P = [y_1, y_2, \ldots, y_m].$$

Clearly $P \subset K$. So we have to show $K(v \geq t) \subset P$. Assume the contrary, then there is a halfspace H_1 with $P \cap H_1 = \emptyset$ whose bounding hyperplane is tangent to $K(v \geq t)$. Note that no y_i is in H_1. Let H_2 be the halfspace whose bounding hyperplane is parallel to that of H_1, and which is tangent to $K(v \geq \tau)$ at the point z. Set $C_j = K \cap H_j$, $j = 1, 2$. Lemma 6.8 says that vol $C_2 \leq d\tau$.

By the maximality of the x_i, $M(z, 1/2)$ intersects some $M(x_i, 1/2)$ and so

$$y_i \in M(x_i, 1) \subset M(z, 5).$$

Here $y_i \in M(x_i, 1)$ follows from the standard position of K. It is not hard to see that the cap C_2^6 contains $M(z, 5)$. Further, the cap C_2^6 is contained in the cap C_1 as their bounding hyperplanes are parallel and vol $C_1 \geq t$ (by Lemma 6.8), while, by the same lemma, vol $C_2^6 \leq 6^d$ vol $C_2 \leq 6^d d\tau = t$. Thus $y_i \in C_1 \subset H_1$, a contradiction.

We turn now to the first type of approximation question. We will consider here the family of all polytopes inscribed in K with at most n s–dimensional faces. Denote this class of polytopes by $\mathcal{P}_n(K, s)$. The usual question of approximation by inscribed polytopes with at most n vertices, the case $\mathcal{P}_n(K, 0)$ in our notation, is well understood, see [25]. Given a smooth enough convex body K, for every polytope in $\mathcal{P}_n(K, 0)$

$$\mathrm{appr}(K, P) \leq c(d, K)n^{-\frac{2}{d-1}}(1 + o(1)),$$

as $n \to \infty$. In the other direction, there exist polytopes P in $\mathcal{P}_n(K, 0)$ with

$$\mathrm{appr}(K, P) \geq c(d, K)n^{-\frac{2}{d-1}}(1 + o(1)).$$

Here even the constant, and its dependence on K and d, are almost completely known. In the same paper, Gruber proves an asymptotic formula for circumscribed polytopes with at most n facets, and in [33], Ludwig gives exact asymptotic formulae for the unrestricted case with n vertices and n facets, respectively. (Approximation is measured as the relative volume of the symmetric difference of P and K.)

Is there a similar estimate for $\mathcal{P}_n(K, s)$ when $0 < s < d - 1$? Or a weaker one, giving the order of magnitude of $\mathrm{appr}(K, P)$? This unusual approximation question has come up in connection with the integer convex hull (cf. [11]).

Again, M-regions and cap coverings are going to help. I present the basic ideas of the proof in the case when $K = B^d$, the unit ball in $\mathbb{R}^d$. This extends without serious difficulty to convex bodies whose Gaussian curvature is bounded away from 0 and ∞. It should be mentioned that K. Böröczky Jr in [15] has worked out several other cases of this type, for instance, inscribed, circumscribed, and unrestricted polytopes with at most n s-dimensional faces (again when $0 < s < d - 1$). His approach is different: it is based on local quadratic approximation of the boundary and uses power diagrams. Here is the result for the unit ball.

Theorem 17.3. *For every polytope* $P \in \mathcal{P}_n(B^d, s)$, *and for large* n

$$\mathrm{appr}(B^d, P) \gg n^{-\frac{2}{d-1}}.$$

The proof below is based on an idea from [11] which is used there when $s = d - 1$. This particular case, when $K = B^d$, was first proved by Rogers [42]. We mention that the theorem holds for smooth convex bodies, not only for the Euclidean ball. But the technique and the arguments are simpler and cleaner in the case of B^d. The interested reader will have no difficulty in extending the proof below to smooth convex bodies.

We may suppose that

$$\mathrm{vol}(B^d \setminus P) \le b_1 n^{-\frac{2}{d-1}},$$

for any particular constant b_1 of our choice (b_1 depending on d), as otherwise there is nothing to prove. We assume further that $s \ge 1$.

Let $F_1, \ldots, F_n$ denote the s-dimensional faces of P and let x_i be the nearest point of F_i to the origin. The minimal cap $C(x_i)$ has width h_i. It is not hard to check that $F_i \subset C(x_i)$. Also, $\mathrm{vol}(C(x_i) \setminus P) \ge \frac{1}{2} \mathrm{vol}\, C(x_i)$. This means that $\mathrm{vol}\, C(x_i)$ must be small, and so h_i must be small. Consequently, $\mathrm{vol}\, C(x_i) \approx h_i^{\frac{d+1}{2}}$, as a quick computation reveals.

Choose next a subsystem $\{x_{i_1}, \ldots, x_{i_m}\}$ from the x_i which is, as we are used to it by now, maximal with respect to the property that the M-regions $M(x_{i_j}, 1/2)$ are pairwise disjoint. To have simple notation set $z_j = x_{i_j}$. By Lemma 6.2, every $C(x_i)$ is contained in some $M(z_j, 5)$. So writing V for the set of vertices of P we clearly have

$$V \subset \cup_1^n F_i \subset \cup_1^n C(z_i) \subset \cup_1^m M(z_j, 5).$$

Fix $\rho = b_2 n^{-\frac{1}{d-1}}$, where b_2 is to be defined later. We want to show that the set $V + \rho B^d$ covers at most half of $S^{d-1} = \mathrm{bd}\, B^d$. We estimate the surface area of this set by that of $S^{d-1} \cap \cup_1^m (M(z_j, 5) + \rho B^d)$, which is clearly

$$\ll \sum_1^m \left(\rho + h_j^{\frac{1}{2}}\right)^{d-1} = \sum_{j=1}^m \sum_{k=0}^{d-1} \binom{d-1}{k} h_j^{\frac{k}{2}} \rho^{d-1-k}$$

$$= \sum_{k=0}^{d-1} \binom{d-1}{k} \rho^{d-1-k} \left(\sum_{j=1}^m h_j^{\frac{k}{2}}\right)$$

$$\leq \sum_{k=0}^{d-1} \binom{d-1}{k} \rho^{d-1-k} m \left(\frac{1}{m} \sum_{j=1}^m h_j^{\frac{d+1}{2}}\right)^{\frac{k}{d+1}}$$

$$= m \left(\rho + \left(\frac{1}{m} \sum_1^m h_j^{\frac{d+1}{2}}\right)^{\frac{1}{d+1}}\right)^{d-1},$$

where we used the inequality between the kth and $(d+1)$st means.

We claim now that the last expression is smaller than half the surface area of S^{d-1} if the constants b_1, b_2 are chosen suitably. Indeed, as $m \leq n$,

$$\rho = b_2 n^{-\frac{1}{d-1}} \leq b_2 m^{-\frac{1}{d-1}}.$$

Next, as the $M(z_j, 1/2)$ are pairwise disjoint and one quarter of their volume is contained in $B^d \setminus P$, $\sum_{j=1}^m h_j^{\frac{d+1}{2}} \ll b_1 n^{-\frac{2}{d-1}}$. This implies

$$\left(\frac{1}{m} \sum h_j^{\frac{d+1}{2}}\right)^{\frac{1}{d+1}} \leq b_1^{\frac{1}{d+1}} m^{-\frac{1}{d-1}}.$$

We just proved that

$$V + \rho B^d \ll m \left(\rho + \left(\frac{1}{m} \sum_1^m h_j^{\frac{d+1}{2}}\right)^{\frac{1}{d+1}}\right)^{d-1}$$

$$\leq m \left((b_2 + b_1^{\frac{1}{d+1}}) m^{-\frac{1}{d-1}}\right)^{d-1} = \left(b_2 + b_1^{\frac{1}{d+1}}\right)^{d-1},$$

where the constant implied by $\ll$ depends only on d. So choosing b_1 and b_2 suitably we can assure that $V + \rho B^d$ misses at least half of S^{d-1}. Consequently,

$$S^{d-1} \setminus (V + \frac{\rho}{2} B^d)$$

contains many pairwise disjoint caps of radius $\rho/2$. This is shown by a greedy algorithm: assume the centres $y_p \in S^{d-1} \setminus (V + \rho B^d)$ of these caps C_p have been chosen for $p = 1, 2, \ldots, q$ and the caps are pairwise disjoint. The caps with centres y_p and radius ρ cover at most

$$q \rho^{d-1} \operatorname{vol}_{d-1} S^{d-1}$$

of S^{d-1}. So as long as this is smaller than the surface area of

$$S^{d-1} \setminus (V + \rho B^d),$$

there is room to choose the next centre y_{q+1}. The algorithm produces as many as $\gg \rho^{-(d-1)} \gg n$ pairwise disjoint caps. They are all disjoint from P, so the volume missed by P is

$$\sum \operatorname{vol} C_p \gg n\rho^{d+1} \gg n^{-\frac{2}{d-1}},$$

finishing the proof of Theorem 17.3.

18 How It All Began: Segments on the Surface of K

The technique of M-regions and cap coverings was invented by Ewald, Larman, and Rogers in their seminal paper [21]. Their aim was to answer a beautiful question of Vic Klee [29], [30]: "Can the boundary of a convex body contain segments in all directions?" (this is the title of [30].) After partial results by McMinn, Besicovitch, Pepe, and Grünbaum and Klee the following basic result was proved in [21].

Theorem 18.1. *Let $S(K)$ denote the set of unit vectors $v \in \mathbb{R}^d$ such that the boundary of the convex body $K \subset \mathbb{R}^d$ contains a segment parallel with v. Then $S(K)$ has σ-finite $(d-2)$-dimensional Hausdorff measure.*

For the proof they invent and develop the technique of cap covering. They prove Lemma 6.4 and Lemma 6.2 which is one of the key steps and lies at the core of the method. It is in this paper where the first economic cap covering is proved and used. The target is to cover the boundary (not the wet part) with caps that have the same width.

Lemma 18.1. *Assume $K \in \mathcal{K}$ contains a ball of radius r and is contained in a ball of radius R. Given a positive $\varepsilon \leq \varepsilon_0$, there are caps $C_1, \ldots, C_m$ with*

(i) $\operatorname{bd} B \subset \cup_1^m C_i$,
(ii) the width of C_i is between 2ε and $36d\varepsilon$,
(iii) $\sum_1^m \operatorname{vol} C_i \ll \varepsilon \operatorname{vol} K$.

Here ε_0 and the constant in $\ll$ depend on d, r, R only.

Thus the caps C_i constitute an "economic cap covering" of $\operatorname{bd} K$ in the sense that each C_i has minimal width $\approx \varepsilon$ and their total volume is $\ll \varepsilon \operatorname{vol} K$. We have seen several versions and strengthenings of this result. It is worth mentioning that the economic cap covering lemma is a relative of the Besicovitch covering theorem.

Theorem 18.1 says that, for every convex body, the set of exceptional directions is small. Precisely, it has σ-finite 1-codimensional Hausdorff measure.

(Exceptional direction means here a direction contained in bd K.) That the exceptional set is small is important and useful in other cases as well, for instance, in integral geometry. It is shown in [21] that the set of exceptional r-flats has σ-finite 1-codimensional Hausdorff measure in the space of all r-flats. (Here $K \in \mathcal{K}$, of course, and an r-flat is exceptional if it intersects bd K in a set of dimension r.)

The method of [21] was simplified by Zalgaller [61], and developed further by Ivanov [27], and Schneider [45], [48]. Ivanov shows that the union of all lines in $\mathbb{R}^d$ that meet bd K in a segment has σ-finite $(d-1)$-dimensional Hausdorff measure. A consequence is that for almost all points $x \notin K$ the shadow boundary of K under central projection from x is sharp. (The reader will have no difficulty stating the analogous consequence of Theorem 18.1.) The following results of Schneider [45] and [48] have applications in integral geometry. The proof method is based on that of [21] but is much more involved.

Theorem 18.2. *Let $K, K' \in \mathcal{K}$. The set of all rotations $\rho \in SO(d)$ for which K and $\rho K'$ contain parallel segments lying in parallel supporting hyperplanes has σ-finite 1-codimensional Hausdorff measure.*

Theorem 18.3. *Let $K, K' \in \mathcal{K}$. The set of all rigid motions ρ for which K and $\rho K'$ have an exceptional common boundary point is of Haar measure zero.*

Here a point x, common to bd K and bd K', is exceptional if the normal cones to K at x and to K' at x contain a common halfline.

Schneider gives a sketch of the proof of Theorem 18.1 in [47], and a full proof, containing Zalgaller's simplification, in his excellent book [46]. The interested reader is advised to consult these references.

References

1. Affentranger, F., Wieacker, J.A.: On the convex hull of uniform random points in a simple d-polytope. Discrete Comput. Geom., **6**, 291–305 (1991)
2. Andrews, G.E.: A lower bound for the volumes of strictly convex bodies with many boundary points. Trans. Amer. Math. Soc., **106**, 270–279 (1963)
3. Arnol'd, V.I.: Statistics of integral convex polytopes (in Russian). Funktsional. Anal. i Prilozhen., **14**, 1–3 (1980). English translation: Funct. Anal. Appl., **14**, 79–84 (1980)
4. Bárány, I.: Intrinsic volumes and f-vectors of random polytopes. Math. Ann., **285**, 671–699 (1989)
5. Bárány, I.: Random polytopes in smooth convex bodies. Mathematika, **39**, 81–92 (1992)
6. Bárány, I.: The technique of M-regions and cap-coverings: a survey. Rend. Circ. Mat. Palermo, Ser. II, Suppl., **65**, 21–38 (1999)
7. Bárány, I.: Sylvester's question: the probability that n points are in convex position. Ann. Probab., **27**, 2020–2034 (1999)
8. Bárány, I.: A note on Sylvester's four-point problem., Studia Sci. Math. Hungar., **38**, 73–77 (2001)

9. Bárány, I., Buchta, Ch.: Random polytopes in a convex polytope, independence of shape, and concentration of vertices. Math. Ann., **297**, 467–497 (1993)

10. Bárány, I., Larman, D.G.: Convex bodies, economic cap coverings, random polytopes. Mathematika, **35**, 274–291 (1988)

11. Bárány, I,. Larman, D.G.: The convex hull of the integer points in a large ball. Math. Ann., **312**, 167–181 (1998)

12. Bárány, I., Reitzner, M.: Central limit theorems for random polytopes in convex polytopes. Manuscript (2005)

13. Bárány, I., Vershik, A. M.: On the number of convex lattice polytopes. Geom. Funct. Anal., **2**, 381–393 (1992)

14. Blaschke, W.: Affine Differentialgeometrie. Springer, Berlin (1923)

15. Böröczky, K. Jr: Polytopal approximation bounding the number of k-faces. J. Approx. Theory, **102**, 263–285 (1999)

16. Bräker, H., Hsing, T., Bingham, N.H.: On the Hausdorff distance between a convex set and an interior random convex hull. Adv. in Appl. Probab., **30**, 295–316 (1998)

17. Buchta, Ch.: An identity relating moments of functionals of convex hulls. Discrete Comput. Geom., **33**, 125–142 (2005)

18. Cabo, A.J., Groeneboom, P.: Limit theorems for functionals of convex hulls. Probab. Theory Related Fields, **100**, 31–55 (1994)

19. Calka, P., Schreiber, T.: Large deviation probabilities for the number of vertices of random polytopes in the ball. Ann. Probab., (to appear)

20. Efron, B.: The convex hull of a random set of points. Biometrika, **52**, 331–343 (1965)

21. Ewald, G., Larman, D.G., Rogers, C.A.: The directions of the line segments and of the r-dimensional balls on the boundary of a convex body in Euclidean space. Mathematika, **17**, 1–20 (1970)

22. Groemer, H.: On the mean value of the volume of a random polytope in a convex set. Arch. Math., **25**, 86–90 (1974)

23. Groeneboom, P.: Limit theorems for convex hulls. Probab. Theory Related Fields, **79**, 327–368 (1988)

24. Gruber, P.M.: In most cases approximation is irregular. Rend. Circ. Mat. Torino, **41**, 19–33 (1983)

25. Gruber, P.M.: Asymptotic estimates for best and stepwise approximation of convex bodies II. Forum Math., **5**, 521–538 (1993)

26. Hsing, T.: On the asymptotic distribution of the area outside a random convex hull in a disk. Ann. Appl. Probab., **4**, 478–493 (1994)

27. Ivanov, B. A.: Exceptional directions for a convex body (in Russian). Mat. Zametki, **20**, 365–371 (1976). English translation: Math. Notes **20**, 763–765 (1976)

28. Kim, J.H., Vu, V.H.: Concentration of multivariate polynomials and its applications. Combinatorica, **20**, 417–434 (2000)

29. Klee, V.: Research problem No. 5. Bull. Amer. Math. Soc., **62**, 419 (1957)

30. Klee, V.: Can the boundary of a d–dimensional convex body contain segment in all directions? Amer. Math. Monthly, **76**, 408–410 (1969)

31. Konyagin, S.B., Sevastyanov, S.V.: Estimation of the number of vertices of a convex integral polyhedron in terms of its volume (in Russian). Funktsional. Anal. i Prilozhen., **18**, 13–15 (1984). English translation: Funct. Anal. Appl., **18**, 11–13 (1984)

32. Leichtweiss, K.: Affine Geometry of Convex Bodies. Barth, Heidelberg (1998)
33. Ludwig, M.: Asymptotic approximation of smooth convex bodies by general polytopes. Mathematika, **46**, 103–125 (1999)
34. Macbeath, A.M.: A theorem on non-homogeneous lattices. Ann. of Math., **56**, 269–293 (1952)
35. Raynaud, H.: Sur l'enveloppe convexe des nuages de points aléatoires dans $\mathbb{R}^n$. J. Appl. Probab., **7**, 35–48 (1970)
36. Reitzner, M.: Random polytopes and the Efron-Stein jackknife inequality. Ann. Probab., **31**, 2136–2166 (2003)
37. Reitzner, M.: The combinatorial structure of random polytopes. Adv. Math., **191**, 178–208 (2005)
38. Reitzner, M.: Central limit theorems for random polytopes. Prob. Theory Appl., (to appear)
39. Rényi, A., Sulanke, R.: Über die konvexe Hülle von n zufällig gewählten Punkten. Z. Wahrsch. Verw. Gebiete, **2**, 75-84 (1963)
40. Rinott, Y.: On normal approximation rates for certain sums of dependent random variables. J. Comput. Appl. Math., **55**, 135–143 (1994)
41. Rockafellar, T.R.: Convex Analysis. Princeton Univ. Press, Princeton, NJ (1970)
42. Rogers, C.A.: The volume of a polyhedron inscribed in a sphere. J. London Math. Soc., **28**, 410–416 (1953)
43. Santaló, L.A.: Integral Geometry and Geometric Probability. Addison-Wesley, Reading, MA (1976)
44. Schmidt, W.: Integral points on surfaces and curves. Monatsh. Math., **99**, 45–82 (1985)
45. Schneider, R.: Kinematic measures for sets of colliding convex bodies. Mathematika, **25**, 1–12 (1978)
46. Schneider, R.: Convex Bodies: the Brunn-Minkowski Theory. Encyclopedia of Mathematics and Its Applications **44**, Cambridge University Press, Cambridge (1993)
47. Schneider, R.: Measures in convex geometry. Rend. Istit. Mat. Univ. Trieste, Suppl., **29**, 215–265 (1999)
48. Schneider, R.: Convex bodies in exceptional relative positions. J. London Math. Soc. (2), **60**, 617–629 (1999)
49. Schneider, R.: Discrete aspects of stochastic geometry. In: Goodman, J.E., O'Rourke, J. (eds.) Handbook of Discrete and Computational Geometry. Second edition, Chapman & Hall/CRC, Boca Raton, FL (2004)
50. Schütt, C.: The convex floating body and polyhedral approximation. Israel J. Math., **73**, 65–77 (1991)
51. Schütt, C.: Random polytopes and affine surface area. Math. Nachr., **170**, 227–249 (1994)
52. Schütt, C.: Floating bodies, illumination bodies, and polytopal approximation. In: Ball, K. M. (ed.) Convex Geometric Analysis. Math. Sci. Res. Inst. Publ., **43**, Cambridge (1999)
53. Schütt, C.: Random polytopes and affine surface area. Math. Nachr., **170**, 227–249 (1994)
54. Schütt, C., Werner, E.: Polytopes with vertices chosen randomly from the boundary of a convex body. Geometric aspects of functional analysis, Lecture Notes in Math., **1807**, 241–422, Springer, Berlin (2003)
55. Sylvester, J.J.: Question 1491. Educational Times, London, April (1864)

56. Vu, V.H.: Concentration of non-Lipschitz functions and applications. Random Structures Algorithms, **20**, 262–316 (2002)
57. Vu, V.H.: Sharp concentration of random polytopes. Geom. Funct. Anal., (to appear)
58. Wendel, J.G.: A problem in geometric probability. Math. Scand., **11**, 109–111 (1962)
59. Wieacker, J.A.: Einige Probleme der polyedrischen Approximation. Diplomarbeit, Albert-Ludwigs-Universität, Freiburg im Breisgau (1978)
60. W. Weil, J. A. Wieacker: Stochastic geometry. In: Gruber, P.M., Wills, J. (eds.) Handbook of Convex Geometry, North-Holland, Amsterdam (1993)
61. Zalgaller, V. A.: k–dimensional directions singular for a convex body F in R^n (in Russian). Zapiski naučn. Sem. Leningrad. Otd. mat. Inst. Steklov, **27**, 67–72 (1972). English translation: J. Soviet Math., **3**, 437–441 (1972)

Integral Geometric Tools for Stochastic Geometry

Rolf Schneider

Mathematisches Institut, Albert-Ludwigs-Universität
Eckerstr. 1, D-79104 Freiburg i. Br., Germany
e-mail: rolf.schneider@math.uni-freiburg.de

Introduction

Integral geometry, as it is understood here, deals with measures on sets of geometric objects, and in particular with the determination of the total measure of various such sets having geometric significance. For example, given two convex bodies in Euclidean space, what is the total invariant measure of the set of all rigid motions which bring the first set into a position where it has nonempty intersection with the second one? Or, what is the total invariant measure of the set of all planes of a fixed dimension having nonempty intersection with a given convex body? Both questions have classical answers, known as the kinematic formula and the Crofton formula, respectively. Results of this type are useful in stochastic geometry. Basic random closed sets, the stationary and isotropic Boolean models with convex grains, are obtained by taking union sets of certain stochastic processes of convex bodies. Simple numerical parameters for the description of such Boolean models are functional densities related to the specific volume, surface area, or Euler characteristic. Kinematic formulae are indispensable tools for the investigation and estimation of such parameters. Weakening the hypotheses on Boolean models, requiring less invariance properties and admitting more general set classes, necessitates the generalization of integral geometric formulae in various directions. An introduction to the needed basic formulae and a discussion of their extensions, analogues and ramifications is the main purpose of the following. The section headings are:

1. From Hitting Probabilities to Kinematic Formulae
2. Localizations and Extensions
3. Translative Integral Geometry
4. Measures on Spaces of Flats

A simplifying aspect of our selection from the realm of integral geometry is, on the side of geometric operations, the restriction to intersections of fixed and variable sets and, on the side of measures, the restriction to Haar measures on groups of rigid motions or translations; only on spaces of flats do we consider non-invariant measures as well.

The following is meant as an introduction to the integral geometry that is relevant for stochastic geometry, with a few glimpses to more recent developments of independent interest. The character of the presentation varies from introductory to survey parts, and corresponding to this, proofs of the stated results are sometimes given in full, occasionally sketched, and often omitted.

1 From Hitting Probabilities to Kinematic Formulae

This chapter gives an introduction to the classical kinematic and Crofton formulae for convex bodies. We start with a deliberately vague question on certain hitting probabilities. This leads us in a natural way to the necessity of calculating certain kinematic measures, as well as to the embryonic idea of a Boolean model.

1.1 A Heuristic Question on Hitting Probabilities

The following question was posed and treated in [4]. Let K and L be two given convex bodies (nonempty compact convex sets) in Euclidean space $\mathbb{R}^d$. We use K to generate a random field of congruent copies of K. That means, countably many congruent copies of K are laid out randomly and independently in space. The bodies may overlap. It is assumed that the random system has a well defined number density, that is, an expected mean number of particles per unit volume. The body L is used as a fixed test body. For a given number $j \in \mathbb{N}_0$, we ask for the probability, p_j, of the event that the test body L is hit by (that is, has nonempty intersection with) exactly j bodies of the random field.

So far, of course, this is only an imprecise heuristic question. It will require several steps to make the question precise. In a first step, we choose a large ball B_r, of radius r and origin 0, that contains L, and we consider only one randomly moving congruent copy of K, under the condition that it hits B_r. What is the probability that it hits also L? To make this a meaningful question, we have to specify the probability distribution of the randomly moving body. The geometrically most natural assumption is that this distribution be induced from the motion invariant measure μ on the group G_d of (orientation preserving) rigid motions of $\mathbb{R}^d$. (We take here for granted that on the locally compact group G_d there exists a Borel measure μ which is finite on compact sets, invariant under left and right multiplications, and not identically zero; it is unique up to a positive factor.) This means that we represent the congruent copies of K in the form gK, where $g \in G_d$ is a rigid motion. We define a probability distribution $\mathbb{P}$ on the space $\mathcal{K}$ of convex bodies (equipped with the Hausdorff metric) in $\mathbb{R}^d$ by

$$\mathbb{P}(\mathcal{A}) := \frac{\mu(\{g \in G_d : gK \cap B_r \neq \emptyset, \; gK \in \mathcal{A}\})}{\mu(\{g \in G_d : gK \cap B_r \neq \emptyset\})}$$

for Borel sets $\mathcal{A} \subset \mathcal{K}$. A random congruent copy of K hitting B_r is then defined as the random variable $\tilde{g}K$, where $\tilde{g}$ is a random variable on some probability space with values in G_d and such that $\tilde{g}K$ has distribution $\mathbb{P}$.

Now it makes sense to ask for the probability, p, of the event that $\tilde{g}K$ meets the body $L \subset B_r$, and this probability is given by

$$p = \frac{\mu(K, L)}{\mu(K, B_r)}, \tag{1}$$

where we have put

$$\mu(K, M) := \mu(\{g \in G_d : gK \cap M \neq \emptyset\})$$

for convex bodies K and M.

How can we compute $\mu(K, M)$? Let us first suppose that K is a ball of radius ρ. Then the measure of all motions g that bring K in a hitting position with M is (under suitable normalization) equal to the measure of all translations which bring the centre of K into the parallel body

$$M + B_\rho := \{m + b : m \in M, \; b \in B_\rho\},$$

and hence to the volume of this body. By a fundamental result of convex geometry, this volume is a polynomial of degree at most d in the parameter ρ, usually written as

$$\lambda_d(M + B_\rho) = \sum_{i=0}^{d} \rho^{d-i} \kappa_{d-i} V_i(M) \tag{2}$$

(λ_d = Lebesgue measure in $\mathbb{R}^d$, κ_j = volume of the j-dimensional unit ball). This result, known as the **Steiner formula**, defines important functionals, the **intrinsic volumes** $V_0, \ldots, V_d$.

We see already from this special case, $K = B_\rho$, that in the computation of the measure $\mu(K, M)$ the intrinsic volumes must play an essential role. It is a remarkable fact that no further functionals are needed for the general case. The **principal kinematic formula** of integral geometry, in its specialization to convex bodies, says that

$$\mu(K, M) = \sum_{i=0}^{d} \alpha_{di} V_i(K) V_{d-i}(M), \tag{3}$$

with certain explicit constants α_{di}. For the moment, we take this formula for granted. A proof will be given in Subsection 1.4.

Recalling that the probability p, of the event that a randomly moving congruent copy of K hitting B_r also hits L, is given by (1), we have now found that

$$p = \frac{\sum_{i=0}^{d} \alpha_{di} V_i(K) V_{d-i}(L)}{\sum_{i=0}^{d} \alpha_{di} V_i(K) V_{d-i}(B_r)}, \qquad (4)$$

which depends (for fixed r) only on the intrinsic volumes of K and L.

In the second step, we consider m randomly chosen congruent copies of K, given in the form $\tilde{g}_1 K, \ldots, \tilde{g}_m K$ with random motions $\tilde{g}_1, \ldots, \tilde{g}_m$. We assume that these random copies are stochastically independent and that they all have the same distribution, as described above. For $j \in \{0, 1, \ldots, m\}$, let p_j denote the probability of the event that the test body L is hit by exactly j of the random congruent copies of K. The assumed independence leads to a binomial distribution, thus

$$p_j = \binom{m}{j} p^j (1-p)^{m-j},$$

with p given by (4).

In the third step, we let the radius r of the ball B_r and the number m of particles tend to ∞, but in such a way that

$$\lim_{\substack{m \to \infty \\ r \to \infty}} \frac{m}{\lambda_d(B_r)} = \gamma$$

with a positive constant γ. From

$$mp = \frac{m}{\lambda_d(B_r)} \frac{\lambda_d(B_r)}{\mu(K, B_r)} \mu(K, L) \qquad \text{and} \qquad \lim_{r \to \infty} \frac{\mu(K, B_r)}{\lambda_d(B_r)} = 1$$

(the latter after a suitable normalization of μ) we get $mp \to \gamma\mu(K, L) =: \theta$ and hence

$$\lim_{r \to \infty} p_j = \frac{\theta^j}{j!} e^{-\theta}$$

with

$$\theta = \gamma\mu(K, L) = \gamma \sum_{i=0}^{d} \alpha_{di} V_i(K) V_{d-i}(L).$$

We have found, not surprisingly, a Poisson distribution. Its parameter, θ, is expressed explicitly in terms of the constant γ, which can be interpreted as the number density of our random system of convex bodies, and the intrinsic volumes of K and L.

This is the answer given in [4]. The answer is explicit and elegant, but still not quite satisfactory. What the authors have computed is a limit of probabilities, and this turned out to be a Poisson distribution. However, this Poisson distribution is not yet interpreted as the distribution of a well-defined random variable. What we would prefer, and what is needed for applications, is a model that allows us to consider from the beginning countably infinite systems of randomly placed convex bodies, with suitable independence properties. This requirement leads us, inevitably and in a natural way, to the notion of a Poisson process of convex particles. More general versions of such particle processes and their union sets, the Boolean models, are one of the topics of Wolfgang Weil's chapter in this volume. Our task is now to provide the integral geometry that is needed for a quantitative treatment of Boolean models. We begin with the Steiner and kinematic formulae.

1.2 Steiner Formula and Intrinsic Volumes

Our first aim is to prove the Steiner formula (2) and to use it for introducing the intrinsic volumes, which are basic functionals of convex bodies. First we collect some notation.

On $\mathbb{R}^d$, we use the standard scalar product $\langle \cdot, \cdot \rangle$ and the Euclidean norm $\| \cdot \|$. The unit ball of $\mathbb{R}^d$ is $B^d := \{x \in \mathbb{R}^d : \|x\| \le 1\}$, and its boundary is the unit sphere S^{d-1}. By λ_k we denote the k-dimensional Lebesgue measure in k-dimensional flats (affine subspaces) of $\mathbb{R}^d$. For convex bodies K we write $\lambda_d(K) =: V_d(K)$. The constant $\kappa_d = \pi^{d/2}/\Gamma(1 + d/2)$ gives the volume of the unit ball B^d. Spherical Lebesgue measure on S^k is denoted by σ_k.

The space $\mathcal{K}$ of convex bodies in $\mathbb{R}^d$ is equipped with the Hausdorff metric. The sum of convex bodies $K, L \in \mathcal{K}$ is defined by

$$K + L := \{x + y : x \in K, y \in L\}.$$

A special case gives the **parallel body** of K at distance $\rho \ge 0$,

$$K_\rho := K + \rho B^d = \{x \in \mathbb{R}^d : d(K, x) \le \rho\},$$

where

$$d(K, x) := \min\{\|x - y\| : y \in K\}$$

is the distance of x from K. Let $K \in \mathcal{K}$ be a convex body. For $x \in \mathbb{R}^d$, there is a unique point $p(K, x)$ in K nearest to x, thus

$$\|p(K, x) - x\| = \min\{\|y - x\| : y \in K\} = d(K, x).$$

This defines a Lipschitz map $p(K, \cdot) : \mathbb{R}^d \to K$, which is called the **nearest-point map** of K, or the **metric projection** onto K.

For the case of a planar convex polygon P, the reader will easily verify, after drawing a picture of the parallel body P_ρ and decomposing P_ρ appropriately, that the area of this parallel body is a quadratic polynomial in ρ. The simple idea showing this extends to higher dimensions, as follows.

A **polyhedral set** in $\mathbb{R}^d$ is a set which can be represented as the intersection of finitely many closed halfspaces. A bounded non-empty polyhedral set is called a **convex polytope** or briefly a **polytope**. Let P be a polytope. If H is a supporting hyperplane of P, then $P \cap H$ is again a polytope. The set $F := P \cap H$ is called a **face** of P, and an **m-face** if dim $F = m$, $m \in \{0, \ldots, d-1\}$. If dim $P = d$, we consider P as a d-face of itself. By $\mathcal{F}_m(P)$

we denote the set of all m-faces of P, and we put $\mathcal{F}(P) := \bigcup_{m=0}^{d} \mathcal{F}_m(P)$. For $m \in \{0, \ldots, d-1\}$, $F \in \mathcal{F}_m(P)$, and a point $x \in \operatorname{relint} F$ (the relative interior of F), let $N(P, F)$ be the **normal cone** of P at F; this is the closed convex cone of outer normal vectors of supporting hyperplanes to P at x, together with the zero vector. It does not depend upon the choice of x. The number

$$\gamma(F, P) := \frac{\lambda_{d-m}(N(P, F) \cap B^d)}{\kappa_{d-m}} = \frac{\sigma_{d-m-1}(N(P, F) \cap S^{d-1})}{(d-m)\kappa_{d-m}} \tag{5}$$

is called the **external angle** of P at its face F. We also put $\gamma(P, P) = 1$ and $\gamma(F, P) = 0$ if either $F = \emptyset$ or F is not a face of P.

Now let a polytope P and a number $\rho > 0$ be given. For $x \in \mathbb{R}^d$, the nearest point $p(P, x)$ lies in the relative interior of a unique face of P. Therefore,

$$P_\rho = \bigcup_{F \in \mathcal{F}(P)} \left[P_\rho \cap p(P, \cdot)^{-1}(\operatorname{relint} F) \right] \tag{6}$$

is a disjoint decomposition of the parallel body P_ρ. For $m \in \{0, \ldots, d-1\}$ and $F \in \mathcal{F}_m$ it follows from the properties of the nearest point map that

$$P_\rho \cap p(P, \cdot)^{-1}(\operatorname{relint} F) = \operatorname{relint} F \oplus (N(P, F) \cap \rho B^d), \tag{7}$$

where $\oplus$ denotes a direct orthogonal sum. An application of Fubini's theorem gives

$$\begin{aligned}\lambda_d(P_\rho \cap p(P, \cdot)^{-1}(\operatorname{relint} F)) &= \lambda_m(F)\lambda_{d-m}(N(P, F) \cap \rho B^d) \\ &= \lambda_m(F)\gamma(F, P)\rho^{d-m}\kappa_{d-m}.\end{aligned}$$

Hence, if we define

$$V_m(P) := \sum_{F \in \mathcal{F}_m(P)} \lambda_m(F)\gamma(F, P), \tag{8}$$

it follows from (6) that

$$V_d(P_\rho) := \sum_{m=0}^{d} \rho^{d-m}\kappa_{d-m}V_m(P). \tag{9}$$

This can be extended to general convex bodies:

Theorem 1.1 (Steiner formula). *There are functionals $V_m : \mathcal{K} \to \mathbb{R}$, $m = 0, \ldots, d$, such that, for $K \in \mathcal{K}$ and $\rho \geq 0$,*

$$\boxed{V_d(K_\rho) = \sum_{m=0}^{d} \rho^{d-m}\kappa_{d-m}V_m(K).} \tag{10}$$

Proof. We use (9) for $\rho = 1, \ldots, d+1$ and solve a system of linear equations with a Vandermonde determinant, to obtain expressions

$$V_m(P) = \sum_{\nu=1}^{d+1} a_{m\nu} V_d(P_\nu), \qquad m = 0, \ldots, d,$$

with coefficients $a_{m\nu}$ independent of the polytope P. With these coefficients, we define

$$V_m(K) := \sum_{\nu=1}^{d+1} a_{m\nu} V_d(K_\nu)$$

for arbitrary convex bodies $K \in \mathcal{K}$. Since, for each fixed $\rho \geq 0$, the mapping $K \mapsto V_d(K + \rho B^d)$ is continuous with respect to the Hausdorff metric, the functional V_m is continuous. Using this and approximating K by a sequence of polytopes, we obtain the asserted result from (9). $\qquad \square$

The polynomial expansion (10) is known as the **Steiner formula**. The functionals $V_0, \ldots, V_d$, uniquely determined by (10), are called the **intrinsic volumes** (also, with different normalizations, the **quermassintegrals** or **Minkowski functionals**). About their geometric meaning, the following can be said. For polytopes, there are the explicit representations (8). They are particularly simple in the cases $m = d, d-1, 0$, and the results carry over to general convex bodies: V_d is the **volume**, $2V_{d-1}$ is the **surface area** (the $(d-1)$-dimensional Hausdorff measure of the boundary, for bodies of dimension d), and V_0 is the constant 1. The functional V_0, although trivial on convex bodies, has its own name and symbol: the **Euler characteristic** χ; the reason will become clear when we consider an extension of V_0 to more general sets. Also the other intrinsic volumes have simple interpretations, either differential geometric, under smoothness assumptions (see Subsection 2.1), or integral geometric (see Subsection 1.4, formula (16)).

It is easily seen from the Steiner formula that the map $V_m : \mathcal{K} \to \mathbb{R}$, $m \in \{0, \ldots, d\}$, has the following properties:

- V_m is invariant under rigid motions and reflections,
- V_m is continuous with respect to the Hausdorff metric,
- V_m is homogeneous of degree m,

that is, it satisfies

$$V_m(\alpha K) = \alpha^m V_m(K) \qquad \text{for } \alpha \geq 0.$$

Using (8) (and approximation), one shows without difficulty that

- V_m does not depend on the dimension of the surrounding space,

that is, if K lies in a Euclidean subspace of $\mathbb{R}^d$, then computation of $V_m(K)$ in that subspace leads to the same result as computation of $V_m(K)$ in $\mathbb{R}^d$. From the integral geometric representation (16) to be proved later it is seen that

- V_m is increasing under set inclusion,

which means that $K \subset L$ for convex bodies K, L implies $V_m(K) \leq V_m(L)$. A very important property is the additivity. A functional $\varphi : \mathcal{K} \to A$ with values in an abelian group is called **additive** or a **valuation** if

$$\varphi(K \cup L) + \varphi(K \cap L) = \varphi(K) + \varphi(L)$$

whenever K, L are convex bodies such that $K \cup L$ is convex (which implies that $K \cap L$ is not empty). If φ is an additive functional, one extends its definition by putting $\varphi(\emptyset) := 0$. As announced, we have:

- V_m is additive.

For a proof, one shows that, for fixed $\rho \geq 0$, the function $\mathbf{1}_\rho$ defined by

$$\mathbf{1}_\rho(K, x) := \begin{cases} 1 \text{ if } x \in K_\rho, \\ 0 \text{ if } x \in \mathbb{R}^d \setminus K_\rho \end{cases}$$

for $K \in \mathcal{K}$ and $x \in \mathbb{R}^d$ satisfies

$$\mathbf{1}_\rho(K \cup L, x) + \mathbf{1}_\rho(K \cap L, x) = \mathbf{1}_\rho(K, x) + \mathbf{1}_\rho(L, x).$$

Integration over x with respect to the Lebesgue measure yields

$$V_d((K \cup L)_\rho) + V_d((K \cap L)_\rho) = V_d(K_\rho) + V_d(L_\rho)$$

for $\rho \geq 0$. Now an application of the Steiner formula and comparison of the coefficients shows that each V_m is additive.

Hints to the literature. For the fundamental facts about convex bodies, we refer to [39], where details of the foregoing can be found.

1.3 Hadwiger's Characterization Theorem for Intrinsic Volumes

Of the properties established above for the intrinsic volumes, already a suitable selection is sufficient for an axiomatic characterization. This is the content of Hadwiger's celebrated characterization theorem:

Theorem 1.2 (Hadwiger's characterization theorem). *Suppose that $\psi : \mathcal{K} \to \mathbb{R}$ is an additive, continuous, motion invariant function. Then there are constants $c_0, \ldots, c_d$ so that*

$$\psi(K) = \sum_{i=0}^{d} c_i V_i(K)$$

for all $K \in \mathcal{K}$.

This result not only throws light on the importance of the intrinsic volumes, showing that they are essentially the only functionals on convex bodies sharing some very natural geometric properties with the volume, it is also useful. Following Hadwiger, we employ it to prove some integral geometric formulae in an elegant way. Whereas Hadwiger's original proof of his characterization theorem was quite long, a shorter proof was published in 1995 by Daniel Klain. It will be presented here, except that a certain extension theorem for additive functionals is postponed to Subsection 2.2, and a certain analytical result will be taken for granted.

The crucial step for a proof of Hadwiger's characterization theorem is the following result.

Proposition 1.1. *Suppose that $\psi : \mathcal{K} \to \mathbb{R}$ is an additive, continuous, motion invariant function satisfying $\psi(K) = 0$ whenever either $\dim K < d$ or K is a unit cube. Then $\psi = 0$.*

Proof. The proof proceeds by induction with respect to the dimension. For $d = 0$, there is nothing to prove. If $d = 1$, ψ vanishes on (closed) segments of unit length, hence on segments of length $1/k$ for $k \in \mathbb{N}$ and therefore on segments of rational length. By continuity, ψ vanishes on all segments and thus on $\mathcal{K}$.

Now let $d > 1$ and suppose that the assertion has been proved in dimensions less than d. Let $H \subset \mathbb{R}^d$ be a hyperplane and I a closed line segment of length 1, orthogonal to H. For convex bodies $K \subset H$ define $\varphi(K) := \psi(K+I)$. Clearly φ has, relative to H, the properties of ψ in the assertion, hence the induction hypothesis yields $\varphi = 0$. For fixed $K \subset H$, we thus have $\psi(K+I) = 0$, and a similar argument as used above for $n = 1$ shows that $\psi(K + S) = 0$ for any closed segment S orthogonal to H. Thus ψ vanishes on right convex cylinders.

Let $K \subset H$ be a convex body again, and let $S = \operatorname{conv}\{0, s\}$ be a segment not parallel to H. If $m \in \mathbb{N}$ is sufficiently large, the cylinder $Z := K + mS$ can be cut by a hyperplane H' orthogonal to S so that the two closed halfspaces H^-, H^+ bounded by H' satisfy $K \subset H^-$ and $K + ms \subset H^+$. Then $\overline{Z} := [(Z \cap H^-) + ms] \cup (Z \cap H^+)$ is a right cylinder, and we deduce that $m\mu(K + S) = \mu(Z) = \mu(\overline{Z}) = 0$. Thus ψ vanishes on arbitrary convex cylinders.

It can be shown that the continuous additive function ψ on $\mathcal{K}$ satisfies the more general additivity property

$$\psi\left(\bigcup_{i=1}^{k} K_i\right) = \sum_{i=1}^{k} \psi(K_i)$$

whenever $K_1, \ldots, K_k$ are convex bodies such that $\dim(K_i \cap K_j) < d$ for $i \neq j$ and that $\bigcup_{i=1}^{k} K_i$ is convex. This follows from Theorem 2.2 and (35) below and the fact that ψ has been assumed to vanish on convex bodies of dimension less than d.

Let P be a polytope and S a segment. The sum $P+S$ has a decomposition

$$P + S = \bigcup_{i=1}^{k} P_i,$$

where $P_1 = P$, the polytope P_i is a convex cylinder for $i > 1$, and $\dim(P_i \cap P_j) < d$ for $i \neq j$. It follows that $\psi(P + S) = \psi(P)$. By induction, we obtain $\psi(P + Z) = \psi(P)$ if Z is a finite sum of segments. Such a body Z is called a **zonotope**, and a convex body which can be approximated by zonotopes is called a **zonoid**. Since the function ψ is continuous, it follows that $\psi(K + Z) = \psi(K)$ for arbitrary convex bodies K and zonoids Z.

We have to use an analytic result for which we do not give a proof. Let K be a centrally symmetric convex body which is sufficiently smooth (say, its support function is of class C^∞). Then there exist zonoids Z_1, Z_2 so that $K + Z_1 = Z_2$ (this can be seen from Section 3.5 in [39], especially Theorem 3.5.3). We conclude that $\psi(K) = \psi(K + Z_1) = \psi(Z_2) = 0$. Since every centrally symmetric convex body K can be approximated by bodies which are centrally symmetric and sufficiently smooth, it follows from the continuity of ψ that $\psi(K) = 0$ for all centrally symmetric convex bodies.

Now let Δ be a simplex, say $\Delta = \operatorname{conv}\{0, v_1, \ldots, v_d\}$, without loss of generality. Let $v := v_1 + \cdots + v_d$ and $\Delta' := \operatorname{conv}\{v, v - v_1, \ldots, v - v_d\}$, then $\Delta' = -\Delta + v$. The vectors $v_1, \ldots, v_d$ span a parallelotope P. It is the union of Δ, Δ' and the part of P, denoted by Q, that lies between the hyperplanes spanned by $v_1, \ldots, v_d$ and $v - v_1, \ldots, v - v_d$, respectively. Now Q is a centrally symmetric polytope, and $\Delta \cap Q$, $\Delta' \cap Q$ are of dimension $d-1$. We deduce that $0 = \psi(P) = \psi(\Delta) + \psi(Q) + \psi(\Delta')$, thus $\psi(-\Delta) = -\psi(\Delta)$. If the dimension d is even, then $-\Delta$ is obtained from Δ by a proper rigid motion, and the motion invariance of ψ yields $\psi(\Delta) = 0$. If the dimension $d > 1$ is odd, we decompose Δ as follows. Let z be the centre of the inscribed ball of Δ, and let p_i be the point where this ball touches the facet F_i of Δ ($i = 1, \ldots, d+1$). For $i \neq j$, let Q_{ij} be the convex hull of the face $F_i \cap F_j$ and the points z, p_i, p_j. The polytope Q_{ij} is invariant under reflection in the hyperplane spanned by $F_i \cap F_j$ and z. If $Q_1, \ldots, Q_m$ are the polytopes Q_{ij} for $1 \leq i < j \leq d+1$ in any order, then $\Delta = \bigcup_{r=1}^{m} Q_r$ and $\dim(Q_r \cap Q_s) < d$ for $r \neq s$. Since $-Q_r$ is the image of Q_r under a proper rigid motion, we have $\psi(-\Delta) = \sum \psi(-Q_r) = \sum \psi(Q_r) = \psi(\Delta)$. Thus $\psi(\Delta) = 0$ for every simplex Δ.

Decomposing a polytope P into simplices, we obtain $\psi(P) = 0$. The continuity of ψ now implies $\psi(K) = 0$ for all convex bodies K. This finishes the induction and hence the proof of the proposition. $\qquad\square$

Proof of Theorem 1.2. We use induction on the dimension. For $d = 0$, the assertion is trivial. Suppose that $d > 0$ and the assertion has been proved in dimensions less than d. Let $H \subset \mathbb{R}^d$ be a hyperplane. The restriction of ψ

to the convex bodies lying in H is additive, continuous and invariant under motions of H into itself. By the induction hypothesis, there are constants $c_0, \ldots, c_{d-1}$ so that $\psi(K) = \sum_{i=0}^{d-1} c_i V_i(K)$ holds for convex bodies $K \subset H$ (here we use the fact that the intrinsic volumes do not depend on the dimension of the surrounding space). By the motion invariance of ψ and V_i, this holds for all $K \in \mathcal{K}$ of dimension less than d. It follows that the function ψ' defined by

$$\psi'(K) := \psi(K) - \sum_{i=0}^{d} c_i V_i(K)$$

for $K \in \mathcal{K}$, where c_d is chosen so that ψ' vanishes at a fixed unit cube, satisfies the assumptions of Proposition 1.2. Hence $\psi' = 0$, which completes the proof of Theorem 1.2.   □

Hints to the literature. Hadwiger's characterization theorem was first proved for dimension three in [13] and for general dimensions in [14]; the proof is reproduced in [16]. The simpler proof as presented here appears in [24]; see also [25].

1.4 Integral Geometric Formulae

We use Hadwiger's characterization theorem to prove some basic integral geometric results for convex bodies. They involve invariant measures on groups of motions or on spaces of flats, where these groups and homogeneous spaces are equipped with their usual topologies. In the following, a measure on a topological space X is always defined on $\mathcal{B}(X)$, its σ-algebra of Borel sets. Let SO_d be the group of proper (i.e., orientation preserving) rotations of $\mathbb{R}^d$. It is a compact group and carries a unique rotation invariant probability measure, which we denote by ν. As before, G_d denotes the group of rigid motions of $\mathbb{R}^d$. Let μ be its invariant (or Haar) measure, normalized so that $\mu(\{g \in G_d : gx \in B^d\}) = \kappa_d$ for $x \in \mathbb{R}^d$. More explicitly, the mapping $\gamma : \mathbb{R}^d \times SO_d \to G_d$ defined by $\gamma(x, \vartheta)y := \vartheta y + x$ for $y \in \mathbb{R}^d$ is a homeomorphism, and μ is the image measure of the product measure $\lambda_d \otimes \nu$ under γ.

By $\mathcal{L}_q^d$ we denote the Grassmannian of q-dimensional linear subspaces of $\mathbb{R}^d$, for $q \in \{0, \ldots, d\}$, and by ν_q its rotation invariant probability measure. Similarly, $\mathcal{E}_q^d$ is the space of q-flats in $\mathbb{R}^d$, and μ_q is its motion invariant measure, normalized so that $\mu_q(\{E \in \mathcal{E}_q^d : E \cap B^d \neq \emptyset\}) = \kappa_{d-q}$. This, too, we make more explicit. We choose a fixed subspace $L \in \mathcal{L}_q^d$ and denote by $L^\perp$ its orthogonal complement. The mappings $\beta_q : SO_d \to \mathcal{L}_q^d$, $\vartheta \mapsto \vartheta L$, and $\gamma_q : L^\perp \times SO_d \to \mathcal{E}_q^d$, $(x, \vartheta) \mapsto \vartheta(L + x)$, are continuous and surjective. Now ν_q is the image measure of the invariant measure ν under β_q, and μ_q is the image measure of the product measure $\lambda_{d-q}^{L^\perp} \otimes \nu$ under γ_q, where $\lambda_{d-q}^{L^\perp}$ denotes Lebesgue measure on $L^\perp$.

Once the invariant measures μ and μ_k are available, it is of interest to determine, for convex bodies $K, M \in \mathcal{K}$, the total measures $\mu(\{g \in G_d :$

$K \cap gM \neq \emptyset\}$) and $\mu_k(\{E \in \mathcal{E}_k^d : K \cap E \neq \emptyset\})$. We write these as the integrals

$$\int_{G_d} \chi(K \cap gM)\,\mu(dg) \qquad \text{and} \qquad \int_{\mathcal{E}_k^d} \chi(K \cap E)\,\mu_k(dE),$$

recalling that $\chi(K) = 1$ for a convex body K and $\chi(\emptyset) = 0$. Since $\chi = V_0$, a more general task is to determine the integrals

$$\int_{G_d} V_j(K \cap gM)\,\mu(dg) \qquad \text{and} \qquad \int_{\mathcal{E}_k^d} V_j(K \cap E)\,\mu_k(dE),$$

for $j = 0, \ldots, d$ (recall that $V_j(\emptyset) = 0$, by convention). For that, we use Hadwiger's characterization theorem. We begin with the latter integral. By

$$\psi(K) := \int_{\mathcal{E}_k^d} V_j(K \cap E)\,\mu_k(dE) \qquad \text{for } K \in \mathcal{K}$$

we define a functional $\psi : \mathcal{K} \to \mathbb{R}$. It is not difficult to show that this functional is additive, motion invariant and continuous (for the continuity, compare the argument used in the proof of Theorem 1.4). Hadwiger's characterization theorem yields a representation

$$\psi(K) = \sum_{r=0}^{d} c_r V_r(K).$$

Here only one coefficient is different from zero. In fact, from

$$\psi(K) = \int_{\mathcal{L}_k^d} \int_{L^\perp} V_j(K \cap (L + y))\,\lambda_{d-k}(dy)\,\nu_k(dL)$$

one sees that ψ has the homogeneity property

$$\psi(\alpha K) = \alpha^{d-k+j}\psi(K)$$

for $\alpha > 0$. Since V_k is homogeneous of degree k, we deduce that $c_r = 0$ for $r \neq d - k + j$. Thus, we have obtained

$$\int_{\mathcal{E}_k^d} V_j(K \cap E)\,\mu_k(dE) = cV_{d-k+j}(K)$$

with some constant c. In order to determine this constant, we choose for K the unit ball B^d. For $\epsilon \geq 0$, the Steiner formula gives

$$\sum_{j=0}^{d} \epsilon^{d-j}\kappa_{d-j}V_j(B^d) = V_d(B^d + \epsilon B^d) = (1+\epsilon)^d\kappa_d = \sum_{j=0}^{d} \epsilon^{d-j}\binom{d}{j}\kappa_d,$$

hence

$$V_j(B^d) = \frac{\binom{d}{j}\kappa_d}{\kappa_{d-j}} \qquad \text{for } j = 0, \ldots, d. \tag{11}$$

Choosing $L \in \mathcal{L}_k^d$, we obtain

$$
\begin{aligned}
cV_{d-k+j}(B^d) &= \int_{\mathcal{E}_k^d} V_j(B^d \cap E)\,\mu_k(\mathrm{d}E) \\
&= \int_{SO_d} \int_{L^\perp} V_j(B^d \cap \vartheta(L+x))\,\lambda_{d-k}(\mathrm{d}x)\,\nu(\mathrm{d}\vartheta) \\
&= \int_{L^\perp \cap B^d} (1 - \|x\|^2)^{j/2} V_j(B^d \cap L)\,\lambda_{d-k}(\mathrm{d}x) \\
&= \frac{\binom{k}{j}\kappa_k}{\kappa_{k-j}} \int_{L^\perp \cap B^d} (1 - \|x\|^2)^{j/2}\,\lambda_{d-k}(\mathrm{d}x).
\end{aligned}
$$

Introducing polar coordinates, we transform the latter integral into a Beta integral and finally obtain

$$c = \frac{\binom{k}{j}\kappa_k \kappa_{d-k+j}}{V_{d-k+j}(B^d)\kappa_{k-j}\kappa_j} = c_{j,d}^{k,d-k+j},$$

where we denote by

$$c_{k,l}^{i,j} := \frac{i!\kappa_i\, j!\kappa_j}{k!\kappa_k\, l!\kappa_l} \tag{12}$$

a frequently occurring constant. By using the identity

$$m!\kappa_m = 2^m \pi^{\frac{m-1}{2}} \Gamma\left(\frac{m+1}{2}\right),$$

this can also be put in the form

$$c_{k,l}^{i,j} = \frac{\Gamma(\frac{i+1}{2})\Gamma(\frac{j+1}{2})}{\Gamma(\frac{k+1}{2})\Gamma(\frac{l+1}{2})}. \tag{13}$$

More generally, we define

$$c_{s_1,\ldots,s_k}^{r_1,\ldots,r_k} := \prod_{i=1}^{k} \frac{r_i!\kappa_{r_i}}{s_i!\kappa_{s_i}}. \tag{14}$$

This notation is only defined with the same number of upper and lower indices; hence, when $c_{s,d,\ldots,d}^{r_1,\ldots,r_k}$ appears, it is clear that the index d is repeated $k-1$ times.

We have obtained the following result.

Theorem 1.3. *Let $K \in \mathcal{K}$ be a convex body. For $k \in \{1, \ldots, d-1\}$ and $j \leq k$ the* **Crofton formula**

$$\boxed{\int_{\mathcal{E}_k^d} V_j(K \cap E)\, \mu_k(\mathrm{d}E) = c_{j,d}^{k,d-k+j} V_{d-k+j}(K)} \tag{15}$$

holds.

The special case $j = 0$ of (15) gives

$$V_m(K) = c_{m,d-m}^{0,d} \int_{\mathcal{E}_{d-m}^d} \chi(K \cap E)\, \mu_{d-m}(\mathrm{d}E) \tag{16}$$

and thus provides an integral geometric interpretation of the intrinsic volumes: $V_m(K)$ *is, up to a normalizing factor, the invariant measure of the set of $(d - m)$-flats intersecting K.*

Using the explicit representation of the measure μ_{d-m} and the fact that the map $L \mapsto L^\perp$ transforms ν_{d-m} into ν_d, we can rewrite the representation (16) as

$$V_m(K) = c_{m,d-m}^{0,d} \int_{\mathcal{L}_m^d} \lambda_m(K|L)\, \nu_m(\mathrm{d}L), \tag{17}$$

where $K|L$ denotes the image of K under orthogonal projection to the subspace L. The special case $m = 1$ shows that V_1, up to a factor, is the **mean width**.

From Hadwiger's characterization theorem, we now deduce a general kinematic formula, involving a functional on convex bodies that need not have any invariance property.

Theorem 1.4 (Hadwiger's general integral geometric theorem).
If $\varphi : \mathcal{K} \to \mathbb{R}$ is an additive continuous function, then

$$\boxed{\int_{G_d} \varphi(K \cap gM)\, \mu(\mathrm{d}g) = \sum_{k=0}^{d} \varphi_{d-k}(K) V_k(M)} \tag{18}$$

for $K, M \in \mathcal{K}$, where the coefficients $\varphi_{d-k}(K)$ are given by

$$\varphi_{d-k}(K) = \int_{\mathcal{E}_k^d} \varphi(K \cap E)\, \mu_k(\mathrm{d}E). \tag{19}$$

Proof. The μ-integrability of the integrand in (18) is seen as follows. For $K, M \in \mathcal{K}$, let $G_d(K, M)$ be the set of all motions $g \in G_d$ for which K and gM touch, that is, $K \cap gM \neq \emptyset$ and K and gM can be separated weakly by a hyperplane. It is not difficult to check that $\gamma(x, \vartheta) \in G_d(K, M)$ if and only if $x \in \partial(K - \vartheta M)$ and, hence, that $\mu(G_d(K, M)) = 0$.

Let $g \in G_d \setminus G_d(K, M)$, and let $(M_j)_{j \in \mathbb{N}}$ be a sequence in $\mathcal{K}$ converging to M. Then $gM_j \to gM$ and hence $K \cap gM_j \to K \cap gM$ (see [51, Hilfssatz 2.1.3]), thus $\varphi(K \cap gM_j) \to \varphi(K \cap gM)$ for $j \to \infty$. It follows that the map $g \mapsto \varphi(K \cap gM)$ is continuous outside a (closed) μ-null set. Moreover, the continuous function φ is bounded on the compact set $\{K' \in \mathcal{K} : K' \subset K\}$, and $\mu(\{g \in G_d : \varphi(K \cap gM) \neq 0\}) \leq \mu(\{g \in G_d : K \cap gM \neq \emptyset\}) < \infty$. This shows the μ-integrability of the function $g \mapsto \varphi(K \cap gM)$.

Now we fix a convex body $K \in \mathcal{K}$ and define

$$\psi(M) := \int_{G_d} \varphi(K \cap gM)\, \mu(dg) \quad \text{for } M \in \mathcal{K}.$$

Then $\psi : \mathcal{K} \to \mathbb{R}$ is obviously additive and motion invariant. The foregoing consideration together with the bounded convergence theorem shows that ψ is continuous. Theorem 1.2 yields the existence of constants $\varphi_0(K), \ldots, \varphi_d(K)$ so that

$$\psi(M) = \sum_{i=0}^{d} \varphi_{d-i}(K) V_i(M)$$

for all $M \in \mathcal{K}$. The constants depend, of course, on the given body K, and we now have to determine them.

Let $k \in \{0, \ldots, d\}$, and choose $L_k \in \mathcal{L}_k^d$. Let $C \subset L_k$ be a k-dimensional unit cube with centre 0, and let $r > 0$. Then

$$\psi(rC) = \sum_{i=0}^{d} \varphi_{d-i}(K) V_i(rC) = \sum_{i=0}^{k} \varphi_{d-i}(K) r^i V_i(C).$$

On the other hand, using the rotation invariance of λ_d,

$$\psi(rC)$$
$$= \int_{G_d} \varphi(K \cap grC)\, \mu(dg)$$
$$= \int_{SO_d} \int_{\mathbb{R}^d} \varphi(K \cap (\vartheta rC + x))\, \lambda_d(dx)\, \nu(d\vartheta)$$
$$= \int_{SO_d} \int_{L_k^\perp} \int_{L_k} \varphi(K \cap (\vartheta rC + \vartheta x_1 + \vartheta x_2))\, \lambda_k(dx_1) \lambda_{d-k}(dx_2)\, \nu(d\vartheta)$$
$$= \int_{SO_d} \int_{L_k^\perp} \int_{L_k} \varphi(K \cap [\vartheta r(C + x_1) + \vartheta x_2]) r^k\, \lambda_k(dx_1) \lambda_{d-k}(dx_2)\, \nu(d\vartheta).$$

Comparison yields

$$\varphi_{d-k}(K)$$
$$= \lim_{r \to \infty} \int_{SO_d} \int_{L_k^\perp} \int_{L_k} \varphi(K \cap [\vartheta r(C + x_1) + \vartheta x_2])\, \lambda_k(dx_1) \lambda_{d-k}(dx_2)\, \nu(d\vartheta).$$

For $r \to \infty$, we have

$$\varphi(K \cap [\vartheta r(C + x_1) + \vartheta x_2]) \to \begin{cases} \varphi(K \cap \vartheta(L_k + x_2)) & \text{if } 0 \in \text{relint}\,(C + x_1), \\ 0 & \text{if } 0 \notin C + x_1. \end{cases}$$

Hence, the bounded convergence theorem gives

$$\varphi_{d-k}(K) = \int_{SO_d} \int_{L_k^\perp} \varphi(K \cap \vartheta(L_k + x_2)) \lambda_k(C) \lambda_{d-k}(dx_2)\, \nu(d\vartheta)$$

$$= \int_{\mathcal{E}_k^d} \varphi(K \cap E)\, \mu_k(dE),$$

as asserted.   □

In Theorem 1.4 we can choose for φ, in particular, the intrinsic volume V_j. In this case, the Crofton formula (15) tells us that

$$(V_j)_{d-k}(K) = \int_{\mathcal{E}_k^d} V_j(K \cap E)\, \mu_k(dE) = c_{j,d}^{k,d-k+j} V_{d-k+j}(K).$$

Hence, we obtain the following result.

Theorem 1.5. *Let $K, M \in \mathcal{K}$ be convex bodies, and let $j \in \{0, \ldots, d\}$. Then the* **principal kinematic formula**

$$\boxed{\int_{G_d} V_j(K \cap gM)\, \mu(dg) = \sum_{k=j}^{d} c_{j,d}^{k,d-k+j} V_k(K) V_{d-k+j}(M)} \qquad (20)$$

holds.

We note that the special case $j = 0$, or

$$\int_{G_d} \chi(K \cap gM)\, \mu(dg) = \sum_{k=0}^{d} c_{0,d}^{k,d-k} V_k(K) V_{d-k}(M), \qquad (21)$$

gives the formula (3) stated in the introduction.

Hadwiger's general formula can be iterated, that is, extended to a finite number of moving convex bodies.

Theorem 1.6. *Let $\varphi : \mathcal{K} \to \mathbb{R}$ be an additive continuous function, and let $K_0, K_1, \ldots, K_k \in \mathcal{K}$, $k \geq 1$, be convex bodies. Then*

$$\int_{G_d} \cdots \int_{G_d} \varphi(K_0 \cap g_1 K_1 \cap \cdots \cap g_k K_k)\, \mu(dg_1) \cdots \mu(dg_k)$$

$$= \sum_{\substack{r_0,\ldots,r_k=0 \\ r_0+\cdots+r_k=kd}}^{d} c_{d-r_0,d,\ldots,d}^{d,r_1,\ldots,r_k}\, \varphi_{r_0}(K_0) V_{r_1}(K_1) \cdots V_{r_k}(K_k),$$

where the coefficients are given by (14).

As before, the specialization $\varphi = V_j$ yields a corollary.

Theorem 1.7 (Iterated kinematic formula). *Let* $K_0, K_1, \ldots, K_k \in \mathcal{K}$, $k \geq 1$, *be convex bodies, and let* $j \in \{0, \ldots, d\}$. *Then*

$$\int_{G_d} \cdots \int_{G_d} V_j(K_0 \cap g_1 K_1 \cap \cdots \cap g_k K_k)\, \mu(dg_1) \cdots \mu(dg_k)$$

$$= \sum_{\substack{m_0,\ldots,m_k=j \\ m_0+\cdots+m_k=kd+j}}^{d} c_{j,d,\ldots,d}^{d,m_0,\ldots,m_k} V_{m_0}(K_0) \cdots V_{m_k}(K_k).$$

Proof of Theorem 1.6. The proof proceeds by induction with respect to k. Theorem 1.4 is the case $k = 1$. Suppose that $k \geq 1$ and that the assertion of Theorem 1.6, and hence that of Theorem 1.7, has been proved for $k+1$ convex bodies. Let $K_0, \ldots, K_{k+1} \in \mathcal{K}$. Using Fubini's theorem twice, the invariance of the measure μ, then Theorem 1.4 followed by Theorem 1.7 for $k+1$ convex bodies, we obtain

$$\int_{G_d} \cdots \int_{G_d} \varphi(K_0 \cap g_1 K_1 \cap \cdots \cap g_{k+1} K_{k+1})\, \mu(dg_1) \cdots \mu(dg_{k+1})$$

$$= \int_{G_d} \cdots \int_{G_d} \varphi(K_0 \cap g_1(K_1 \cap g_2 K_2 \cap \cdots \cap g_{k+1} K_{k+1}))\, \mu(dg_1)$$

$$\times\, \mu(dg_2) \cdots \mu(dg_{k+1})$$

$$= \int_{G_d} \cdots \int_{G_d} \sum_{j=0}^{d} \varphi_{d-j}(K_0) V_j(K_1 \cap g_2 K_2 \cap \cdots \cap g_{k+1} K_{k+1})$$

$$\times\, \mu(dg_2) \cdots \mu(dg_{k+1})$$

$$= \sum_{j=0}^{d} \varphi_{d-j}(K_0) \sum_{\substack{m_0,\ldots,m_k=j \\ m_0+\cdots+m_k=kd+j}}^{d} c_{j,d,\ldots,d}^{d,m_0,\ldots,m_k} V_{m_0}(K_1) \cdots V_{m_k}(K_{k+1})$$

$$= \sum_{\substack{r_0,\ldots,r_{k+1}=0 \\ r_0+\cdots+r_{k+1}=(k+1)d}}^{d} c_{d-r_0,d,\ldots,d}^{d,r_1,\ldots,r_{k+1}}\, \varphi_{r_0}(K_0) V_{r_1}(K_1) \cdots V_{r_{k+1}}(K_{k+1}).$$

This completes the proof. □

Hints to the literature. The idea of deducing integral geometric formulae for convex bodies from a characterization of the intrinsic volumes essentially goes back to W. Blaschke. It was put on a solid basis by Hadwiger [12]. Hadwiger's general integral geometric theorem and its deduction from the characterization theorem appear in [15] and [16]. The iteration of Theorem 1.6 (for general φ) is formulated here for the first time.

The standard source for integral geometry is [37]. An introduction to integral geometry in the spirit of these lectures, with a special view to applications in stochastic geometry, is given in [51].

2 Localizations and Extensions

The envisaged applications of the kinematic formula (20) to stochastic geometry require its extension in several directions. In this section, we first treat a local version. It involves local versions of the intrinsic volumes, in the form of curvature measures and their generalizations. The rest of the section deals with extensions to non-convex sets.

2.1 The Kinematic Formula for Curvature Measures

The notion of the parallel body of a convex body K can be generalized, by taking only those points x into account for which the nearest point $p(K, x)$ in K belongs to some specified set of points, and/or the normalized vector from $p(K, x)$ to x belongs to some specified set of directions. Again there is a Steiner formula, and the coefficients define the curvature measures. In this section we show how the kinematic and Crofton formulae can be extended to these curvature measures. First we introduce a general version of curvature measures.

By a **support element** of the convex body $K \in \mathcal{K}$ we understand a pair (x, u) where x is a boundary point of K and u is an outer unit normal vector of K at x. The set of all support elements of K is called the **generalized normal bundle** of K and is denoted by Nor K. It is a subset of the product space $\Sigma := \mathbb{R}^d \times S^{d-1}$.

Recall that $p(K, x)$ is the point in the convex body K nearest to $x \in \mathbb{R}^d$ and that $d(K, x) := \|x - p(K, x)\|$ is the distance of x from K. For $x \notin K$ we have $d(K, x) > 0$, and we put $u(K, x) := (x - p(K, x))/d(K, x)$; then $(p(K, x), u(K, x)) \in$ Nor K. For $\rho > 0$ and each Borel set $S \subset \Sigma$, a **local parallel set** is now defined by

$$M_\rho(K, S) := \{x \in K_\rho \setminus K : (p(K, x), u(K, x)) \in S\}.$$

The Steiner formula extends as follows.

Theorem 2.1 (Local Steiner formula). *For $K \in \mathcal{K}$, there are finite measures $\Xi_0(K, \cdot), \ldots, \Xi_{d-1}(K, \cdot)$ on Σ such that, for $\rho \geq 0$ and every $S \in \mathcal{B}(\Sigma)$,*

$$\lambda_d(M_\rho(K, S)) = \sum_{m=0}^{d-1} \rho^{d-m} \kappa_{d-m} \Xi_m(K, S). \tag{22}$$

The measures $\Xi_0(K, \cdot), \ldots, \Xi_{d-1}(K, \cdot)$ are called the **support measures** or **generalized curvature measures** of K.

The principle of the proof is clear from the proof of Theorem 1.1, the Steiner formula; only the details are slightly more technical. We sketch here the main ideas. We write $\mu_\rho(K, S) := \lambda_d(M_\rho(K, S))$ for $S \in \mathcal{B}(\Sigma)$. First, for a polytope P one obtains

$$\mu_\rho(P, S) = \sum_{m=0}^{d-1} \rho^{d-m} \kappa_{d-m} \Xi_m(P, S)$$

if one puts

$$\Xi_m(P, S) := \sum_{F \in \mathcal{F}_m(P)} \int_F \frac{\sigma_{d-1-m}(N(P, F) \cap \{u \in S^{d-1} : (y, u) \in S\})}{(d-m)\kappa_{d-m}} \lambda_m(dy).$$

Next, one shows that $\mu_\rho(K, \cdot)$ is a measure on Σ and that $\lim_{j \to \infty} K_j = K$ for convex bodies K_j, K implies that the sequence $(\mu_\rho(K_j, \cdot))_{j \in \mathbb{N}}$ converges weakly to $\mu_\rho(K, \cdot)$. Thus, the map $K \mapsto \mu_\rho(K, \cdot)$ is weakly continuous. Moreover, this map is additive. For each fixed $S \in \mathcal{B}(\Sigma)$, the function $\mu_\rho(\cdot, S)$ is measurable. One can now follow the arguments of Section 1.2 and establish the existence of the measures $\Xi_0(K, \cdot), \ldots, \Xi_{d-1}(K, \cdot)$ for general $K \in \mathcal{K}$ so that (22) holds. Proceeding essentially as for the intrinsic volumes (which correspond to the case $S = \Sigma$), one then shows that the maps $\Xi_m : \mathcal{K} \times \mathcal{B}(\Sigma) \to \mathbb{R}$ have the following properties:

- **Motion covariance**: $\Xi_m(gK, g.S) = \Xi_m(K, S)$ for $g \in G_d$, where $g.S := \{(gx, g_0 u) : (x, u) \in S\}$, g_0 denoting the rotation part of g,
- **Homogeneity**: $\Xi_m(\alpha K, \alpha \cdot S) = \Xi_m(K, S)$ for $\alpha \geq 0$, where $\alpha \cdot S := \{(\alpha x, u) : (x, u) \in S\}$,
- **Weak continuity**: $K_j \to K$ implies $\Xi_m(K_j, \cdot) \to \Xi_m(K, \cdot)$ weakly,
- $\Xi_m(\cdot, S)$ is **additive**, for each fixed $S \in \mathcal{B}(\Sigma)$,
- $\Xi_m(\cdot, S)$ is **measurable**, for each fixed $S \in \mathcal{B}(\Sigma)$.

In the following, we will mainly use the first of two natural specializations of the support measures, which are defined by

$$\Phi_m(K, A) := \Xi_m(K, A \times S^{d-1}) \qquad \text{for } A \in \mathcal{B}(\mathbb{R}^d),$$
$$\Psi_m(K, B) := \Xi_m(K, \mathbb{R}^d \times B) \qquad \text{for } B \in \mathcal{B}(S^{d-1}).$$

Thus, $\Phi_m(K, \cdot)$ is the image measure of $\Xi_m(K, \cdot)$ under the projection $(x, u) \mapsto x$, and $\Psi_m(K, \cdot)$ is the image measure of $\Xi_m(K, \cdot)$ under the projection $(x, u) \mapsto u$. The measure $\Phi_m(K, \cdot)$ is called the mth **curvature measure** of K, and $\Psi_m(K, \cdot)$ is called the mth **area measure** of K, but the reader should be warned that often the same terminology is used for differently normalized measures. In particular, the measure $S_{d-1}(K, \cdot) = 2\Psi(K, \cdot)$ is commonly known as the **area measure** of K.

The defining Steiner formula for the curvature measures can be written in the form

$$\lambda_d(\{x \in K_\rho : p(K, x) \in A\}) = \sum_{m=0}^{d} \rho^{d-m} \kappa_{d-m} \Phi_m(K, A) \qquad (23)$$

for $A \in \mathcal{B}(\mathbb{R}^d)$. Here we have admitted all $x \in K_\rho$ with $p(K, x) \in A$ on the left side; therefore, the right side contains the term

$$\Phi_d(K, A) := \lambda_d(K \cap A).$$

The reason for the name 'curvature measure' becomes clear if one considers a convex body K whose boundary ∂K is a regular hypersurface of class C^2 with positive positive Gauss-Kronecker curvature. In that case, the local parallel volume can be computed by differential-geometric means, and one obtains for $m = 0, \ldots, d-1$ the representation

$$\Phi_m(K, A) = \frac{\binom{d}{m}}{d\kappa_{d-m}} \int_{A \cap \partial K} H_{d-1-m} \, dS. \qquad (24)$$

Here, H_k denotes the kth normalized elementary symmetric function of the principal curvatures of ∂K, and dS is the volume form on ∂K. Thus the curvature measures are (up to normalizing factors) indefinite integrals of curvature functions, and they replace the latter in the non-smooth case. The corresponding representation for the area measures is

$$\Psi_m(K, B) = \frac{\binom{d}{m}}{d\kappa_{d-m}} \int_B s_m \, d\sigma_{d-1} \qquad (25)$$

for $B \in \mathcal{B}(S^{d-1})$. Here, s_m is the mth normalized elementary symmetric function of the principal radii of curvature of ∂K, as a function of the outer unit normal vector.

For a polytope P, the explicit representation given above for the support measure $\Xi_m(P, \cdot)$ specializes to

$$\Phi_m(P, A) = \sum_{F \in \mathcal{F}_m(P)} \gamma(F, P) \lambda_m(F \cap A) \qquad (26)$$

for $A \in \mathcal{B}(\mathbb{R}^d)$ and

$$\Psi_m(P, B) = \sum_{F \in \mathcal{F}_m(P)} \frac{\sigma_{d-1-m}(N(P,F) \cap B)\lambda_m(F)}{(d-m)\kappa_{d-m}} \tag{27}$$

for $B \in \mathcal{B}(S^{d-1})$.

For arbitrary $K \in \mathcal{K}$, it is clear from (23) that the curvature measures $\Phi_0(K, \cdot), \ldots, \Phi_{d-1}(K, \cdot)$ are concentrated on the boundary of K. We mention without proof that the measures $\Phi_0(K, \cdot)$ and $\Phi_{d-1}(K, \cdot)$ have simple intuitive interpretations. Let $\mathcal{H}^{d-1}$ denote $(d-1)$-dimensional Hausdorff measure. If $\dim K \neq d-1$, then

$$\Phi_{d-1}(K, A) = \frac{1}{2}\mathcal{H}^{d-1}(A \cap \partial K).$$

For $\dim K = d-1$, one trivially has $\Phi_{d-1}(K, A) = \mathcal{H}^{d-1}(A \cap \partial K)$. The measure Φ_0 is the normalized area of the spherical image. Let $\sigma(K, A) \subset S^{d-1}$ denote the set of all outer unit normal vectors of K at points of $A \cap \partial K$, then

$$\Phi_0(K, A) = \frac{1}{d\kappa_d}\mathcal{H}^{d-1}(\sigma(K, A)).$$

Now we state the local versions of Theorems 1.6 and 1.3.

Theorem 2.2. *Let* $K, M \in \mathcal{K}$ *be convex bodies, let* $j \in \{0, \ldots, d\}$, *and let* $A, B \in \mathcal{B}(\mathbb{R}^d)$ *be Borel sets. Then the* **local principal kinematic formula**

$$\int_{G_d} \Phi_j(K \cap gM, A \cap gB) \, \mu(\mathrm{d}g) = \sum_{k=j}^{d} c_{j,d}^{k,d-k+j} \Phi_k(K, A)\Phi_{d-k+j}(M, B) \tag{28}$$

holds. For $k \in \{1, \ldots, d-1\}$ *and* $j \leq k$ *the* **local Crofton formula**

$$\int_{\mathcal{E}_k^d} \Phi_j(K \cap E, A \cap E) \, \mu_k(\mathrm{d}E) = c_{j,d}^{k,d-k+j}\Phi_{d-k+j}(K, A) \tag{29}$$

holds. The coefficients $c_{j,d}^{k,d-k+j}$ *are those given by* (12).

We will describe the main ideas of a proof for the case of polytopes. The result for general convex bodies is then obtained by approximation, using the weak continuity of the curvature measures. We omit the details of this approximation, as well as all arguments concerning measurability, null sets, and integration techniques. Due to the explicit representation of the Haar measure μ, the integral over G_d in (28) can be split in the form

$$\int_{G_d} \Phi_j(K \cap gM, A \cap gB) \, \mu(\mathrm{d}g)$$

$$= \int_{SO_d} \int_{\mathbb{R}^d} \Phi_j(K \cap (\vartheta M + x), A \cap (\vartheta B + x)) \, \lambda_d(\mathrm{d}x) \, \nu(\mathrm{d}\vartheta).$$

First we treat only the inner integral (without loss of generality, for $\vartheta = $ identity). This gives us the opportunity to introduce some notions and results of translative integral geometry, which will be elaborated upon in Section 3.

Let $K, M \in \mathcal{K}$ be d-dimensional polytopes, and let $A, B \in \mathcal{B}(\mathbb{R}^d)$ be Borel sets. We have to investigate the integral

$$I := \int_{\mathbb{R}^d} \Phi_j(K \cap (M + x), A \cap (B + x)) \, \lambda_d(dx).$$

By (26), the jth curvature measure of a polytope P is given by

$$\Phi_j(P, \cdot) = \sum_{F \in \mathcal{F}_j(P)} \gamma(F, P)\lambda_F,$$

where we have introduced the abbreviation

$$\lambda_F(\cdot) := \lambda_{\dim F}(F \cap \cdot). \tag{30}$$

It follows that

$$I = \int_{\mathbb{R}^d} \sum_{F' \in \mathcal{F}_j(K \cap (M + x))} \gamma(F', K \cap (M + x))\lambda_{F'}(A \cap (B + x)) \, \lambda_d(dx). \tag{31}$$

The faces $F' \in \mathcal{F}_j(K \cap (M + x))$ are precisely the j-dimensional sets of the form $F' = F \cap (G + x)$ with a face $F \in \mathcal{F}_k(K)$ and a face $G \in \mathcal{F}_i(M)$, where $k, i \in \{j, \ldots, d\}$. We may assume that $k + i = d + j$, since only such pairs F, G contribute to the integral. Therefore, we obtain

$$I = \sum_{k=j}^{d} \sum_{F \in \mathcal{F}_k(K)} \sum_{G \in \mathcal{F}_{d-k+j}(M)}$$
$$\int_{\mathbb{R}^d} \gamma(F \cap (G + x), K \cap (M + x))\lambda_{F \cap (G+x)}(A \cap (B + x)) \, \lambda_d(dx).$$

In the integrand, we may assume that relint $F \cap$ relint $(G + x) \neq \emptyset$, since other vectors x do not contribute to the integral, and in this case the **common exterior angle**

$$\gamma(F, G; K, M) := \gamma(F \cap (G + x), K \cap (M + x)) \tag{32}$$

does not depend on x. Putting

$$J(F, G) := \int_{\mathbb{R}^d} \lambda_{F \cap (G+x)}(A \cap (B + x)) \, \lambda_d(dx),$$

we thus have

$$I = \sum_{k=j}^{d} \sum_{F \in \mathcal{F}_k(K)} \sum_{G \in \mathcal{F}_{d-k+j}(M)} \gamma(F, G; K, M)J(F, G).$$

To compute the integral $J(F, G)$ for given faces $F \in \mathcal{F}_k(K)$ and $G \in \mathcal{F}_{d-k+j}(M)$, we decompose the space $\mathbb{R}^d$ in a way adapted to these faces and apply Fubini's theorem. The result is

$$J(F, G) = [F, G] \lambda_F(A) \lambda_G(B),$$

where the 'generalized sine function' $[F, G]$ is defined as follows.

Let $L, L' \subset \mathbb{R}^d$ be two linear subspaces. We choose an orthonormal basis of $L \cap L'$ and extend it to an orthonormal basis of L and also to an orthonormal basis of L'. Let P denote the parallelepiped that is spanned by the vectors obtained in this way. We define $[L, L'] := \lambda_d(P)$. Then $[L, L']$ depends only on the subspaces L and L'. If $L + L' \neq \mathbb{R}^d$, then $[L, L'] = 0$. We extend this definition to faces F, G of polytopes by putting $[F, G] := [L, L']$, where L and L' are the linear subspaces which are translates of the affine hulls of F and G, respectively.

Inserting the expression for $J(F, G)$ in the integral I, we end up with the following **principal translative formula for polytopes**.

Theorem 2.3. *If $K, M \in \mathcal{K}$ are polytopes and $A, B \in \mathcal{B}(\mathbb{R}^d)$, then for $j \in \{0, \ldots, d\}$,*

$$\int_{\mathbb{R}^d} \Phi_j(K \cap (M + x), A \cap (B + x)) \, \lambda_d(\mathrm{d}x)$$

$$= \sum_{k=j}^{d} \sum_{F \in \mathcal{F}_k(K)} \sum_{G \in \mathcal{F}_{d-k+j}(M)} \gamma(F, G; K, M)[F, G] \lambda_F(A) \lambda_G(B).$$

The kinematic formula at which we are aiming requires, for polytopes, the computation of

$$\int_{G_d} \Phi_j(K \cap gM, A \cap gB) \, \mu(\mathrm{d}g)$$

$$= \int_{SO_d} \int_{\mathbb{R}^d} \Phi_j(K \cap (\vartheta M + x), A \cap (\vartheta B + x)) \, \lambda_d(\mathrm{d}x) \, \nu(\mathrm{d}\vartheta)$$

$$= \sum_{k=j}^{d} \sum_{F \in \mathcal{F}_k(K)} \sum_{G \in \mathcal{F}_{d-k+j}(M)} \lambda_F(A) \lambda_G(B)$$

$$\times \int_{SO_d} \gamma(F, \vartheta G; K, \vartheta M)[F, \vartheta G] \, \nu(\mathrm{d}\vartheta).$$

Here we have used the fact that $\lambda_{\vartheta G}(\vartheta B) = \lambda_G(B)$. The summands with $k = j$ or $k = d$ are easily determined; we get $\Phi_j(K, A)\Phi_d(M, B)$ for $k = j$ and $\Phi_d(K, A)\Phi_j(M, B)$ for $k = d$. The remaining integrals over the rotation group are determined in the following theorem.

Theorem 2.4. *Let* $K, M \in \mathcal{K}$ *be polytopes, let* $j \in \{0, \ldots, d-2\}$, $k \in \{j+1, \ldots, d-1\}$, $F \in \mathcal{F}_k(K)$ *and* $G \in \mathcal{F}_{d-k+j}(M)$. *Then*

$$\int_{SO_d} \gamma(F, \vartheta G; K, \vartheta M)[F, \vartheta G]\, \nu(\mathrm{d}\vartheta) = c_{j,d}^{k,d-k+j}\gamma(F, K)\gamma(G, M),$$

where $c_{j,d}^{k,d-k+j}$ *is as in* (12).

For this formula, a proof can be given which uses the fact that the spherical Lebesgue measure is, up to a constant factor, the only rotation invariant finite Borel measure on the sphere. In this way, one obtains (28), but with unknown coefficients instead of $c_{j,d}^{k,d-k+j}$. The values of these coefficients then follow from the fact that in the case $A = B = \mathbb{R}^d$ the result must coincide with (20). Thus (28) is obtained. The local Crofton formula (29) can be deduced from (28) by an argument similar to that used in the proof of Theorem 1.4.

There is also a version of the local kinematic and Crofton formulae for support measures. Such a variant, which we mention only briefly, is possible if the intersection of Borel sets in $\mathbb{R}^d$ is replaced by a suitable law of composition for subsets of Σ, which is adapted to intersections of convex bodies. For $S, S' \subset \Sigma$, let

$$S \wedge S' := \{(x, u) \in \Sigma : \text{ there are } u_1, u_2 \in S^{d-1} \text{ with}$$
$$(x, u_1) \in S, (x, u_2) \in S', u \in \text{pos}\,\{u_1, u_2\}\},$$

where $\text{pos}\,\{u_1, u_2\} := \{\lambda_1 u_1 + \lambda_2 u_2 : \lambda_1, \lambda_2 \geq 0\}$ is the positive hull of $\{u_1, u_2\}$. For a q-flat $E \in \mathcal{E}_q^d$, $q \in \{1, \ldots, d-1\}$, one defines

$$S \wedge E := \{(x, u) \in \Sigma : \text{ there are } u_1, u_2 \in S^{d-1} \text{ with}$$
$$(x, u_1) \in S, x \in E, u_2 \in E^{\perp}, u \in \text{pos}\,\{u_1, u_2\}\},$$

where $E^{\perp}$ is the linear subspace totally orthogonal to E. Now for given convex bodies $K, K' \in \mathcal{K}$, Borel sets $S \subset \text{Nor}\,K$ and $S' \subset \text{Nor}\,K'$, and for $j \in \{0, \ldots, d-2\}$, the formula

$$\boxed{\int_{G_d} \Xi_j(K \cap gK', S \wedge gS')\, \mu(\mathrm{d}g) = \sum_{k=j+1}^{d-1} c_{j,d}^{k,d-k+j}\Xi_k(K, S)\Xi_{d-k+j}(K', S')}$$

$$\tag{33}$$

holds (for $j = d-1$, both sides would give 0). The local Crofton formula has the following extension. Let $K \in \mathcal{K}$ be a convex body, $k \in \{1, \ldots, d-1\}$, $j \in \{0, \ldots, k-1\}$, and let $S \subset \text{Nor}\,K$ be a Borel set. Then

$$\boxed{\int_{\mathcal{E}_k^d} \Xi_j(K \cap E, S \wedge E)\, \mu_k(\mathrm{d}E) = c_{j,d}^{k,d-k+j}\Xi_{d-k+j}(K, S).} \tag{34}$$

Hints to the literature. For a more thorough introduction to support and curvature measures we refer to [39]. Detailed proofs of the kinematic and Crofton formulae for curvature measures of convex bodies are found in [39] and [51]. Formulae (33) and (34) are due to Glasauer [6], under an additional assumption. This assumption was removed by Schneider [42]. An analogue of Hadwiger's general integral geometric theorem for measure valued valuations on convex bodies was proved in [40]. A simpler proof was given in [5], and a generalization in [7].

2.2 Additive Extension to Polyconvex Sets

So far, our integral geometric investigations were confined to convex bodies. In view of applications, this is a too narrow class of sets. The additivity of the intrinsic volumes and curvature measures permits us to extend these and the pertinent integral geometric intersection formulae to finite unions of convex bodies. This additive extension will be achieved in the present section.

By $\mathcal{R}$ we denote the system of all finite unions of convex bodies in $\mathbb{R}^d$ (including the empty set). The system $\mathcal{R}$, which is closed under finite unions and intersections, is called the **convex ring** (a questionable translation of the German 'Konvexring'). The elements of $\mathcal{R}$ will be called **polyconvex** sets.

Let φ be a function on $\mathcal{R}$ with values in some abelian group. The function φ is called **additive** or a **valuation** if

$$\varphi(K \cup L) + \varphi(K \cap L) = \varphi(K) + \varphi(L)$$

for $K, L \in \mathcal{R}$ and $\varphi(\emptyset) = 0$. Every such valuation satisfies the **inclusion-exclusion principle**

$$\varphi(K_1 \cup \cdots \cup K_m) = \sum_{r=1}^{m} (-1)^{r-1} \sum_{i_1 < \cdots < i_r} \varphi(K_{i_1} \cap \cdots \cap K_{i_r}) \qquad (35)$$

for $K_1, \ldots, K_m \in \mathcal{R}$, as follows by induction.

If φ is a valuation on $\mathcal{K}$, one may ask whether it has an extension to a valuation on $\mathcal{R}$. If such an extension exists and is also denoted by φ, then its values on $\mathcal{R}$ are given by (35), thus the extension is unique. Conversely, however, one cannot just employ (35) for the definition of such an extension, since the representation of an element of $\mathcal{R}$ in the form $K_1 \cup \cdots \cup K_m$ with $K_i \in \mathcal{K}$ is in general not unique. Hence, the existence of an additive extension, if there is one, must be proved in a different way.

A simple example of a valuation on $\mathcal{R}$ is given by the **indicator function**. For $K \in \mathcal{R}$, let

$$\mathbf{1}_K(x) := \begin{cases} 1 & \text{for } x \in K, \\ 0 & \text{for } x \in \mathbb{R}^d \setminus K. \end{cases}$$

For $K, L \in \mathcal{R}$ we trivially have $\mathbf{1}_{K \cup L}(x) + \mathbf{1}_{K \cap L}(x) = \mathbf{1}_K(x) + \mathbf{1}_L(x)$ for $x \in \mathbb{R}^d$. Hence, the mapping $\varphi : \mathcal{R} \to V$, $K \mapsto \mathbf{1}_K$, is an additive function on

$\mathcal{R}$ with values in the vector space V of finite linear combinations of indicator functions of polyconvex sets. By (35), V consists of all linear combinations of indicator functions of convex bodies.

We will now prove a general extension theorem for valuations on $\mathcal{K}$.

Theorem 2.5. *Let X be a topological vector space, and let $\varphi : \mathcal{K} \to X$ be a continuous valuation. Then φ has an additive extension to the convex ring $\mathcal{R}$.*

Proof. Let $\varphi : \mathcal{K} \to X$ be additive and continuous. An essential part of the proof is the following claim.

Proposition 2.1. *The equality*

$$\sum_{i=1}^{m} \alpha_i \mathbf{1}_{K_i} = 0$$

with $m \in \mathbb{N}$, $\alpha_i \in \mathbb{R}$, $K_i \in \mathcal{K}$ implies

$$\sum_{i=1}^{m} \alpha_i \varphi(K_i) = 0.$$

Assume this proposition were false. Then there is a smallest number $m \in \mathbb{N}$, necessarily $m \geq 2$, for which there exist numbers $\alpha_1, \ldots, \alpha_m \in \mathbb{R}$ and convex bodies $K_1, \ldots, K_m \in \mathcal{K}$ such that

$$\sum_{i=1}^{m} \alpha_i \mathbf{1}_{K_i} = 0, \tag{36}$$

but

$$\sum_{i=1}^{m} \alpha_i \varphi(K_i) =: a \neq 0. \tag{37}$$

Let $H \subset \mathbb{R}^d$ be a hyperplane with $K_1 \subset \operatorname{int} H^+$, where H^+, H^- are the two closed halfspaces bounded by H. By (36) we have

$$\sum_{i=1}^{m} \alpha_i \mathbf{1}_{K_i \cap H^-} = 0, \quad \sum_{i=1}^{m} \alpha_i \mathbf{1}_{K_i \cap H} = 0.$$

Since $K_1 \cap H^- = \emptyset$ and $K_1 \cap H = \emptyset$, each of these two sums has at most $m-1$ non-zero summands. From the minimality of m (and from $\varphi(\emptyset) = 0$) we get

$$\sum_{i=1}^{m} \alpha_i \varphi(K_i \cap H^-) = 0, \quad \sum_{i=1}^{m} \alpha_i \varphi(K_i \cap H) = 0.$$

Since $K_i = (K_i \cap H^+) \cup (K_i \cap H^-)$ and $(K_i \cap H^+) \cap (K_i \cap H^-) = K_i \cap H$, the additivity of φ on $\mathcal{K}$ yields

$$\sum_{i=1}^{m} \alpha_i \varphi(K_i \cap H^+) = a, \tag{38}$$

whereas (36) gives

$$\sum_{i=1}^{m} \alpha_i 1_{K_i \cap H^+} = 0. \tag{39}$$

A standard separation theorem for convex bodies implies the existence of a sequence $(H_j)_{j \in \mathbb{N}}$ of hyperplanes with $K_1 \subset \operatorname{int} H_j^+$ for $j \in \mathbb{N}$ and

$$K_1 = \bigcap_{j=1}^{\infty} H_j^+.$$

If the argument that has led us from (36), (37) to (39), (38) is applied k-times, we obtain

$$\sum_{i=1}^{m} \alpha_i \varphi \left(K_i \cap \bigcap_{j=1}^{k} H_j^+ \right) = a.$$

For $k \to \infty$ this yields

$$\sum_{i=1}^{m} \alpha_i \varphi(K_i \cap K_1) = a, \tag{40}$$

since

$$\lim_{k \to \infty} K_i \cap \bigcap_{j=1}^{k} H_j^+ = K_i \cap K_1$$

in the sense of the Hausdorff metric (if $K_i \cap K_1 \neq \emptyset$, otherwise use $\varphi(\emptyset) = 0$) and φ is continuous. Equality (36) implies

$$\sum_{i=1}^{m} \alpha_i 1_{K_i \cap K_1} = 0. \tag{41}$$

The procedure leading from (36) and (37) to (41) and (40) can be repeated, replacing the bodies K_i and K_1 by $K_i \cap K_1$ and K_2, then by $K_i \cap K_1 \cap K_2$ and K_3, and so on. Finally one obtains

$$\sum_{i=1}^{m} \alpha_i 1_{K_1 \cap \cdots \cap K_m} = 0$$

and

$$\sum_{i=1}^{m} \alpha_i \varphi(K_1 \cap \cdots \cap K_m) = a$$

(because of $K_i \cap K_1 \cap \cdots \cap K_m = K_1 \cap \cdots \cap K_m$). Now $a \neq 0$ implies $\sum_{i=1}^{m} \alpha_i \neq 0$ and hence $\mathbf{1}_{K_1 \cap \cdots \cap K_m} = 0$ by the first relation, but this yields $\varphi(K_1 \cap \cdots \cap K_m) = 0$, contradicting the second relation. This completes the proof of the proposition.

Now we consider the real vector space V of all finite linear combinations of indicator functions of elements of $\mathcal{K}$. For $K \in \mathcal{R}$ we have $\mathbf{1}_K \in V$, as noted earlier. For given $f \in V$ we choose a representation

$$f = \sum_{i=1}^{m} \alpha_i \mathbf{1}_{K_i}$$

with $m \in \mathbb{N}$, $\alpha_i \in \mathbb{R}$, $K_i \in \mathcal{K}$ and define

$$\tilde{\varphi}(f) := \sum_{i=1}^{m} \alpha_i \varphi(K_i).$$

The proposition proved above shows that this definition is possible, since the right-hand side does not depend on the special representation chosen for f. Evidently, $\tilde{\varphi} : V \to X$ is a linear map satisfying $\tilde{\varphi}(\mathbf{1}_K) = \varphi(K)$ for $K \in \mathcal{K}$. We can now extend φ from $\mathcal{K}$ to $\mathcal{R}$ by defining

$$\varphi(K) := \tilde{\varphi}(\mathbf{1}_K) \qquad \text{for } K \in \mathcal{R}.$$

By the linearity of $\tilde{\varphi}$ and the additivity of the map $K \mapsto \mathbf{1}_K$ we obtain, for $K, M \in \mathcal{R}$,

$$\varphi(K \cup M) + \varphi(K \cap M) = \tilde{\varphi}(\mathbf{1}_{K \cup M}) + \tilde{\varphi}(\mathbf{1}_{K \cap M}) = \tilde{\varphi}(\mathbf{1}_{K \cup M} + \mathbf{1}_{K \cap M})$$
$$= \tilde{\varphi}(\mathbf{1}_K + \mathbf{1}_M) = \tilde{\varphi}(\mathbf{1}_K) + \tilde{\varphi}(\mathbf{1}_M)$$
$$= \varphi(K) + \varphi(M).$$

Thus φ is additive on $\mathcal{R}$. $\qquad\qquad\qquad\qquad\qquad\qquad\qquad\qquad\qquad$ $\square$

The extension theorem can be applied to the map $K \mapsto \Phi_m(K, \cdot)$ from $\mathcal{K}$ into the vector space of finite signed measures on $\mathcal{B}(\mathbb{R}^d)$ with the weak topology, since this map is additive and continuous. Hence, the curvature measures have unique additive extensions, as finite signed measures, to the convex ring $\mathcal{R}$. The values of the extension can be obtained from the inclusion-exclusion principle (35), which we now write in a more concise form.

For $m \in \mathbb{N}$, let $S(m)$ denote the set of all non-empty subsets of $\{1, \ldots, m\}$. For $v \in S(m)$, let $|v| := \operatorname{card} v$. If $K_1, \ldots, K_m$ are given, we write

$$K_v := K_{i_1} \cap \cdots \cap K_{i_m} \qquad \text{for } v = \{i_1, \ldots, i_r\} \in S(m).$$

With these conventions, the inclusion-exclusion principle (35) for an additive function φ can be written in the form

$$\varphi(K_1 \cup \cdots \cup K_m) = \sum_{v \in S(m)} (-1)^{|v|-1} \varphi(K_v). \tag{42}$$

We can now easily extend our integral geometric formulae for curvature measures (which includes the case of intrinsic volumes) to polyconvex sets. Let $K \in \mathcal{R}$. We choose a representation

$$K = \bigcup_{i=1}^{m} K_i$$

with convex bodies $K_1, \ldots, K_m$. Since $\varPhi_k$ is additive on $\mathcal{R}$, the inclusion-exclusion principle gives

$$\varPhi_k(K, \cdot) = \sum_{v \in S(m)} (-1)^{|v|-1} \varPhi_k(K_v, \cdot).$$

Now let $M \in \mathcal{K}$ be a convex body, and let $A, B \in \mathcal{B}(\mathbb{R}^d)$. Since the principal kinematic formula holds for convex bodies, we obtain

$$\int_{G_d} \varPhi_j(K \cap gM, A \cap gB)\, \mu(\mathrm{d}g)$$

$$= \int_{G_d} \varPhi_j\left(\bigcup_{i=1}^{m}(K_i \cap gM), A \cap gB\right) \mu(\mathrm{d}g)$$

$$= \int_{G_d} \sum_{v \in S(m)} (-1)^{|v|-1} \varPhi_j(K_v \cap gM, A \cap gB)\, \mu(\mathrm{d}g)$$

$$= \sum_{v \in S(m)} (-1)^{|v|-1} \sum_{k=j}^{d} c_{j,d}^{k,d-k+j} \varPhi_k(K_v, A)\varPhi_{d-k+j}(M, B)$$

$$= \sum_{k=j}^{d} c_{j,d}^{k,d-k+j} \varPhi_k(K, A)\varPhi_{d-k+j}(M, B).$$

Hence, the kinematic formula holds for $K \in \mathcal{R}$ and $M \in \mathcal{K}$. In a similar way, it can now be extended to $K \in \mathcal{R}$ and $M \in \mathcal{R}$. An analogous extension is possible for the Crofton formulae.

Hints to the literature. In adopting the name 'polyconvex' for the elements of the convex ring, we followed Klain and Rota [25], who in turn followed E. de Giorgi. The extension theorem 2.5 and its proof reproduced here are due to Groemer [10]. For the support measures, and thus for the curvature measures, a more explicit construction of an additive extension to polyconvex sets is found in Section 4.4 of [39]. It is based on an extension of the Steiner formula for polyconvex sets, with the Lebesgue measure replaced by the integral of the multiplicity function that arises from additive extension of the indicator function of a parallel set. An application of this extension in stochastic geometry appears in [38].

2.3 Curvature Measures for More General Sets

The class of sets for which the curvature measures Φ_j have been defined, and the local principal kinematic formula (28) has been proved is, up to now, the convex ring, which consists of the finite unions of convex bodies. These polyconvex sets may seem sufficiently general for the simpler purposes of applied stochastic geometry, since sets consisting of very many very small convex bodies can be considered as sufficiently good models for real materials. Nevertheless, from theoretical as well as practical viewpoints, it seems desirable to extend the definition of curvature measures and the validity of kinematic formulae beyond the domain of the convex ring. In this subsection we briefly describe such extensions.

Our approach to the curvature measures involved two steps: the definition via the local Steiner formula (23) (or (22), more generally) and additive extension. The local Steiner formula expresses the Lebesgue measure of a local parallel set, and the definition of the latter makes essential use of the nearest point map $p(K, \cdot)$ of a convex body K, and thus of the fact that to each point $x \in \mathbb{R}^d$ there is a unique nearest point in K. It can be proved that a closed set $A \subset \mathbb{R}^d$ with the property that to each point of $\mathbb{R}^d$ there is a unique nearest point in A, must necessarily be convex ([39], Theorem 1.2.4). At first glance, this seems to indicate that the Steiner formula approach is restricted to convex sets. However, this is not the case. In fact, it is sufficient to have the Steiner formula (23) for small positive values of the distance ρ. This leads us to the sets with positive reach.

Let $K \subset \mathbb{R}^d$ be a nonempty closed set. The **reach** of K, denoted by reach(K), is the largest number ρ (or ∞) such that to each $x \in \mathbb{R}^d$ with distance $d(K, x)$ from x to K smaller than ρ, there is a unique point in K nearest to x; this point is then denoted by $p(K, x)$. If reach(K) > 0, then K is called a **set with positive reach**.

Let $K \subset \mathbb{R}^d$ be a compact set with positive reach. For every Borel set $A \subset \mathbb{R}^d$ and for $0 \leq \rho < $ reach(K) one has a polynomial expansion

$$\lambda_d(\{x \in \mathbb{R}^d : d(K, x) \leq \rho \text{ and } p(K, x) \in A\}) = \sum_{m=0}^{d} \rho^{d-m} \kappa_{d-m} \Phi_m(K, A),$$

and this defines the **curvature measures** $\Phi_0(K, \cdot), \ldots, \Phi_d(K, \cdot)$ of K. If K is convex, these are the known positive measures; in the general case, they are finite signed measures on $\mathcal{B}(\mathbb{R}^d)$. Several of the properties of the curvature measures of convex bodies carry over to the curvature measures of compact sets with positive reach. We mention here only the extension of the principal kinematic formula: If $K, M \subset \mathbb{R}^d$ are compact sets with positive reach, then $K \cap gM$ is a set with positive reach for μ-almost all rigid motions $g \in G_d$, and the kinematic formula (28) holds for Borel sets A, B.

A Steiner type formula for arbitrary closed sets, involving a generalized version of the support measures, is treated in [19]. Since it is applicable in stochastic geometry, we will describe it here, without proofs. To explain the principal ideas, we start with a convex body K which has a regular boundary of class C^2 with positive Gauss-Kronecker curvature. First we state a more general version of the local Steiner formula. Instead of the volume of a local parallel set, we consider the integral $\int_{\mathbb{R}^d \setminus K} f \, d\lambda_d$ of a bounded measurable function f with compact support. It is intuitively clear, and can be proved, that

$$\int_{\mathbb{R}^d \setminus K} f \, d\lambda_d = \int_0^\infty \int_{\partial K_t} f \, dS_t \, dt, \tag{43}$$

where dS_t denotes the volume form of the boundary of the parallel body K_t. We transform the inner integral into an integral over the sphere S^{d-1}. For $u \in S^{d-1}$, let x_u be the (unique) boundary point of K with outer unit normal vector u. Then

$$\int_{\partial K_t} f \, dS_t = \int_{S^{d-1}} f(x_u + tu) \prod_{i=1}^{d-1} r_i^{(t)}(u) \, \sigma_{d-1}(du),$$

where $r_i^{(t)}(u)$, $i = 1, \ldots, d-1$, are the principal radii of curvature of ∂K_t at $x_u + tu$. They are given by $r_i^{(t)}(u) = r_i^{(0)}(u) + t$, hence

$$\prod_{i=1}^{d-1} r_i^{(t)}(u) = \sum_{i=0}^{d-1} \binom{d-1}{i} t^{d-i-1} s_i(u),$$

with s_i as in (25). The result

$$\int_{\mathbb{R}^d \setminus K} f \, d\lambda_d = \sum_{i=0}^{d-1} \binom{d-1}{i} \int_0^\infty \int_{S^{d-1}} t^{d-i-1} f(x_u + tu) s_i(u) \, \sigma_{d-1}(du) \, dt$$

can be written as an integral over the generalized normal bundle $\mathrm{Nor}\, K$ with respect to the support measures, using (25). Setting $\omega_m := m\kappa_m$, we get

$$\int_{\mathbb{R}^d \setminus K} f \, d\lambda_d = \sum_{i=0}^{d-1} \omega_{d-i} \int_0^\infty \int_{\mathrm{Nor}\, K} t^{d-i-1} f(x + tu) \, \Xi_i(K, d(x, u)) \, dt. \tag{44}$$

There is nothing in this formula which refers to the smoothness assumptions made for K. In fact, (44) can be extended to general convex bodies. The special choice

$$f(x) := \mathbf{1}_S(p(K, x), u(K, x)) \mathbf{1}_{K_\rho \setminus K}(x)$$

then gives the local Steiner formula (22).

Heuristically, one would expect that (44) remains true for a non-convex closed set A, provided that the function f has its support in a region where the nearest point in A is uniquely determined. Fortunately, the points where the latter does not hold can be neglected. So we assume now only that $A \subset \mathbb{R}^d$ is a nonempty closed set. As for convex bodies, we define $d(A, x) := \min\{\|a - x\| : a \in A\}$, and we let $p(A, x) := a$ whenever a is a uniquely determined point in A nearest to x. If $d(A, x) > 0$ and $p(A, x)$ exists, we define $u(A, x) := (x - p(A, x))/d(A, x)$. The **exoskeleton** $\mathrm{exo}(A)$ of A is defined as the set of all points x of $\mathbb{R}^d \setminus A$ for which $p(A, x)$ does not exist (because there is more than one point in A nearest to x). One can prove that $\mathrm{exo}(A)$ has d-dimensional Lebesgue measure zero. The **generalized normal bundle** of A is defined by

$$\mathrm{Nor}\, A := \{(p(A, x), u(A, x)) : x \notin A \cup \mathrm{exo}(A)\},$$

and the **reach function** $\delta(A, \cdot) : \mathbb{R}^d \times S^{d-1} \to [0, \infty]$ of A by

$$\delta(A, x, u) := \inf\{t \geq 0 : x + tu \in \mathrm{exo}(A)\} \qquad \text{if } (x, u) \in \mathrm{Nor}\, A,$$

(with $\inf \emptyset = \infty$) and $\delta(A, x, u) := 0$ if $(x, u) \notin \mathrm{Nor}\, A$.

In [19], the **support measures** $\Theta_0(A, \cdot), \ldots, \Theta_{d-1}(A, \cdot)$ of A are introduced as real-valued, σ-additive set functions on the system of all Borel sets in Σ which are contained in

$$(\Sigma \setminus \mathrm{Nor}\, A) \cup \{(x, u) : x \in B, \ \delta(A, x, u) \geq s\}$$

for some $s > 0$ and some compact $B \subset \mathbb{R}^d$. These signed measures vanish on every Borel subset of $\Sigma \setminus \mathrm{Nor}\, A$. Denoting by $|\Theta_i|$ the total variation measure of Θ_i and putting $a \wedge b := \min\{a, b\}$ for $a, b \in \mathbb{R}$, we formulate the following result from [19], which is a far-reaching generalization of the Steiner formula.

Theorem 2.6. *The support measures* $\Theta_0(A, \cdot), \ldots, \Theta_{d-1}(A, \cdot)$ *of a nonempty closed set* $A \subset \mathbb{R}^d$ *satisfy*

$$\int_{\mathrm{Nor}\, A} \mathbf{1}_B(x)(\delta(A, x, u) \wedge r)^{d-j} |\Theta_j|(A, \mathrm{d}(x, u)) < \infty$$

for all compact sets $B \subset \mathbb{R}^d$ *and all* $r > 0$ $(j = 0, \ldots, d-1)$, *and*

$$\int_{\mathbb{R}^d \setminus A} f \, \mathrm{d}\lambda_d = \sum_{i=0}^{d-1} \omega_{d-i} \int_0^\infty \int_{\mathrm{Nor}\, A} t^{d-i-1} \mathbf{1}\{t < \delta(A, x, u)\}$$
$$\times f(x + tu)\, \Theta_i(A, \mathrm{d}(x, u))\, \mathrm{d}t$$

for every measurable bounded function $f : \mathbb{R}^d \to \mathbb{R}$ *with compact support.*

The proof, for which we must refer to [19], makes essential use of the fact that for a given compact set A there exists a sequence $(A_k)_{k \in \mathbb{N}}$ of sets of positive reach such that

$$\text{Nor } A \subset \bigcup_{k \in \mathbb{N}} \text{Nor } A_k$$

and, for $(x, u) \in \text{Nor } A$,

$$\delta(A, x, u) \leq \sup\{\text{reach}(A_k) : (x, u) \in \text{Nor } A_k, \, k \in \mathbb{N}\}.$$

In particular, Nor A is countably $(d-1)$-rectifiable, so that generalized curvatures, as in [61], can be defined.

Hints to the literature. Sets with positive reach and their curvature measures were introduced by Federer [3], who also obtained kinematic and Crofton formulae for these measures. His theory has been further developed in the work of Martina Zähle and co-authors. This work treats current representations of Federer's curvature measures in [61], a short proof of the principal kinematic formula for sets with positive reach in [36], and extensions to certain finite unions of sets with positive reach in [62], [34].

There have been a number of successful attempts to define curvature measures, and to obtain kinematic formulae, for very general and quite abstract classes of sets. We refer to the brief survey in [22] (Subsection 2.1) and the references given there.

3 Translative Integral Geometry

The simple and elegant form of the kinematic formulae (20) and (28), in particular the separation of the two involved convex bodies on the right-hand sides, is due to the fact that the integrals are with respect to the invariant measure on the motion group. The stochastic geometry of stationary but not necessarily isotropic random sets requires analogous investigations with respect to the translation group, for example, the determination of the integrals

$$\int_{\mathbb{R}^d} V_j(K \cap (M + t)) \, \lambda_d(\mathrm{d}t). \tag{45}$$

Convention. In order to achieve a more concise form of translative formulae, we use in the following an operator notation for translations, namely

$$xM := M + x \quad \text{for } M \subset \mathbb{R}^d \text{ and } x \in \mathbb{R}^d.$$

Further, in integrations with respect to Lebesgue measure, we omit the measure, thus $\int f(x) \, \lambda_d(\mathrm{d}x)$ is written as $\int f(x) \, \mathrm{d}x$.

Recall that the indicator function of a set $A \subset \mathbb{R}^d$ is denoted by $\mathbf{1}_A$. We write $M^* := \{y \in \mathbb{R}^d : -y \in M\}$ for the reflection of a set M in 0.

The cases $j = d$ and $j = d - 1$ of the integral (45) are still simple. We have $\mathbf{1}_{tM}(x) = \mathbf{1}_{xM^*}(t)$ and hence, by Fubini's theorem,

$$\int_{\mathbb{R}^d} V_d(K \cap tM) \, \mathrm{d}t$$

$$= \int_{\mathbb{R}^d} \int_{\mathbb{R}^d} 1_{K \cap tM}(x) \, \mathrm{d}x \, \mathrm{d}t = \int_{\mathbb{R}^d} \int_{\mathbb{R}^d} 1_K(x) 1_{xM^*}(t) \, \mathrm{d}t \, \mathrm{d}x$$

$$= \int_{\mathbb{R}^d} 1_K(x) V_d(xM^*) \, \mathrm{d}x = V_d(M^*) \int_{\mathbb{R}^d} 1_K(x) \, \mathrm{d}x,$$

thus

$$\int_{\mathbb{R}^d} V_d(K \cap tM) \, \mathrm{d}t = V_d(K) V_d(M). \tag{46}$$

With only slightly more effort one can show that

$$\int_{\mathbb{R}^d} V_{d-1}(K \cap tM) \, \mathrm{d}t = V_{d-1}(K) V_d(M) + V_d(K) V_{d-1}(M).$$

However, for the functional $V_0 = \chi$, a separation of K and M on the right-hand side does not take place. Since $\chi(K \cap tM) = 1_{K+M^*}(t)$, we have

$$\int_{\mathbb{R}^d} \chi(K \cap tM) \, \mathrm{d}t = V_d(K + M^*). \tag{47}$$

Convex geometry tells us that

$$V_d(K + M^*) = \sum_{i=0}^{d} \binom{d}{i} V_i(K, M^*)$$

with

$$V_i(K, M^*) := V(\underbrace{K, \ldots, K}_{i}, \underbrace{M^*, \ldots, M^*}_{d-i}),$$

where the function $V : \mathcal{K}^d \to \mathbb{R}$ is the **mixed volume**. The essential observation is that the obtained expression cannot be simplified further. Thus, in translative integral geometry we must live with more complicated functionals, depending on several convex bodies simultaneously. A translative formula for curvature measures, which we will now study, necessarily involves new measures depending on several convex bodies.

3.1 The Principal Translative Formula for Curvature Measures

We have already obtained a translative formula for the curvature measures of polytopes, namely Theorem 2.3. We rewrite this result in a form that is convenient for the following. Let $K, M \in \mathcal{K}$ be polytopes, let $A, B \subset \mathbb{R}^d$ be Borel sets, and let $j \in \{0, \ldots, d\}$; then

$$\int_{\mathbb{R}^d} \Phi_j(K \cap xM, A \cap xB) \, \mathrm{d}x = \sum_{m=j}^{d} \Phi_{m,d-m+j}^{(j)}(K, M; A \times B),$$

where for $m, k \in \{j, \dots, d\}$ with $m + k = d + j$ we have introduced measures $\Phi_{m,k}^{(j)}(K, M; \cdot)$ on $(\mathbb{R}^d)^2$ by

$$\Phi_{m,k}^{(j)}(K, M; \cdot) := \sum_{F \in \mathcal{F}_m(K)} \sum_{G \in \mathcal{F}_k(M)} \gamma(F, G; K, M)[F, G]\lambda_F \otimes \lambda_G.$$

To extend this to more than two polytopes, we first extend the notation. For a polyhedral set P (a nonempty intersection of finitely many closed half-spaces) in $\mathbb{R}^d$ and for a face F of P, let $\gamma(F, P)$ be the (normalized) exterior angle of P at F, defined by (5). If $K_1, \dots, K_k$ are polyhedral sets and F_i is a face of K_i for $i = 1, \dots, k$, we choose points $x_i \in \operatorname{relint} F_i$ for $i = 1, \dots, k$ and define the **common exterior angle** $\gamma(F_1, \dots, F_k; K_1, \dots, K_k)$ by

$$\gamma(F_1, \dots, F_k; K_1, \dots, K_k) := \gamma\left(\bigcap_{i=1}^k (F_i - x_i), \bigcap_{i=1}^k (K_i - x_i)\right).$$

This definition does not depend on the choice of the points x_i.

Further, we need the notion of the determinant of subspaces, extending the definition of the generalized sine function $[\cdot, \cdot]$ given in Subsection 2.1. Let $L_1, \dots, L_k$ be linear subspaces of $\mathbb{R}^d$ with $\sum_{i=1}^k \dim L_i =: m \leq d$. Choose an orthonormal basis in each L_i (the empty set if $\dim L_i = 0$) and let $\det(L_1, \dots, L_k)$ be the m-dimensional volume of the parallelepiped spanned by the union of these bases (1, by definition, if $\dim L_i = 0$ for $i = 1, \dots, k$). Then one defines

$$[L_1, \dots, L_k] := \det(L_1^\perp, \dots, L_k^\perp) \quad \text{if } \sum_{i=1}^k \dim L_i \geq (k - 1)d$$

and $[L_1, \dots, L_k] := 0$ if $\sum_{i=1}^k \dim L_i < (k-1)d$. Obviously, any d-dimensional argument of $[L_1, \dots, L_k]$ can be deleted without changing the value. We also note that $[L] = 1$ and that $[L_1, \dots, L_k] = 0$ if $L_1, \dots, L_k$ are not in general relative position (the subspaces $L_1, \dots, L_k$ are in general relative position if $L_1 \cap \dots \cap L_k$ has dimension $\max\{0, \dim L_1 + \dots + \dim L_k - (k-1)d\}$).

For nonempty subsets $F_1, \dots, F_k \subset \mathbb{R}^d$ we set

$$[F_1, \dots, F_k] := [L(F_1), \dots, L(F_k)],$$

where $L(F_i)$ is the linear subspace parallel to the affine hull of F_i $(i = 1, \dots, k)$.

Now let polytopes $K_1, \dots, K_k \in \mathcal{K}$ $(k \in \mathbb{N})$ be given. For indices $m_1, \dots, m_k \in \{1, \dots, d\}$ satisfying $j := \sum_{i=1}^k m_i - (k-1)d \geq 0$ we introduce measures on $(\mathbb{R}^d)^k$, the **mixed measures** $\Phi_{m_1, \dots, m_k}^{(j)}(K_1, \dots, K_k; \cdot)$, by

$$\Phi_{m_1, \dots, m_k}^{(j)}(K_1, \dots, K_k; \cdot)$$
$$:= \sum_{F_1 \in \mathcal{F}_{m_1}(K_1)} \cdots \sum_{F_k \in \mathcal{F}_{m_k}(K_k)} \gamma(F_1, \dots, F_k; K_1, \dots, K_k)[F_1, \dots, F_k]$$
$$\times \lambda_{F_1} \otimes \cdots \otimes \lambda_{F_k}.$$

Note that for $k = 1$ we have $j = m_1$ and hence $\Phi_m^{(j)}(K; \cdot) = \Phi_m(K; \cdot)$.

We can use essentially the same integration technique as sketched for the proof of Theorem 2.3, or that theorem combined with induction, to obtain the following iterated translative formulae for polytopes $K_1, \ldots, K_k \in \mathcal{K}$ $(k \geq 2)$ and Borel sets $A_1, \ldots, A_k \in \mathcal{B}(\mathbb{R}^d)$:

$$
\int_{\mathbb{R}^d} \cdots \int_{\mathbb{R}^d} \Phi_j(K_1 \cap x_2 K_2 \cap \cdots \cap x_k K_k, A_1 \cap x_2 A_2 \cap \cdots \cap x_k A_k) \, dx_2 \cdots dx_k
$$
$$
= \sum_{\substack{m_1, \ldots, m_k = j \\ m_1 + \cdots + m_k = (k-1)d + j}} \Phi_{m_1, \ldots, m_k}^{(j)}(K_1, \ldots, K_k; A_1 \times \cdots \times A_k). \tag{48}
$$

This formula is equivalent to the validity of the equation

$$
\int_{(\mathbb{R}^d)^{k-1}} \int_{\mathbb{R}^d} g(y, y - x_2, \ldots, y - x_k) \, \Phi_j(K_1 \cap x_2 K_2 \cap \cdots \cap x_k K_k, dy)
$$
$$
\times d(x_2, \ldots, x_k)
$$
$$
= \sum_{\substack{m_1, \ldots, m_k = j \\ m_1 + \cdots + m_k = (k-1)d + j}} \int_{(\mathbb{R}^d)^k} g(x_1, x_2, \ldots, x_k)
$$
$$
\times \Phi_{m_1, \ldots, m_k}^{(j)}(K_1, \ldots, K_k; d(x_1, \ldots, x_k)) \tag{49}
$$

for all nonnegative measurable functions g on $(\mathbb{R}^d)^k$. In fact, if the first formula holds, then the second holds for $g = 1_{A_1 \times \cdots \times A_k}$ and thus for elementary functions, hence, by the standard extension, it holds for nonnegative measurable functions. Conversely, if the second formula holds, then the first is true for compact sets $A_1, \ldots, A_k$, since $1_{A_1 \times \cdots \times A_k}$ is then the limit of a decreasing sequence of continuous functions. Since both sides of the equation are measures in $A_1, \ldots, A_k$, the equation holds for Borel sets.

Formula (48) can be extended to general convex bodies, that is, for $K_1, \ldots, K_k \in \mathcal{K}$ and numbers $k, m_1, \ldots, m_k, j$ as above, there exist finite measures $\Phi_{m_1, \ldots, m_k}^{(j)}(K_1, \ldots, K_k; \cdot)$ on $(\mathbb{R}^d)^k$, the **mixed measures**, so that (48) holds. The proof uses approximation of general convex bodies by polytopes, formula (48) for polytopes, and the weak continuity of the curvature measures.

Theorem 3.1 (Iterated translative formula). *Let $K_1, \ldots, K_k \in \mathcal{K}$, where $k \geq 2$, be convex bodies. For $m_1, \ldots, m_k \in \{1, \ldots, d\}$ with $j := \sum_{i=1}^{k} m_i - (k-1)d \geq 0$, there are finite measures $\Phi_{m_1, \ldots, m_k}^{(j)}(K_1, \ldots, K_k; \cdot)$ on $(\mathbb{R}^d)^k$ such that*

$$
\int_{\mathbb{R}^d} \cdots \int_{\mathbb{R}^d} \Phi_j(K_1 \cap x_2 K_2 \cap \cdots \cap x_k K_k, A_1 \cap x_2 A_2 \cap \cdots \cap x_k A_k) \, dx_2 \cdots dx_k
$$
$$
= \sum_{\substack{m_1, \ldots, m_k = j \\ m_1 + \cdots + m_k = (k-1)d + j}} \Phi_{m_1, \ldots, m_k}^{(j)}(K_1, \ldots, K_k; A_1 \times \cdots \times A_k) \tag{50}
$$

for Borel sets $A_1, \ldots, A_k \in \mathcal{B}(\mathbb{R}^d)$.

We collect some properties of the mixed measures. Most of them can be deduced without difficulty from the corresponding properties of the mixed measures of polytopes, which are obvious.

- **Symmetry:**

$$\Phi^{(j)}_{m_1,\ldots,m_k}(K_1,\ldots,K_k; A_1 \times \cdots \times A_k)$$
$$= \Phi^{(j)}_{m_{\pi(1)},\ldots,m_{\pi(k)}}(K_{\pi(1)},\ldots,K_{\pi(k)}; A_{\pi(1)} \times \cdots \times A_{\pi(k)})$$

for each permutation π of $\{1,\ldots,k\}$.
- **Support property**: the support of $\Phi^{(j)}_{m_1,\ldots,m_k}(K_1,\ldots,K_k;\cdot)$ is contained in $S_1 \times \cdots \times S_k$, where $S_i = K_i$ if $m_i = d$ and $S_i = \partial K_i$ if $m_i < d$.
- **Translation covariance**: for $t_1,\ldots,t_k \in \mathbb{R}^d$,

$$\Phi^{(j)}_{m_1,\ldots,m_k}(t_1 K_1,\ldots,t_k K_k; t_1 A_1 \times \cdots \times t_k A_k)$$
$$= \Phi^{(j)}_{m_1,\ldots,m_k}(K_1,\ldots,K_k; A_1 \times \cdots \times A_k).$$

- **Homogeneity**: for $\alpha_1,\ldots,\alpha_k \geq 0$,

$$\Phi^{(j)}_{m_1,\ldots,m_k}(\alpha_1 K_1,\ldots,\alpha_k K_k; \alpha_1 A_1 \times \cdots \times \alpha_k A_k)$$
$$= \alpha_1^{m_1} \cdots \alpha_k^{m_k} \Phi^{(j)}_{m_1,\ldots,m_k}(K_1,\ldots,K_k; A_1 \times \cdots \times A_k).$$

- **Weak continuity**: the map $(K_1,\ldots,K_k) \mapsto \Phi^{(j)}_{m_1,\ldots,m_k}(K_1,\ldots,K_k;\cdot)$ from $\mathcal{K}^k$ into the space of finite signed measures on $(\mathbb{R}^d)^k$ with the weak topology is continuous.
- **Additivity**: $\Phi^{(j)}_{m_1,\ldots,m_k}(K_1,\ldots,K_k; A_1 \times \cdots \times A_k)$ is additive in each of its first k arguments.
- **Decomposition property:**

$$\Phi^{(j)}_{m_1,\ldots,m_{k-1},d}(K_1,\ldots,K_k; A_1 \times \cdots \times A_k)$$
$$= \Phi^{(j)}_{m_1,\ldots,m_{k-1}}(K_1,\ldots,K_{k-1}; A_1 \times \cdots \times A_{k-1}) \lambda_d(K_k \cap A_k)$$

(and similarly for the other arguments, by symmetry).
- **Reduction property:**

$$\Phi^{(j)}_{m_1,\ldots,m_k}(K_1,\ldots,K_k; A_1 \times \cdots \times A_k)$$
$$= \left(\frac{2}{\kappa_{d-1}}\right)^j \frac{1}{j!\kappa_j} \Phi^{(0)}_{m_1,\ldots,m_k,d-1,\ldots,d-1}(K_1,\ldots,K_k,\underbrace{B^d,\ldots,B^d}_{j};$$
$$A_1 \times \cdots \times A_k \times (\mathbb{R}^d)^j).$$

- **Local determination**: If $K_1', \ldots, K_k' \in \mathcal{K}$ and $A_i \subset \operatorname{int} K_i'$ for $i = 1, \ldots, k$, then

$$\Phi_{m_1,\ldots,m_k}^{(j)}(K_1 \cap K_1', \ldots, K_k \cap K_k'; A_1 \times \cdots \times A_k)$$
$$= \Phi_{m_1,\ldots,m_k}^{(j)}(K_1, \ldots, K_k; A_1 \times \cdots \times A_k).$$

The decomposition property has a useful consequence. The condition $m_1 + \cdots + m_k = (k-1)d + j$ implies that at most $d - j$ of the indices $m_1, \ldots, m_k$ can be smaller than d. Hence, all mixed measures with upper index (j) can be expressed in terms of Lebesgue measure and the finitely many mixed measures $\Phi_{m_1,\ldots,m_r}^{(j)}$, where $r \in \{1, \ldots, d-j\}$. By the reduction property, all mixed measures can further be reduced to the measures $\Phi_{m_1,\ldots,m_k}^{(0)}$ with $k \in \{1, \ldots, d\}$ and $m_1, \ldots, m_k \in \{1, \ldots, d-1\}$ satisfying $m_1 + \cdots + m_k = (k-1)d$. The mixed measures with upper index (0) will therefore be considered as **basic**.

The last of the listed properties, local determination, can be used to extend the definition of the mixed measures, in an obvious way, to closed convex sets that are not necessarily bounded. The iterated translative formula (50) remains valid if the Borel sets A_i corresponding to unbounded convex sets K_i are bounded.

The total measures

$$\Phi_{m_1,\ldots,m_k}^{(j)}(K_1, \ldots, K_k; (\mathbb{R}^d)^k) =: V_{m_1,\ldots,m_k}^{(j)}(K_1, \ldots, K_k)$$

are called the **mixed functionals**, and those with upper index (0) the **basic mixed functionals**. In the case of polytopes $K_1, \ldots, K_d$, the mixed functionals are explicitly given by

$$V_{m_1,\ldots,m_k}^{(j)}(K_1, \ldots, K_k)$$
$$= \sum_{F_1 \in \mathcal{F}_{m_1}(K_1)} \cdots \sum_{F_k \in \mathcal{F}_{m_k}(K_k)} \gamma(F_1, \ldots, F_k; K_1, \ldots, K_k)[F_1, \ldots, F_k]$$
$$\times V_{m_1}(F_1) \cdots V_{m_k}(F_k).$$

The mixed measures, and therefore also the mixed functionals, satisfy various integral geometric relations, among them translative formulae, rotation formulae, and Crofton type formulae.

Hints to the literature. The technical details omitted in this subsection can be found in [50] and [58]. A thorough investigation of the mixed measures, including integral geometric relations and special representations, appears in [60]. In [8], representations of mixed measures in terms of support measures are applied to Boolean models.

Using methods of geometric measure theory, Rataj and Zähle [33] have obtained a general translative formula for support measures of sets with positive reach. An iterated version is proved in [28]. Various extensions and supplements appear in [29], [17], [63], [34], [35]. Translative Crofton formulae for support measures are treated in [30].

3.2 Basic Mixed Functionals and Support Functions

In this subsection, we will study some special cases of the mixed measures in greater detail and under particular aspects. First, we consider the basic mixed functionals $V_{m_1,\ldots,m_k}^{(0)}$. They are uniquely determined as the coefficients in the polynomial expansion

$$
\int_{\mathbb{R}^d} \cdots \int_{\mathbb{R}^d} \chi(\alpha_1 K_1 \cap x_2\alpha_2 K_2 \cap \cdots \cap x_k\alpha_k K_k)\, dx_2 \cdots dx_k
$$

$$
= \sum_{\substack{m_1,\ldots,m_k=0 \\ m_1+\cdots+m_k=(k-1)d}}^{d} \alpha_1^{m_1} \cdots \alpha_k^{m_k} V_{m_1,\ldots,m_k}^{(0)}(K_1,\ldots,K_k) \tag{51}
$$

for $K_1,\ldots,K_k \in \mathcal{K}$, $\alpha_1 \ldots, \alpha_k \geq 0$, $k \geq 2$ (a special case of (50)). Our first aim is to show that the notion of basic mixed functionals can, in a certain sense, be viewed as dual to the notion of mixed volumes, which constitute an important set of functionals in the classical theory of convex bodies. It suffices to consider polytopes (the extension to general convex bodies is achieved by approximation). For these, we will obtain a class of representations of the basic mixed measures of greater generality than their original definition, which is the representation

$$
V_{m_1,\ldots,m_k}^{(0)}(K_1,\ldots,K_k)
$$

$$
= \sum_{F_1 \in \mathcal{F}_{m_1}(K_1)} \cdots \sum_{F_k \in \mathcal{F}_{m_k}(K_k)} \gamma(F_1,\ldots,F_k; K_1,\ldots,K_k)[F_1,\ldots,F_k]
$$

$$
\times V_{m_1}(F_1) \cdots V_{m_k}(F_k). \tag{52}
$$

We describe a general construction leading to different types of mixed functionals. For $k \geq 2$, we consider the Euclidean vector space $X := (\mathbb{R}^d)^k$ (with the scalar product induced from that of $\mathbb{R}^d$), together with a surjective linear map $f : X \to \mathbb{R}^q$ onto a Euclidean space $\mathbb{R}^q$, where $q < kd$. For $L := (\ker f)^\perp$, the orthogonal complement of the kernel of f, let $\pi_L : X \to L$ denote the orthogonal projection. If $\tilde{f} := f|L$, then $f = \tilde{f} \circ \pi_L$, and $\tilde{f} : L \to \mathbb{R}^q$ is an isomorphism.

For a given polytope $P \subset X$, let $\mathcal{G}_v(P)$ be a set of q-dimensional faces of P with the property that, for the induced Lebesgue measure Λ_L in L,

$$
\Lambda_L(\pi_L P) = \sum_{F \in \mathcal{G}_v(P)} \Lambda_L(\pi_L F) \tag{53}
$$

and hence

$$
V_q(fP) = \sum_{F \in \mathcal{G}_v(P)} V_q(fF). \tag{54}
$$

Such sets $\mathcal{G}_v(P)$, depending on a parameter v, can be obtained as follows. We choose a vector $v \in L^\perp \setminus \{0\}$ satisfying

$$\dim F(P, u) \le q \qquad \text{for all } u \in L_v^+ := L + \{\alpha v : \alpha > 0\} \tag{55}$$

($F(P, u)$ denotes the face of P with exterior normal vector u). The condition (55) excludes only vectors from finitely many proper linear subspaces of $L^\perp$. With this choice, the set

$$\mathcal{G}_v(P) := \{F \in \mathcal{F}_q(P) : N(P, F) \cap L_v^+ \ne \emptyset\}$$

will satisfy (53). This is seen if one decomposes the projection π_L into the orthogonal projection $\pi_{L,v}$ onto $\mathrm{lin}(L \cup \{v\})$, followed by the orthogonal projection from this space to L. In fact, let $\mathcal{S}_q$ be the set of q-faces of $\pi_{L,v}P$ having a normal vector in L_v^+. The images of these q-faces under projection to L cover πP without overlappings. On the other hand, under $\pi_{L,v}$, the set $\mathcal{G}_v(P)$ is in one-to-one correspondence with the set $\mathcal{S}_q$. This proves (53).

Now let $P_1, \ldots, P_k \subset \mathbb{R}^d$ be polytopes, and let $P := P_1 \times \cdots \times P_k$. Every q-face F of P is of the form $F = F_1 \times \cdots \times F_k$ with faces $F_i \in \mathcal{F}_{m_i}(P_i)$ $(i = 1, \ldots, k)$ for suitable $m_1, \ldots, m_k \in \{0, \ldots, d\}$ satisfying $m_1 + \cdots + m_k = q$. For $\alpha_1, \ldots, \alpha_k \ge 0$, we clearly have

$$\begin{aligned}
&V_q(f(\alpha_1 F_1 \times \cdots \times \alpha_k F_k)) \\
&= \alpha_1^{m_1} \cdots \alpha_k^{m_k} c(f, F_1, \ldots, F_k) V_{m_1}(F_1) \cdots V_{m_k}(F_k),
\end{aligned}$$

where $c(f, F_1, \ldots, F_k)$ is the factor by which the linear map $f | \mathrm{aff}(F_1 \times \cdots \times F_k)$ changes the q-dimensional volume. Together with (54), this gives

$$\begin{aligned}
&V_q(f(\alpha_1 P_1 \times \cdots \times \alpha_k P_k)) \\
&= \sum_{\substack{m_1, \ldots, m_k = 0 \\ m_1 + \cdots + m_k = q}}^{d} \alpha_1^{m_1} \cdots \alpha_k^{m_k} V_{m_1, \ldots, m_k}^f(P_1, \ldots, P_k)
\end{aligned} \tag{56}$$

with

$$\begin{aligned}
&V_{m_1, \ldots, m_k}^f(P_1, \ldots, P_k) \\
&= \sum_{\substack{F_i \in \mathcal{F}_{m_i}(P_i), i=1, \ldots, k \\ F_1 \times \cdots \times F_k \in \mathcal{G}_v(P_1 \times \cdots \times P_k)}} c(f, F_1, \ldots, F_k) V_{m_1}(F_1) \cdots V_{m_k}(F_k).
\end{aligned} \tag{57}$$

Here v is chosen according to (55), for $P = P_1 \times \cdots \times P_k$. The expansion (56) determines the coefficients $V_{m_1, \ldots, m_k}^f(P_1, \ldots, P_k)$ uniquely; hence they do not depend on the choice of the set $\mathcal{G}_v(P_1 \times \cdots \times P_k)$. These coefficients represent a **general type of mixed functionals**, depending on the choice of the number k and the linear map f.

For a concrete example, let $f : (\mathbb{R}^d)^k \to \mathbb{R}^d$ be defined by $f(x_1, \ldots, x_d) := x_1 + \cdots + x_d$. Then

$$f(\alpha_1 P_1 \times \cdots \times \alpha_k P_k) = \alpha_1 P_1 + \cdots + \alpha_k P_k,$$

and (56) gives

$$V_d(\alpha_1 P_1 + \cdots + \alpha_k P_k)$$

$$= \sum_{\substack{m_1,\ldots,m_k=0 \\ m_1+\cdots+m_k=d}}^{d} \alpha_1^{m_1} \cdots \alpha_k^{m_k} V_{m_1,\ldots,m_k}^{f}(P_1,\ldots,P_k).$$

This shows that

$$V_{m_1,\ldots,m_k}^{f}(P_1,\ldots,P_k) = \binom{d}{m_1,\ldots,m_k} V(\underbrace{P_1,\ldots,P_1}_{m_1},\ldots,\underbrace{P_k,\ldots,P_k}_{m_k}),$$

where $V : \mathcal{K}^d \to \mathbb{R}$ denotes the **mixed volume**, known from the theory of convex bodies. Equation (57) provides a class of special representations for the mixed volumes of polytopes.

In the second example, let $g : (\mathbb{R}^d)^k \to (\mathbb{R}^d)^{k-1}$ be defined by $g(y_1,\ldots,y_k)$ $:= (y_1 - y_2, \ldots, y_1 - y_k)$. Then

$$g(P_1 \times \cdots \times P_k) = \{(x_2,\ldots,x_k) \in (\mathbb{R}^d)^{k-1} : P_1 \cap x_2 P_2 \cap \cdots \cap x_k P_k \neq \emptyset\},$$

hence

$$V_{(k-1)d}(g(\alpha_1 P_1 \times \cdots \times \alpha_k P_k))$$

$$= \int_{\mathbb{R}^d} \cdots \int_{\mathbb{R}^d} \chi(\alpha_1 P_1 \cap x_2 \alpha_2 P_2 \cap \cdots \cap x_k \alpha_k P_k) dx_2 \cdots dx_k.$$

Now (56) shows that

$$V_{m_1,\ldots,m_k}^{g}(P_1,\ldots,P_k) = V_{m_1,\ldots,m_k}^{(0)}(P_1,\ldots,P_k),$$

thus we obtain the basic mixed functionals. The constructions leading to mixed volumes and to basic mixed functionals can be considered as duals of each other, since *the kernels of the employed maps f and g are complementary orthogonal subspaces of* $(\mathbb{R}^d)^k$.

In order to make the representation (57) more explicit in the concrete cases, one has to translate condition (55) to the actual situation and to determine the factors $c(f, F_1, \ldots, F_k)$. We give here the result only for the second example. In that case, we say that a vector $w \in \mathbb{R}^d$ is **admissible for the polytopes** $P_1,\ldots,P_k$ if

$$w \notin N(P_1, F_1) + \cdots + N(P_k, F_k)$$

whenever $F_1 \in \mathcal{F}_{m_1}(P_1), \ldots, F_k \in \mathcal{F}_{m_k}(P_k)$ with numbers $m_1,\ldots,m_k \in \{1,\ldots,d\}$ satisfying $m_1 + \cdots + m_k = (k-1)d + 1$. Then we get the following result. If w is admissible for $P_1,\ldots,P_k$, then

$$V_{m_1,\ldots,m_k}^{(0)}(P_1,\ldots,P_k)$$

$$= \sum_{F_1 \in \mathcal{F}_{m_1}(P_1)} \cdots \sum_{F_k \in \mathcal{F}_{m_k}(P_k)} \mathbf{1}_{N(P_1,F_1)+\cdots+N(P_k,F_k)}(w)$$

$$\times [F_1,\ldots,F_k] V_{m_1}(F_1) \cdots V_{m_k}(F_k). \tag{58}$$

This representation is more general than (52). Equation (58) holds for all vectors $w \in S^{d-1}$, with the exception of those in finitely many great subspheres (depending on $P_1, \ldots, P_k$). Integration over S^{d-1} with respect to spherical Lebesgue measure σ_{d-1} yields (52).

Our second aim in this subsection is the presentation of a set-valued analogue of the iterated translative formula (51), of the form

$$\int_{\mathbb{R}^d} \cdots \int_{\mathbb{R}^d} (\alpha_1 K_1 \cap x_2 \alpha_2 K_2 \cap \cdots \cap x_k \alpha_k K_k) \, \mathrm{d}x_2 \cdots \mathrm{d}x_k$$

$$= \sum_{\substack{m_1, \ldots, m_k = 0 \\ m_1 + \cdots + m_k = (k-1)d+1}}^{d} \alpha_1^{m_1} \cdots \alpha_k^{m_k} T_{m_1, \ldots, m_k}(K_1, \ldots, K_k) + t, \qquad (59)$$

with convex bodies $T_{m_1, \ldots, m_k}(K_1, \ldots, K_k)$ and a translation vector t. The integral of a function with values in $\mathcal{K}$ is defined via support functions, so that this formula is equivalent to a relation for support functions.

It is convenient in the following to use the **centred support function** h^*. This is the support function with respect to the Steiner point s, thus

$$h^*(K, u) = h(K - s(K), u) = h(K, u) - \langle s(K), u \rangle,$$

where the **Steiner point** of the convex body K is defined by

$$s(K) := \frac{1}{\kappa_d} \int_{S^{d-1}} h(K, u) u \, \sigma_{d-1}(\mathrm{d}u).$$

The centred support function is related to a special case of the mixed measures. We recall from Subsection 3.1 that the mixed measures can be extended to unbounded closed convex sets. The special representations for polytopes extend to polyhedral sets. We use this for the sets

$$u^+ := \{x \in \mathbb{R}^d : \langle x, u \rangle \geq 0\}, \qquad u^\perp := \{x \in \mathbb{R}^d : \langle x, u \rangle = 0\},$$

where $u \in S^{d-1}$. Choosing a Borel set $A(u) \subset u^\perp$ with $\lambda_{d-1}(A(u)) = 1$, we have, for polytopes P,

$$h^*(P, u) = \sum_{F \in \mathcal{F}_1(P)} \gamma(F, u^\perp; P, u^+)[F, u^\perp] V_1(F) \qquad (60)$$

$$= \Phi_{1,d-1}^{(0)}(P, u^+; \mathbb{R}^d \times A(u)). \qquad (61)$$

To prove (60), we first remark that, for a polytope P,

$$\sum_{e \in \mathcal{F}_0(P)} \gamma(e, P) e = s(P), \qquad \sum_{e \in \mathcal{F}_0(P)} \gamma(e, P) = \chi(P) = 1, \qquad (62)$$

where, for the ease of notation, we do not distinguish between a vertex e and the corresponding 0-face $\{e\}$ of P. Let

$$H_{u,t}^+ := \{x \in \mathbb{R}^d : \langle x, u \rangle \geq t\}$$

for $u \in S^{d-1}$ and $t \in \mathbb{R}$, and choose a number c satisfying $P \subset H_{u,c}^+$. Then

$$h(P, u) - c = \int_c^\infty \chi\left(P \cap H_{u,t}^+\right) \mathrm{d}t. \qquad (63)$$

Using (62), we get

$$\int_c^\infty \chi\left(P \cap H_{u,t}^+\right) \mathrm{d}t$$

$$= \int_c^\infty \sum_{e \in \mathcal{F}_0(P \cap H_{u,t}^+)} \gamma\left(e, P \cap H_{u,t}^+\right) \mathrm{d}t$$

$$= \int_c^\infty \sum_{e \in \mathcal{F}_0(P)} \gamma(e, P) \mathbf{1}\{\langle e, u \rangle \geq t\} \, \mathrm{d}t$$

$$+ \int_c^\infty \sum_{F \in \mathcal{F}_1(P)} \gamma(F, u^\perp; P, u^+)\chi(F \cap H_{u,t}) \, \mathrm{d}t$$

$$= \sum_{e \in \mathcal{F}_0(P)} \gamma(e, P)(\langle e, u \rangle - c) + \sum_{F \in \mathcal{F}_1(P)} \gamma(F, u^\perp; P, u^+)[F, u^\perp]V_1(F)$$

$$= \langle s(P), u \rangle - c + \sum_{F \in \mathcal{F}_1(P)} \gamma(F, u^\perp; P, u^+)[F, u^\perp]V_1(F),$$

which together with (63) proves (60).

As mentioned earlier, the mixed measures, which arose from an iterated translative formula, satisfy themselves an iterated formula. We can apply this to the mixed measure (61), or else we can use essentially the same method that led to the iterated formula (48). In either way, the following theorem can be obtained.

Theorem 3.2. *If $P_1, \ldots, P_k \subset \mathbb{R}^d$ are polytopes and $\alpha_1, \ldots, \alpha_k \geq 0$ $(k \geq 2)$, then*

$$\int_{\mathbb{R}^d} \cdots \int_{\mathbb{R}^d} h^*(\alpha_1 P_1 \cap x_2\alpha_2 P_2 \cap \cdots \cap x_k\alpha_k P_k, \cdot) \, \mathrm{d}x_2 \cdots \mathrm{d}x_k$$

$$= \sum_{\substack{m_1, \ldots, m_k = 0 \\ m_1 + \cdots + m_k = (k-1)d+1}}^{d} \alpha_1^{m_1} \cdots \alpha_k^{m_k} h_{m_1, \ldots, m_k}^*(P_1, \ldots, P_k; \cdot) \qquad (64)$$

with

$$h_{m_1, \ldots, m_k}^*(P_1, \ldots, P_k; u) \qquad (65)$$

$$:= \sum_{F_1 \in \mathcal{F}_{m_1}(P_1)} \cdots \sum_{F_k \in \mathcal{F}_{m_k}(P_k)} \gamma\left(F_1, \ldots, F_k, u^\perp; P_1, \ldots, P_k, u^+\right)$$

$$\times [F_1, \ldots, F_k, u^\perp]V_{m_1}(F_1) \cdots V_{m_k}(F_k) \qquad (66)$$

for $u \in S^{d-1}$.

The left-hand side of (64) is clearly a support function. It is, at the moment, not clear that each summand on the right-hand side is also a support function. To prove that this holds true, we need a class of more general representations of the functions $h^*_{m_1,\dots,m_k}(P_1,\dots,P_k;\cdot)$. These can be obtained by methods similar to those employed previously in this subsection. Let $u \in \mathbb{R}^d \setminus \{0\}$. We say that the vector $w \in \mathbb{R}^d$ is **admissible for** $(P_1,\dots,P_k,u)$ if it is admissible for $(P_1,\dots,P_k)$ and if

$$w \notin N(P_1,F_1) + \cdots + N(P_k,F_k) + \text{pos}\{-u\}$$

(where pos denotes the positive hull) whenever $F_1 \in \mathcal{F}_{m_1}(P_1),\dots,F_k \in \mathcal{F}_{m_k}(P_k)$ for numbers $m_1,\dots,m_k \in \{1,\dots,d\}$ satisfying $m_1 + \cdots + m_k = (k-1)d + 2$. If $w \in \mathbb{R}^d$ is admissible for $(P_1,\dots,P_k,u)$, it can be deduced from (60) and (58) that

$$
\begin{aligned}
h^*_{m_1,\dots,m_k}&(P_1,\dots,P_k;u) \\
&= \langle v_{m_1,\dots,m_k}(P_1,\dots,P_k,w),u\rangle \\
&\quad + \sum_{F_1 \in \mathcal{F}_{m_1}(P_1)} \cdots \sum_{F_k \in \mathcal{F}_{m_k}(P_k)} \mathbf{1}_{N(P_1,F_1)+\cdots+N(P_k,F_k)+\text{pos}\{-u\}}(w) \\
&\quad \times \|u\|[F_1,\dots,F_k,u^\perp] V_{m_1}(F_1) \cdots V_{m_k}(F_k)
\end{aligned}
\tag{67}
$$

for $u \in \mathbb{R}^d$. By $v_{m_1,\dots,m_k}(P_1,\dots,P_k,w)$ we have denoted a vector which could be given explicitly. The point now is that every $u_0 \in \mathbb{R}^d \setminus \{0\}$ has a convex neighbourhood U with the following properties: there is a vector w that is admissible for $(P_1,\dots,P_k,u)$, for all $u \in U$, and the function

$$u \mapsto \mathbf{1}_{N(P_1,F_1)+\cdots+N(P_k,F_k)+\text{pos}\{-u\}}(w)$$

is constant for $u \in U$. This implies that the right-hand side of (67) defines a convex function of $u \in U$. Hence, the function $h^*_{m_1,\dots,m_k}(P_1,\dots,P_k;\cdot)$ is locally a support function and therefore also globally. The convex body determined by this support function is denoted by $T_{m_1,\dots,m_k}(P_1,\dots,P_k)$. With this definition, (59) holds for polytopes, and an extension to general convex bodies can be achieved by approximation.

Hints to the literature. The general type of mixed functionals, of which the mixed volumes and the basic mixed functionals are special cases, was introduced in [43]; the representation (58) is found there. Theorem 3.2 was first obtained in [59]. Also formula (60) appears there. The simpler proof of the latter, as given here, and the general representation (67), are in [47]. The convexity of the functions $h^*_{m_1,\dots,m_k}(P_1,\dots,P_k;\cdot)$, and thus formula (59) with convex bodies $T_{m_1,\dots,m_k}(K_1,\dots,K_k)$, was first proved in [9] (for $k = 2$); the simpler proof sketched here is carried out in [47]. McMullen [26] has developed a general theory of 'mixed fibre polytopes', of which the mixed polytopes $T_{m_1,\dots,m_k}(P_1,\dots,P_k)$ are a special case.

3.3 Further Topics of Translative Integral Geometry

We briefly mention further results of translative integral geometry. First, the elementary formula (47),

$$\int_{\mathbb{R}^d} \chi(K \cap tM)\,\mathrm{d}t = \sum_{i=0}^{d} \binom{d}{i} V_i(K, M^*),$$

has counterparts where one or both of the bodies are replaced by their boundaries. This fact is highly non-elementary. One reason is that the boundary of a general convex body need not even be a set with positive reach, another is the observation that the intersection of the boundary of a convex body with another convex body need not have finite Euler characteristic (for which now a topological definition is needed). Moreover, approximation by polytopes (for which the results are easy) cannot be applied. Nevertheless, the following has been proved. If $K, M \in \mathcal{K}$ are d-dimensional convex bodies, then

$$\int_{\mathbb{R}^d} \chi(\partial K \cap tM)\,\mathrm{d}t$$
$$= \sum_{i=0}^{d-1} \binom{d}{i} \left\{ V_i(K, M^*) + (-1)^{d-i-1} V_i(K, M) \right\} \tag{68}$$

and

$$\int_{\mathbb{R}^d} \chi(\partial K \cap t\partial M)\,\mathrm{d}t$$
$$= (1 + (-1)^d) \sum_{i=0}^{d-1} \binom{d}{i} \left\{ V_i(K, M^*) + (-1)^{i-1} V_i(K, M) \right\}. \tag{69}$$

From these formulae one can deduce, by additional integrations over the rotation group, the kinematic formulae

$$\int_{G_d} \chi(\partial K \cap gM)\,\mu(\mathrm{d}g)$$
$$= \sum_{k=0}^{d-1} (1 - (-1)^{d-k}) c_{0,d}^{k,d-k} V_k(K) V_{d-k}(M) \tag{70}$$

and

$$\int_{G_d} \chi(\partial K \cap g\partial M)\,\mu(\mathrm{d}g)$$
$$= (1 + (-1)^d) \sum_{k=0}^{d-1} (1 - (-1)^k) c_{0,d}^{k,d-k} V_k(K) V_{d-k}(M). \tag{71}$$

Our second topic in this subsection is motivated by Hadwiger's general integral geometric theorem (Theorem 1.4). It provides an abstract version of the principal kinematic formula, holding for arbitrary continuous valuations (additive functions). One may ask whether a similarly general result holds in translative integral geometry. However, this can hardly be expected. What can be achieved, is an analogous result for continuous valuations that are simple. A valuation on $\mathcal{K}$ is **simple** if it is zero on convex bodies of dimension less than d.

The following theorem involves, besides the support function $h(K, \cdot)$ of a convex body K, also its area measure $S_{d-1}(K, \cdot) = 2\Xi_{d-1}(K, \mathbb{R}^d \times \cdot)$. The geometric meaning of this measure is as follows. For a Borel subset A of the unit sphere S^{d-1}, the value $S_{d-1}(K, A)$ is the area (the $(d-1)$-dimensional Hausdorff measure) of the set of boundary points of K at which there exists an outer unit normal vector in A.

Theorem 3.3. *Let $\varphi : \mathcal{K} \to \mathbb{R}$ be a continuous simple valuation. Then*

$$\int_{\mathbb{R}^d} \varphi(K \cap xM) \, \mathrm{d}x = \varphi(K)V_d(M) + \int_{S^{d-1}} f_{K,\varphi}(u) S_{d-1}(M, \mathrm{d}u)$$

for $K, M \in \mathcal{K}$, where $f_{K,\varphi} : S^{d-1} \to \mathbb{R}$ is the odd function given by

$$f_{K,\varphi}(u) = -\varphi(K)h(K, u) + \int_{-h(K,-u)}^{h(K,u)} \varphi(K \cap H^-(u, t)) \, \mathrm{d}t.$$

Hints to the literature. The kinematic formulae (70) and (71) were conjectured by Firey (see Problem 18 in the collection [11]). A proof of the more general formulae (68) and (69), and thus of (70) and (71), was given in [21]. In [20], these integral geometric results were extended to lower-dimensional sets, and iterated formulae were established; these were applied to stochastic geometry. Theorem 3.3 was proved in [48].

4 Measures on Spaces of Flats

If $M \in \mathcal{K}$ is a k-dimensional convex body, $k \in \{1, \ldots, d-1\}$, the case $j = 0$ of the Crofton formula (15) reduces to

$$\int_{\mathcal{E}_{d-k}^d} \operatorname{card}(M \cap E) \, \mu_{d-k}(\mathrm{d}E) = c_{0,d}^{k,d-k} \lambda_k(M), \tag{72}$$

since $V_0(M \cap E) = \operatorname{card}(M \cap E)$ for μ_{d-k}-almost all $(d-k)$-planes E (here card denotes the number of elements, possibly ∞), and $V_k(M) = \lambda_k(M)$. Formula (72) remains true if M is a k-dimensional compact C^1 submanifold of $\mathbb{R}^d$ (or, more generally, a $(\mathcal{H}^k, k)$-rectifiable Borel set) and λ_k denotes the

k-dimensional Euclidean surface area. Relations of this type are also known as **Crofton formulae**. They provide beautiful integral geometric interpretations of k-dimensional areas.

In this section, we study such Crofton formulae from a more general and 'reverse' point of view: given a notion of k-dimensional area that replaces λ_k, does there exist a measure on the space $\mathcal{E}^d_{d-k}$ of $(d-k)$-planes so that a counterpart to (72) is valid for a sufficiently large class of k-dimensional surfaces M? The measures on $\mathcal{E}^d_{d-k}$ we are seeking are always locally finite Borel measures. We shall admit signed measures as well, but we distinguish clearly between the cases of positive and of signed measures. Invariance properties of these measures are only postulated to the extent that the considered areas are themselves invariant. In the first subsection, areas and measures will be translation invariant, but no invariance property is assumed in the second subsection.

The third subsection is devoted to a special topic from stochastic geometry, the Poisson hyperplane processes. The connection with the Crofton formulae treated before will not be that of an application, but will rather consist in a common structural background. A basic feature in both studies is a relation between certain measures on the space of hyperplanes and special convex bodies, the (generalized) zonoids. Another common feature is the generation of lower dimensional flats as intersections of hyperplanes, and a corresponding generation of measures.

4.1 Minkowski Spaces

Our first topic are Crofton formulae in Minkowski geometry. A **Minkowski space** is (here) a finite dimensional normed space, say $(\mathbb{R}^d, \|\cdot\|)$. The unit ball of this space,

$$B := \{x \in \mathbb{R}^d : \|x\| \leq 1\},$$

is a convex body with 0 as interior point and centre of symmetry. The space of such convex bodies will be denoted by $\mathcal{C}^d$. The norm $\|\cdot\|$ induces a metric d by $d(x, y) := \|x - y\|$ for $x, y \in \mathbb{R}^d$.

For the subsequent computations, it is convenient to retain also the Euclidean structure on $\mathbb{R}^d$ given by the scalar product $\langle \cdot, \cdot \rangle$, although we work in a Minkowski space with norm $\|\cdot\|$. This 'impure' procedure simplifies calculations and presentation.

The metric d induces, in a well-known way, a notion of curve length for rectifiable curves. This curve length is invariant under translations. We denote the Minkowskian length by vol_1; thus, in particular, $\mathrm{vol}_1(S) = \|a - b\|$ if S is the segment with endpoints a, b.

In contrast to the case $k = 1$, where the metric induces a natural notion of curve length, for $k > 1$ there is no canonical notion of a k-dimensional area in Minkowski spaces, but rather a variety of options. The principal ambiguity can be made clear in the case $k = d$, the case of a notion of volume.

A reasonable notion of volume in the Minkowski space $(\mathbb{R}^d, \|\cdot\|)$ should be a Borel measure, assigning a positive finite value to every nonempty bounded open set, and determined solely by the Minkowskian metric. For the latter reason, it should be invariant under Minkowskian isometries and thus, in particular, under translations. The theory of Haar measures tells us that such a measure is unique up to a positive constant factor, and thus it is a constant multiple of the Lebesgue measure λ_d induced by the chosen Euclidean structure. Thus, a Minkowskian notion of volume in $(\mathbb{R}^d, \|\cdot\|)$ is fixed if we assign a value of this volume to the unit ball B. Since the notion of volume should be the same in isometric Minkowski spaces, the value assigned to B should depend only on the equivalence class of B under linear transformations. Thus, choosing a notion of Minkowskian volume for d-dimensional Minkowski spaces is equivalent to choosing a positive real function α_d on the space $\mathcal{C}^d$ which is invariant under linear transformations of $\mathbb{R}^d$. If α_d is chosen, then in the Minkowski space with unit ball B the induced volume $\alpha_{d,B}$ is given by

$$\alpha_{d,B}(M) = \frac{\alpha_d(B)}{\lambda_d(B)}\lambda_d(M)$$

for $M \in \mathcal{B}(\mathbb{R}^d)$ (recall that $\mathcal{B}(X)$ is the σ-algebra of Borel sets of the topological space X). This convenient representation does not depend on the choice of the Euclidean structure.

The choice just of a normalizing factor may seem rather unimportant, but it is not. Different choices make essential differences if we now employ this procedure to the definition of lower dimensional areas. Let $k \in \{1, \ldots, d\}$. For a k-dimensional convex body M, we denote by L_M the linear subspace of $\mathbb{R}^d$ parallel to the affine hull of M. Since a Minkowskian k-area is assumed to be translation invariant, its value at M should depend only on the Minkowski metric in the subspace L_M. The unit ball of this Minkowski space is $B \cap L_M$. This leads us to the following axioms for a Minkowskian k-area. Let $\mathcal{C}^k$ denote the set of k-dimensional convex bodies in $\mathbb{R}^d$ which are centrally symmetric with respect to 0. A k-**normalizing function** is a function $\alpha_k : \mathcal{C}^k \to \mathbb{R}^+$ which is (i) continuous, (ii) invariant under linear transformations, and (iii) satisfies $\alpha_k(E^k) = \kappa_k$ if E^k is a k-dimensional ellipsoid. Such a function induces, in a Minkowski space $(\mathbb{R}^d, \|\cdot\|)$ with unit ball B, a **Minkowskian k-area** $\alpha_{k,B}$ by

$$\alpha_{k,B}(M) := \frac{\alpha_k(B \cap L_M)}{\lambda_k(B \cap L_M)}\lambda_k(M)$$

for every k-dimensional convex body M. Again, this is independent of the Euclidean structure. The axiom (i) for a k-normalizing function seems reasonable, (ii) ensures the invariance of the k-area under Minkowskian isometries, and (iii) is assumed in order to obtain the standard k-area if the space is Euclidean. The Minkowskian k-area can be extended to more general sets, for example to k-dimensional C^1-submanifolds M, by

$$\alpha_{k,B}(M) = \int_M \frac{\alpha_k(B \cap T_x M)}{\lambda_k(B \cap T_x M)} \lambda_k(\mathrm{d}x),$$

where $T_x M$ denotes the tangent space of M at x, considered as a subspace of $\mathbb{R}^d$.

For the quotient appearing in the integrand, we use the notation

$$\frac{\alpha_k(B \cap L)}{\lambda_k(B \cap L)} =: \sigma_{k,B}(L) \qquad \text{for } L \in \mathcal{L}_k^d$$

(which depends on the Euclidean structure) and call $\sigma_{k,B}$ the **scaling function**; then

$$\alpha_{k,B}(M) = \sigma_{k,B}(L_M)\lambda_k(M).$$

A 1-normalizing function is uniquely determined, hence $\sigma_{1,B} = 1$, thus for $k = 1$ we get the Minkowskian curve length vol_1 again.

Now we can study the existence of Crofton formulae. We assume that a k-normalizing function α_k and a Minkowski space $(\mathbb{R}^d, \|\cdot\|)$ with unit ball B are given.

Definition. *A* **Crofton measure** *for* $\alpha_{k,B}$ *is a translation invariant signed measure* η_{d-k} *on* $\mathcal{E}_{d-k}^d$ *for which*

$$\int_{\mathcal{E}_{d-k}^d} \mathrm{card}\,(M \cap E)\,\eta_{d-k}(\mathrm{d}E) = \alpha_{k,B}(M) \tag{73}$$

holds for every k-dimensional convex body M.

It is no essential restriction of the generality to consider here only convex k-dimensional sets M. If (73) holds for these, the formula can be extended to more general k-dimensional surfaces.

Let us suppose that a Crofton measure η_{d-k} for $\alpha_{k,B}$ exists. Since it is translation invariant and locally finite, there is a finite signed measure φ on $\mathcal{L}_{d-k}^d$ such that

$$\int_{\mathcal{E}_{d-k}^d} f \, \mathrm{d}\eta_{d-k} = \int_{\mathcal{L}_{d-k}^d} \int_{L^\perp} f(L + x)\, \lambda_k(\mathrm{d}x)\, \varphi(\mathrm{d}L) \tag{74}$$

holds for every nonnegative measurable function f on $\mathcal{E}_{d-k}^d$ (see, e.g., [52, 4.1.1]). Let M be a k-dimensional convex body. For a subspace L, we denote by $|L$ the orthogonal projection to L. With the function $[\cdot, \cdot]$ introduced in Subsection 2.1, we have $\lambda_k(M|L^\perp) = \lambda_k(M)[L_M, L]$. Now we obtain

$$\alpha_{k,B}(M) = \int_{\mathcal{E}_{d-k}^d} \mathrm{card}\,(M \cap E)\,\eta_{d-k}(\mathrm{d}E)$$

$$= \int_{\mathcal{E}_{d-k}^d} \mathbf{1}\{M \cap E \neq \emptyset\}\,\eta_{d-k}(\mathrm{d}E)$$

$$= \int_{\mathcal{L}_{d-k}^d} \int_{L^\perp} \mathbf{1}\{M \cap (L+x) \neq \emptyset\}\, \lambda_k(\mathrm{d}x)\, \varphi(\mathrm{d}L)$$

$$= \int_{\mathcal{L}_{d-k}^d} \lambda_k(M|L^\perp)\, \varphi(\mathrm{d}L)$$

$$= \lambda_k(M) \int_{\mathcal{L}_{d-k}^d} [L_M, L]\, \varphi(\mathrm{d}L).$$

This yields

$$\sigma_{k,B}(E) = \int_{\mathcal{L}_{d-k}^d} [E, L]\, \varphi(\mathrm{d}L) \qquad \text{for } E \in \mathcal{L}_k^d. \tag{75}$$

Conversely, if (75) is satisfied with a finite signed measure φ, then we can use (74) to define a signed measure η_{d-k} on $\mathcal{E}_{d-k}^d$, and this is a Crofton measure for $\alpha_{k,B}$.

The crucial integral equation (75) is now first considered for $k = 1$. Choosing $v \in S^{d-1}$ and for M the segment with endpoints v and $-v$, for which $\alpha_{1,B}(M) = \mathrm{vol}_1(M) = 2\|v\|$, and representing $(d-1)$-dimensional linear subspaces by their Euclidean unit normal vectors, we see that (75) is equivalent to

$$\|v\| = \int_{S^{d-1}} |\langle u, v \rangle|\, \rho(\mathrm{d}u)$$

with an even finite signed measure ρ on S^{d-1}.

Introducing the polar unit ball (the dual body of B, where $\mathbb{R}^d$ and its dual space have been identified via the scalar product),

$$B^o := \{x \in \mathbb{R}^d : \langle x, y \rangle \leq 1 \text{ for all } y \in B\},$$

we have $\|v\| = h(B^o, v)$ (e.g., [39, p. 44]); hence (75), for $k = 1$, is equivalent to

$$h(B^o, v) = \int_{S^{d-1}} |\langle u, v \rangle|\, \rho(\mathrm{d}u) \qquad \text{for } v \in \mathbb{R}^d. \tag{76}$$

A convex body B^o whose support function has a representation (76) with a finite signed measure ρ is called a **generalized zonoid**, and it is a zonoid (as defined in Subsection 1.3, i.e., a convex body which can be approximated by finite vector sums of line segments) if there is such a representation with a positive measure ρ. Every body in $\mathcal{C}^d$ with sufficiently smooth support function is a generalized zonoid. Therefore, the generalized zonoids are dense in the space $\mathcal{C}^d$, whereas the zonoids are nowhere dense in $\mathcal{C}^d$. The crosspolytope is an example of a centrally symmetric convex body which is not a generalized zonoid. Hence, in the Minkowski space ℓ_∞^d, no Crofton measure for vol_1 exists.

Now we suppose that a *positive* Crofton measure η_{d-1} for vol_1 exists, and we draw a second conclusion. For this, we choose m points $p_1, \ldots, p_m \in \mathbb{R}^d$ and integers $N_1, \ldots, N_m$ with

$$\sum_{i=1}^m N_i = 1. \tag{77}$$

Let H be a hyperplane not incident with one of the points $p_1, \ldots, p_m$, and let H^+, H^- be the two closed halfspaces bounded by H. Then, denoting the segment with endpoints p_i, p_j by $\overline{p_i p_j}$, we have

$$\sum_{i<j} 1\{H \cap \overline{p_i p_j} \neq \emptyset\} N_i N_j = \left(\sum_{p_i \in H^+} N_i \right) \left(\sum_{p_j \in H^-} N_j \right)$$

$$= \left(\sum_{p_i \in H^+} N_i \right) \left(1 - \sum_{p_i \in H^+} N_i \right)$$

$$\leq 0,$$

where we have used (77) and the fact that $z(1-z) \leq 0$ for every integer z. Integrating the obtained inequality over all $H \in \mathcal{E}_{d-1}^d$ with respect to the positive measure η_{d-1} (and observing that the set of hyperplanes through one of the points $p_1, \ldots, p_m$ has measure zero), we obtain

$$\sum_{i<j} d(p_i, p_j) N_i N_j \leq 0. \tag{78}$$

Generally, a metric d satisfying (78) for all m-tuples $(N_1, \ldots, N_m)$ of integers with (77) and all $m \in \mathbb{N}$ is called a **hypermetric**. We say that our Minkowski space $(\mathbb{R}^d, \| \cdot \|)$ is **hypermetric** if its induced metric d is a hypermetric. Now we can formulate a theorem.

Theorem 4.1. *In the Minkowski space $(\mathbb{R}^d, \| \cdot \|)$ with unit ball B, a Crofton measure for vol_1 exists if and only if the polar unit ball B^o is a generalized zonoid. The following conditions are equivalent:*

(a) *There exists a positive Crofton measure for vol_1.*

(b) *The polar unit ball B^o is a zonoid.*

(c) *The Minkowski space $(\mathbb{R}^d, \| \cdot \|)$ is hypermetric.*

For the implication (c) $\Rightarrow$ (b), which we do not prove here, we refer to [1] and the references given there.

We turn to k-areas for $k > 1$ and first introduce two special cases of such areas, which play a prominent role. The **Busemann k-area**, denoted by β_k, is defined by the constant k-normalizing function, $\alpha_k(C) = \kappa_k$ for $C \in \mathcal{C}^k$. The **Holmes-Thompson k-area** is defined by the k-normalizing function $\alpha_k(C) := \kappa_k^{-1} \mathrm{vp}(C)$ for $C \in \mathcal{C}^k$, where

$$\mathrm{vp}(C) := \lambda_k(C) \lambda_k(C^o)$$

is the **volume product**; here the polar body C^o is taken with respect to the k-dimensional linear subspace containing C. The volume product is invariant under linear transformations and therefore independent of the Euclidean structure. The scaling function of the Busemann k-area is given by

$$\sigma^{\mathrm{Bus}}_{k,B}(L) = \frac{\kappa_k}{\lambda_k(B \cap L)} \qquad \text{for } L \in \mathcal{L}^d_k,$$

and hence the Busemann k-area of a k-dimensional C^1 submanifold M by

$$\beta_k(M) = \int_M \frac{\kappa_k}{\lambda_k(B \cap T_x M)} \, \lambda_k(\mathrm{d}x). \tag{79}$$

For the scaling function of the Holmes-Thompson k-area we obtain $\sigma^{\mathrm{HT}}_{k,B}(L) = \kappa^{-1}\lambda_k((B \cap L)^o)$. Convex geometry tells us that $(B \cap L)^o = B^o|L$, hence

$$\sigma^{\mathrm{HT}}_{k,B}(L) = \frac{\lambda_k(B^o|L)}{\kappa_k} \qquad \text{for } L \in \mathcal{L}^d_k. \tag{80}$$

We denote the Holmes-Thompson k-area by vol_k; then

$$\mathrm{vol}_k(M) = \int_M \frac{\lambda_k(B^o|T_x M)}{\kappa_k} \, \lambda_k(\mathrm{d}x) \tag{81}$$

for a k-dimensional C^1 submanifold M.

Let $\mathcal{H}^k_B$ denote the k-dimensional Hausdorff measure that is induced by the metric d of our Minkowski space. It can be shown that the Busemann k-area $\beta_k(M)$ is nothing but $\mathcal{H}^k_B(M)$ (if the Hausdorff measure is suitably normalized). Using this Hausdorff measure instead of the Euclidean area measure λ_k, we can write the Holmes-Thompson area in the form

$$\mathrm{vol}_k(M) = \frac{1}{\kappa^2_k} \int_M \mathrm{vp}(B \cap T_x M) \, \mathcal{H}^k_B(\mathrm{d}x), \tag{82}$$

which does not use the auxiliary Euclidean structure any more.

The Holmes-Thompson k-area of M is equal to the symplectic volume of M as defined in the theory of Finsler spaces.

Now we study the existence of Crofton measures for a general Minkowskian $(d-1)$-area $\alpha_{d-1,B}$. Writing

$$\sigma(u) := \langle u, u \rangle^{1/2} \sigma_{d-1,B}(u^\perp) \qquad \text{for } u \in \mathbb{R}^d \setminus \{0\},$$

we see that (75) for $k = d - 1$ is equivalent to

$$\sigma(u) = \int_{S^{d-1}} |\langle u, v \rangle| \rho(\mathrm{d}v), \qquad u \in S^{d-1}, \tag{83}$$

with an even finite signed measure ρ on S^{d-1}.

Suppose that a **positive** Crofton measure exists for $\alpha_{d-1,B}$. Then ρ in (83) is a positive measure, hence σ is the support function of a zonoid. This zonoid is denoted by $\mathbf{I}_{\alpha,B}$ and called the **isoperimetrix**. The name comes from the isoperimetric problem: it can be shown that a convex body in $\mathbb{R}^d$

with given positive volume has smallest $(d-1)$-area $\alpha_{d-1,B}$ of its boundary if and only if it is homothetic to the isoperimetrix.

Conversely, if (83) holds with a positive measure ρ, then there exists a positive Crofton measure for the $(d-1)$-area $\alpha_{d-1,B}$. For the Holmes-Thompson $(d-1)$-area this is always the case, in any Minkowski space. In fact, from a well-known formula for projection volumes of convex bodies, we get for the Holmes-Thompson area

$$\sigma(u) = \frac{1}{\kappa_{d-1}}\lambda_{d-1}(B^\circ|u^\perp) = \frac{1}{2\kappa_{d-1}}\int_{S^{d-1}}|\langle u,v\rangle|\,S_{d-1}(B^\circ,dv)$$

for $u \in S^{d-1}$, and here the area measure $S_{d-1}(B^\circ,\cdot)$ is positive.

For the Busemann $(d-1)$-area, it can be shown that the function σ is again a support function, but not necessarily of a zonoid or a generalized zonoid. As a consequence, there need not exist a Crofton measure for the Busemann $(d-1)$-area. More precise information is contained in the following theorem, whose proof is based on properties of zonoids and generalized zonoids. Here, C^d denotes a cube in $\mathbb{R}^d$ with centre 0.

Theorem 4.2. (a) *There exist Minkowski spaces, with unit ball arbitrarily close to B^d, in which there is no Crofton measure for the Busemann $(d-1)$-area. There also exist Minkowski spaces with unit ball arbitrarily close to B^d, but different from an ellipsoid, in which there is a positive Crofton measure for the Busemann $(d-1)$-area.*

(b) *In every Minkowski space of sufficiently large dimension d and with unit ball sufficiently close to C^d, there is no positive Crofton measure for the Busemann $(d-1)$-area.*

(c) *There exist Minkowski spaces, for example ℓ^d_∞ and ℓ^d_1, in which there is no positive Crofton measure for any general Minkowskian $(d-1)$-area, except for the multiples of the Holmes-Thompson $(d-1)$-area.*

(d) *In every Minkowski space, there is a positive Crofton measure for the Holmes-Thompson $(d-1)$-area.*

The preceding theorem is sufficient reason for us to concentrate, from now on, on the Holmes-Thompson area. This is even more justified in view of the following theorem.

Theorem 4.3. *Let $(\mathbb{R}^d, \|\cdot\|)$ be a Minkowski space. If in this space there exists a Crofton measure (a positive Crofton measure) for vol_1, then there also exists a Crofton measure (a positive Crofton measure) for vol_k, $k \in \{2,\dots,d-2\}$.*

Proof. Suppose that in the Minkowski space with unit ball B there is a Crofton measure for vol_1. Then, by (76), there is an even finite signed measure ρ on S^{d-1} such that

$$h(B^o, u) = \int_{S^{d-1}} |\langle u, v \rangle| \rho(dv) \qquad \text{for } u \in \mathbb{R}^d. \tag{84}$$

We employ a result from the theory of generalized zonoids. For vectors $u_1, \ldots, u_k \in S^{d-1}$, we denote by $L(u_1, \ldots, u_k)$ the linear subspace spanned by these vectors and by $[u_1, \ldots, u_k]$ the k-dimensional Euclidean volume of the parallelepiped spanned by them. Then for $E \in \mathcal{L}_k^d$ we have

$$\lambda_k(B^o | E) \tag{85}$$

$$= \frac{2^k}{k!} \int_{S^{d-1}} \cdots \int_{S^{d-1}} [E, L(u_1, \ldots, u_k)^{\perp}][u_1, \ldots, u_k] \, \rho(du_1) \cdots \rho(du_k)$$

(see [57], and observe that the proof is valid if ρ is a signed measure). We define a signed measure $\rho^{(k)}$ on $\mathcal{L}_k^d$ by

$$\rho^{(k)}(A) := c_k \int_{S^{d-1}} \cdots \int_{S^{d-1}} 1_A(L(u_1, \ldots, u_k))[u_1, \ldots, u_k] \, \rho(du_1) \cdots \rho(du_k)$$

for Borel sets $A \subset \mathcal{L}_k^d$, where

$$c_k = \frac{2^k}{k! \kappa_k}. \tag{86}$$

Then we can write (80) in the form

$$\sigma_{k,B}^{HT}(E) = \int_{\mathcal{L}_k^d} [E, L^{\perp}] \rho^{(k)}(dL) = \int_{\mathcal{L}_{d-k}^d} [E, L] \rho_{(k)}(dL)$$

for $E \in \mathcal{L}_k^d$; here $\rho_{(k)}$ is the image measure of $\rho^{(k)}$ under the map $L \mapsto L^{\perp}$ from $\mathcal{L}_k^d$ to $\mathcal{L}_{d-k}^d$. We see that the integral equation (75) has a solution for $\sigma_{k,B}^{HT}$, hence there is a Crofton measure for vol_k.

If there is a positive Crofton measure for vol_1, then ρ is a positive measure, which implies that $\rho_{(k)}$ is a positive measure, hence there is a positive Crofton measure for vol_k. □

In the proof of Theorem 4.3 we started with a Crofton measure for vol_1, say η, and constructed a Crofton measure for vol_k, say η_{d-k}. This construction has a nice geometric interpretation. Given is a measure η on the space of hyperplanes, and we need a measure on the space of $(d-k)$-planes. It turns out that η_{d-k}, as constructed, is the image measure of $c_k \eta^{\otimes k} \llcorner \mathcal{H}_k^*$ under the intersection map

$$(H_1, \ldots, H_k) \mapsto H_1 \cap \cdots \cap H_k$$

from $\mathcal{H}_k^*$ to $\mathcal{E}_{d-k}^d$, where $\mathcal{H}_k^*$ denotes the set of all k-tuples $(H_1, \ldots, H_k)$ of hyperplanes with $\dim(H_1 \cap \cdots \cap H_k) = d-k$ and $\llcorner$ denotes restriction. More explicitly, for $\mathcal{A} \in \mathcal{B}(\mathcal{E}_{d-k}^d)$ we have

$$\eta_{d-k}(\mathcal{A}) = c_k \underbrace{\eta \otimes \cdots \otimes \eta}_{k}(\{(H_1, \ldots, H_k) \in (\mathcal{E}_{d-1}^d)^k : H_1 \cap \cdots \cap H_k \in \mathcal{A}\}).$$

There are two main cases where the assumption of Theorem 4.3 is satisfied. If the norm $\| \cdot \|$, which is equal to $h(B^o, \cdot)$, is sufficiently smooth, then B^o is a generalized zonoid, hence a Crofton measure exists for vol_1. If the Minkowski space $(\mathbb{R}^d, \| \cdot \|)$ is hypermetric, then Theorem 4.1 says that a positive Crofton measure for vol_1 exists. Hence, in either of these two cases, the Holmes-Thompson area of any dimension satisfies an intersection formula of Crofton type.

Hints to the literature. Motivated by earlier work of Busemann, a study of integral geometric formulae for areas in affine spaces, and particularly in Minkowski spaces, was made in [53]. Much of the material exposed here, including the proof of Theorem 4.3, is found there, together with additional information. Parts (a) and (b) of Theorem 4.2 are proved in [44], and part (c) in [41]. For general information about geometry in Minkowski spaces, we refer to [55].

4.2 Projective Finsler Spaces

The main result of the previous subsection, the existence of Crofton measures for Holmes-Thompson areas in smooth or hypermetric Minkowski spaces, can be extended to certain spaces where one has no longer any nontrivial invariance under a transformation group. We sketch here, without proofs, the main ideas of such an extension. It takes place in a natural generalization of Minkowski spaces, the projective Finsler spaces. Generally speaking, a Finsler space is a differentiable manifold together with a norm in each tangent space, satisfying certain smoothness assumptions. Here we consider only $\mathbb{R}^d$ as the underlying manifold (where we canonically identify each tangent space of $\mathbb{R}^d$ with $\mathbb{R}^d$ itself), and we consider Finsler metrics on $\mathbb{R}^d$ which are compatible with the affine structure of $\mathbb{R}^d$, in the sense we shall now make precise.

By a **Finsler metric** on $\mathbb{R}^d$ we understand here a continuous function $F : \mathbb{R}^d \times \mathbb{R}^d \to [0, \infty)$ such that $F(x, \cdot) =: \| \cdot \|_x$ is a norm on $\mathbb{R}^d$ for each $x \in \mathbb{R}^d$. If this holds, the pair $(\mathbb{R}^d, F)$ is called a **Finsler space**. (We should rather speak of a **generalized** Finsler space, since the common definition of a Finsler space includes smoothness assumptions and a stronger convexity property of the norms $\| \cdot \|_x$, but we will delete 'generalized' in the following.) In such a Finsler space, the **length** of a parameterized piecewise C^1 curve $\gamma : [a, b] \to \mathbb{R}^d$ is defined by

$$L_F(\gamma) := \int_a^b F(\gamma(t), \gamma'(t)) \, dt;$$

this is independent of the parameterization since $F(\gamma(t), \cdot)$ is homogeneous of degree one. For $p, q \in \mathbb{R}^d$, the **distance** $d_F(p, q)$ is defined as the infimum of the lengths of all piecewise C^1 curves connecting p and q. Then d_F is a metric, called the **metric induced by** F. The Finsler space $(\mathbb{R}^d, F)$ is called **projective** if line segments are shortest curves connecting their endpoints. If this holds, the segment with endpoints p, q has length $d_F(p, q)$.

Let $(\mathbb{R}^d, F)$ be a Finsler space. For $x \in \mathbb{R}^d$, the unit ball of the Minkowski space $(\mathbb{R}^d, \|\cdot\|_x)$ is denoted by B_x (recall that we have identified the tangent space $T_x\mathbb{R}^d$ of $\mathbb{R}^d$ at x with $\mathbb{R}^d$). As in the previous subsection, we use a fixed auxiliary scalar product $\langle \cdot, \cdot \rangle$ on $\mathbb{R}^d$. With its aid, we define the polar body of B_x,

$$B_x^o := \{v \in \mathbb{R}^d : \langle u, v \rangle \leq 1 \text{ for all } u \in B_x\}.$$

(Without our simplifying conventions, B_x would be a convex body in the tangent space $T_x\mathbb{R}^d$, and B_x^o would be a convex body in the dual tangent space $T_x^*\mathbb{R}^d$.)

Extending the definition given for Minkowski spaces, one can define the **Holmes-Thompson k-area** of a k-dimensional C^1-submanifold M in $(\mathbb{R}^d, F)$ by

$$\mathrm{vol}_k(M) = \frac{1}{\kappa_k} \int_M \lambda_k(B_x^o | T_x M) \, \lambda_k(\mathrm{d}x).$$

This definition uses the Euclidean structure in several ways, but is, in fact, independent of its choice. Formula (82), which does not use the Euclidean structure, can be extended as follows. If $\mathcal{H}_F^k$ denotes the k-dimensional Hausdorff measure induced by the metric d_F, then

$$\mathrm{vol}_k(M) = \frac{1}{\kappa_k^2} \int_M \mathrm{vp}(B_x \cap T_x M) \, \mathcal{H}_F^k(\mathrm{d}x). \tag{87}$$

As in the special case of Minkowski spaces, the existence of Crofton measures (which we define in the same way) for Holmes-Thompson areas is closely related to the theory of generalized zonoids. First, we study this connection under a smoothness assumption. We assume that the Finsler space $(\mathbb{R}^d, F)$ is **smooth**, meaning that the function F is of class C^∞ (weaker assumptions would be sufficient, but this is not an issue here).

Let $x \in \mathbb{R}^d$ be given. Due to the smoothness assumption, the integral equation

$$F(x, v) = \int_{S^{d-1}} |\langle u, v \rangle| \gamma_x(u) \, \sigma_{d-1}(\mathrm{d}u)$$

has a continuous even solution γ_x on S^{d-1} (see, e.g., [39, Theorem 3.5.3]). Now the assumption that the Finsler space $(\mathbb{R}^d, F)$ is projective has a strong implication on $\gamma_x(u)$, as a function of its two variables x and u: there exists a continuous function $g : S^{d-1} \times \mathbb{R} \to \mathbb{R}$ such that $g(u, \tau) = g(-u, -\tau)$ and

$$\gamma_x(u) = g(u, \langle x, u \rangle) \qquad \text{for } (x, u) \in \mathbb{R}^d \times \mathbb{R}^d.$$

This follows from Pogorelov's [27] work on Hilbert's Fourth Problem (see [45] for a brief sketch of the main ideas). Recalling that $F(x, \cdot) = \|\cdot\|_x = h(B_x^o, \cdot)$, which is the support function of the polar unit ball at x, we now have

$$h(B_x^o, v) = \int_{S^{d-1}} |\langle u, v \rangle| g(u, \langle x, u \rangle) \, \sigma_{d-1}(\mathrm{d}u) \tag{88}$$

for $v \in \mathbb{R}^d$. This representation is of the form of (76) and can be employed in a similar way. We use it for the construction of signed measures η_{d-k} on the space $\mathcal{E}_{d-k}^d$ of $(d-k)$-flats, as in the proof of Theorem 4.3 and the subsequent remark, in the following way. The function $x \mapsto g(u, \langle x, u \rangle)$ is constant on the hyperplane H through x with normal vector u, let $h(H)$ be its value on H. This defines a function h on $\mathcal{E}_{d-1}^d$. Let η be the signed measure on $\mathcal{E}_{d-1}^d$ which has density h with respect to the rigid motion invariant measure on $\mathcal{E}_{d-1}^d$. Explicitly, this comes down to the following. We parameterize hyperplanes of $\mathbb{R}^d$ by

$$H_{u,t} = \{ y \in \mathbb{R}^d : \langle y, u \rangle = t \}, \qquad u \in S^{d-1}, \, t \in \mathbb{R},$$

and then define a signed measure η on $\mathcal{E}_{d-1}^d$ by

$$\eta(\mathcal{A}) := \int_{S^{d-1}} \int_{\mathbb{R}} \mathbf{1}_{\mathcal{A}}(H_{u,t}) g(u, t) \, \mathrm{d}t \, \sigma_{d-1}(\mathrm{d}u) \tag{89}$$

for $\mathcal{A} \in \mathcal{B}(\mathcal{E}_{d-1}^d)$. Next, we define η_{d-k} as the image measure of $c_k \eta^{\otimes k}$ under the intersection map $(H_1, \ldots, H_k) \mapsto H_1 \cap \cdots \cap H_k$ (with c_k given by (86)). Explicitly, this means that

$$\eta_{d-k}(\mathcal{A}) := c_k \int_{S^{d-1}} \cdots \int_{S^{d-1}} \int_{\mathbb{R}} \cdots \int_{\mathbb{R}} \mathbf{1}_{\mathcal{A}}(H_{u_1,t_1} \cap \cdots \cap H_{u_k,t_k})$$
$$g(u_1, t_1) \cdots g(u_k, t_k) \, \mathrm{d}t_1 \cdots \mathrm{d}t_k \, \sigma_{d-1}(\mathrm{d}u_1) \cdots \sigma_{d-1}(\mathrm{d}u_k) \tag{90}$$

for $\mathcal{A} \in \mathcal{B}(\mathcal{E}_{d-k}^d)$.

With these measures, we can now state Crofton formulae in a very general version. A set $M \subset \mathbb{R}^d$ is called $(\mathcal{H}^k, k)$-**rectifiable**, for $k \in \{1, \ldots, d\}$, if $\mathcal{H}^k(M) < \infty$ and there are Lipschitz maps $f_i : \mathbb{R}^k \to \mathbb{R}^d$ ($i \in \mathbb{N}$) such that $\mathcal{H}^k(M \setminus \bigcup_{i \in \mathbb{N}} f_i(\mathbb{R}^k)) = 0$. Here the k-dimensional Hausdorff measure $\mathcal{H}^k$ and the notion of Lipschitz map refer to a Euclidean structure, but the class of $(\mathcal{H}^k, k)$-rectifiable sets is independent of the choice of this structure. The definition of the Holmes-Thompson k-area can be extended to $(\mathcal{H}^k, k)$-rectifiable Borel sets, for example by (87).

Theorem 4.4. *Let $(\mathbb{R}^d, F)$ be a smooth projective Finsler space. Then, for $k \in \{1, \ldots, d-1\}$ and every $(\mathcal{H}^k, k)$-rectifiable Borel set M in $\mathbb{R}^d$,*

$$\int_{\mathcal{E}_{d-k}^d} \mathrm{card}\,(M \cap E) \, \eta_{d-k}(\mathrm{d}E) = \mathrm{vol}_k(M). \tag{91}$$

An even more general version holds. This refers to the case where M and the intersecting flats are not necessarily of complementary dimensions. For $j \in \{1, \ldots, d-1\}$ and $k \in \{d-j, \ldots, d\}$, and for $(\mathcal{H}^k, k)$-rectifiable Borel sets M,

$$\int_{\mathcal{E}_j^d} \text{vol}_{k+j-d}(M \cap E)\, \eta_j(dE) = \frac{c_{k+j-d}c_{d-j}}{c_k}\text{vol}_k(M). \tag{92}$$

We turn to the existence of positive Crofton measures. A projective Finsler space $(\mathbb{R}^d, F)$ is called **hypermetric** if its induced metric d_F is a hypermetric. We assume now that the Finsler space $(\mathbb{R}^d, F)$ is smooth, projective, and hypermetric. Then every polar local unit ball B_x^o is not only a generalized zonoid (which it is by (88)), but even a zonoid (as proved in [1]), which means that the function $g(\cdot, \langle x, \cdot \rangle)$ in (88), and hence g, is nonnegative. It follows that each η_{d-k} is a positive measure.

The existence of positive Crofton measures, as just established, can be extended to general (i.e., not necessarily smooth) hypermetric projective Finsler spaces. Such an extension can be based on the following approximation result, which Pogorelov [27] and Szabó [54] established in their work on Hilbert's Fourth Problem. For every $\epsilon > 0$, there is a smooth projective Finsler space $(\mathbb{R}^d, F_\epsilon)$ such that $\lim_{\epsilon \to 0} F_\epsilon = F$, uniformly on every compact set. Moreover, each $(\mathbb{R}^d, F_\epsilon)$ is hypermetric (see [1]). Therefore, to every ϵ, there exists a positive Crofton measure η_{d-1} for the Holmes-Thompson 1-area, as constructed above. This measure depends on ϵ; we denote it by $\eta_{(\epsilon)}$. Making essential use of the positivity of these measures, one can show that the family $(\eta_{(\epsilon)})_{\epsilon \in (0,1)}$ of measures is relatively compact in the vague topology. Hence, there is a sequence $(\epsilon_i)_{i \in \mathbb{N}}$ tending to zero such that the sequence $(\eta_{(\epsilon_i)})_{i \in \mathbb{N}}$ converges vaguely to a measure η on $\mathcal{E}_{d-1}^d$. With this measure, we repeat the earlier construction: we define the measure η_{d-k} on $\mathcal{E}_{d-k}^d$ as the image measure of $c_k \eta^{\otimes k} \llcorner \mathcal{H}_k^*$ under the map $(H_1, \ldots, H_k) \mapsto H_1 \cap \cdots \cap H_k$ from $\mathcal{H}_k^*$ to $\mathcal{E}_{d-k}^d$. Using the vague convergence of $(\eta_{(\epsilon_i)})_{i \in \mathbb{N}}$ to η, it can be proved that

$$\int_{\mathcal{E}_{d-k}^d} \text{card}\,(M \cap E)\, \eta_{d-k}(dE) = \text{vol}_k(M) \tag{93}$$

for $k \in \{1, \ldots, d-1\}$ and every k-dimensional convex body M. By definition, η_{d-k} is a positive Crofton measure for the Holmes-Thompson k-area in $(\mathbb{R}^d, F)$. However, it has not been investigated, in this case of a general Finsler metric, whether (93) can be extended to $(\mathcal{H}^k, k)$-rectifiable Borel sets M, nor whether (92) has a counterpart.

We remark that a measure η_{d-k} satisfying (93) for all k-dimensional convex bodies M is uniquely determined if either $k = 1$ or $k = d-1$, but not in the intermediate cases.

Theorem 4.2(d) can be extended from Minkowski spaces to projective Finsler spaces, thus a positive Crofton measure for the Holmes-Thompson $(d-1)$-area in the projective Finsler space $(\mathbb{R}^d, F)$ exists even if the space is not hypermetric. The clue for a proof of this fact is again a formula from the theory of generalized zonoids. First we assume that $(\mathbb{R}^d, F)$ is a smooth projective Finsler space. Let $s_{d-1}(B_x^o, u)$ denote the product of the principal radii of curvature of the boundary of B_x^o at the point with outer normal vector

$u \in S^{d-1}$. From the representation (88), one has an explicit integral representation of $s_{d-1}(B_x^o, u)$ in terms of the function g; see [56, Satz 7]. With its aid, one can show that a function δ on $\mathcal{E}_1^d$ can be defined by

$$\delta(\lin\{u\} + x) := s_{d-1}(B_x^o, u) \qquad \text{for } u \in S^{d-1}, \; x \in \mathbb{R}^d$$

and that this function is a density of the signed measure η_1 with respect to the suitably normalized rigid motion invariant measure on the space $\mathcal{E}_1^d$ of lines. Since this density is always nonnegative, the Crofton measure η_1 is positive. The existence of a positive Crofton measure for the Holmes-Thompson $(d-1)$-area in non-smooth projective Finsler spaces can then again be obtained by approximation.

We collect the stated results in a theorem.

Theorem 4.5. *Let $(\mathbb{R}^d, F)$ be a projective Finsler space. In this space, there exists a positive Crofton measure for the Holmes-Thompson $(d-1)$-area. If the space is hypermetric, there exists a positive Crofton measure for the Holmes-Thompson k-area, for each $k \in \{1, \ldots, d-1\}$.*

Hints to the literature. For smooth projective Finsler spaces and smooth submanifolds, a version of the Crofton formula (91) was proved in [2], using the symplectic structure on the space of geodesics of a projective Finsler space. The general formula (92), together with its proof based on the theory of generalized zonoids, appears in [46]. Theorem 4.5 was proved in [45].

4.3 Nonstationary Hyperplane Processes

Finally, we treat a special topic from stochastic geometry which is closely related to the preceding subsection. The relation to Crofton measures in projective Finsler spaces is not one of application, but consists in the similarity of the underlying structures. We study stochastic processes of k-planes in $\mathbb{R}^d$, and in particular of hyperplanes.

First we need some explanations. Let S be an arbitrary locally compact space with a countable base. A subset $F \subset S$ is called **locally finite** if $F \cap C$ is finite for every compact subset C of S. Let $\mathcal{F}_{lf}$ be the system of all locally finite subsets of S. One equips $\mathcal{F}_{lf}$ with the smallest σ-algebra for which all counting functions

$$F \mapsto \text{card}\,(F \cap A), \qquad A \in \mathcal{B}(S),$$

are measurable. A **simple point process** in S is a random variable X on some probability space $(\Omega, \mathsf{A}, \mathbb{P})$ with values in $\mathcal{F}_{lf}$. The expectation

$$\Theta(A) := \mathbb{E}\,\text{card}\,(X \cap A), \qquad A \in \mathcal{B}(S),$$

defines the **intensity measure** of the point process X. The point process X with intensity measure Θ is called a **Poisson process** if Θ is finite on compact sets and if, for every Borel set $A \subset S$ with $\Theta(A) < \infty$ and all $j \in \mathbb{N}_0$ one has

$$\mathbb{P}(\text{card}\,(X \cap A) = j) = \frac{\Theta(A)^j}{j!} e^{-\Theta(A)}.$$

This is now applied to the space $S = \mathcal{E}_k^d$, where $k \in \{0, \ldots, d-1\}$. We assume that X is a simple point process in $\mathcal{E}_k^d$, with an intensity measure $\Theta \not\equiv 0$ which is finite on compact sets. We call X a k-**flat process**. The process X is **stationary (isotropic)** if its distribution is invariant under translations (rotations). Stationary k-flat processes have been thoroughly studied in stochastic geometry. We work here with a weaker assumption.

Definition. *The k-flat process X is* **regular** *if its intensity measure has a continuous density with respect to some translation invariant, locally finite measure on $\mathcal{E}_k^d$.*

Let X be a regular k-flat process, with intensity measure Θ. Thus, there exist a locally finite, translation invariant measure μ on $\mathcal{E}_k^d$ and a nonnegative, continuous function h on $\mathcal{E}_k^d$ such that

$$\Theta(A) = \int_A h \, d\mu \qquad \text{for } A \in \mathcal{B}(\mathcal{E}_k^d).$$

By the decomposition result (74), there is a finite measure Φ on $\mathcal{L}_k^d$ such that

$$\int_{\mathcal{E}_k^d} f \, d\mu = \int_{\mathcal{L}_k^d} \int_{L^\perp} f(L + x)\,\lambda_k(dx)\,\Phi(dL)$$

for every nonnegative, measurable function f on $\mathcal{E}_k^d$. This gives

$$\Theta(A) = \int_{\mathcal{L}_k^d} \int_{L^\perp} \mathbf{1}_A(L + x) h(L + x)\,\lambda_k(dx)\,\Phi(dL) \tag{94}$$

for $A \in \mathcal{B}(\mathcal{E}_k^d)$.

To measure the local 'denseness' of the process X, we consider the expectations

$$\mathbb{E} \sum_{E \in X} \lambda_k(E \cap B)$$

for $B \in \mathcal{B}(\mathbb{R}^d)$ with $\lambda_d(B) < \infty$. One obtains

$$\mathbb{E} \sum_{E \in X} \lambda_k(E \cap B) = \int_B \int_{\mathcal{L}_k^d} h(L + x)\,\Phi(dL)\,\lambda_d(dx).$$

Thus, the measure $\mathbb{E} \sum_{E \in X} \lambda_k(E \cap \cdot)$ has a continuous density with respect to Lebesgue measure λ_d, given by

$$\gamma(x) := \int_{\mathcal{L}_k^d} h(L + x)\,\Phi(dL).$$

This function γ is called the **intensity function** of the process X. If X is stationary, this function is a constant, called the **intensity** of X. The following more intuitive interpretation of the intensity function can be proved:

$$\gamma(x) = \lim_{r \to 0} \frac{1}{\kappa_{d-k} r^{d-k}} \mathbb{E} \, \text{card} \left(\{ E \in X : E \cap (rB^d + x) \neq \emptyset \} \right).$$

From now on we assume that $k = d - 1$, so that X is a regular hyperplane process. The density h is now defined on the space $\mathcal{E}^d_{d-1}$ of hyperplanes, hence we can define a function $g : S^{d-1} \times \mathbb{R} \to (0, \infty)$ by

$$g(u, t) := h(H_{u,t}).$$

Then g is continuous and satisfies $g(u, t) = g(-u, -t)$. The measure Φ on $\mathcal{L}^d_{d-1}$ defines an even measure $\tilde{\Phi}$ on the sphere S^{d-1} satisfying

$$\tilde{\Phi}(A) = \frac{1}{2} \Phi(\{ H_{u,0} : u \in A \})$$

for Borel sets $A \subset S^{d-1}$ without antipodal points. Now (94) can be written in the form

$$\Theta(\mathcal{A}) = \int_{S^{d-1}} \int_{\mathbb{R}} \mathbf{1}_{\mathcal{A}}(H_{u,t}) g(u, t) \, dt \, \tilde{\Phi}(du) \tag{95}$$

for $\mathcal{A} \in \mathcal{B}(\mathcal{E}^d_{d-1})$. Note that this representation is of the type (89), but is more general. In the preceding subsection, the function g was derived from the local unit balls of a given projective Finsler metric. We will now, reversely, use the present function g to construct 'local unit balls' and exhibit their relevance for the geometry of the hyperplane process X.

For each $x \in \mathbb{R}^d$, we define a finite even measure ρ_x on S^{d-1} by

$$\rho_x(A) := \int_A g(u, \langle u, x \rangle) \, \tilde{\Phi}(du), \qquad A \in \mathcal{B}(S^{d-1}).$$

Then we define the **local associated zonoid** Π_x of X at x as the convex body with support function

$$h(\Pi_x, u) = \int_{S^{d-1}} |\langle u, v \rangle| \, \rho_x(dv), \qquad u \in \mathbb{R}^d. \tag{96}$$

The main results of this subsection are two examples showing how these local associated zonoids enter the discussion of natural geometric questions about the hyperplane process X.

First, let $M \subset \mathbb{R}^d$ be a closed line segment. We ask for the expected number of intersection points of M with the hyperplanes of X. Putting

$$\mathcal{H}_M := \{ H \in \mathcal{E}^d_{d-1} : H \cap M \neq \emptyset \},$$

this expected number is given by

$$\mathbb{E}\operatorname{card}(X \cap \mathcal{H}_M) = \Theta(\mathcal{H}_M) = \int_{S^{d-1}} \int_{\mathbb{R}} \mathbf{1}_{\mathcal{H}_M}(H_{u,t}) g(u,t) \, dt \, \tilde{\Phi}(du).$$

For the computation of the inner integral, we choose a C^1 parameterization $y : [a,b] \to \mathbb{R}^d$ of M with $y' \neq 0$. Let $u \in S^{d-1}$ be given, without loss of generality not orthogonal to M, and let $\mathbf{1}_{\mathcal{H}_M}(H_{u,t}) = 1$. Then there is a unique $s \in [a,b]$ with $M \cap H_{u,t} = \{y(s)\}$, hence $t = \langle u, y(s) \rangle$. Substituting t by $s \cdot \operatorname{sgn} \langle u, y' \rangle$, we get

$$\mathbb{E}\operatorname{card}(X \cap \mathcal{H}_M) = \int_{S^{d-1}} \int_a^b g(u, \langle u, y(s) \rangle) \, |\langle u, y'(s) \rangle| \, ds \, \tilde{\Phi}(du)$$
$$= \int_a^b h(\Pi_{y(s)}, y'(s)) \, ds.$$

Defining a Finsler metric F by

$$F(x,u) := h(\Pi_x, u),$$

we see that

$$\mathbb{E}\operatorname{card}(X \cap \mathcal{H}_M) = \int_a^b F(y(s), y'(s)) \, ds,$$

hence we can formulate the following result.

Theorem 4.6. *The expected number of hyperplanes in the regular hyperplane process X hitting the segment M is equal to the Finsler length of M, for the Finsler metric defined by the support function of the local associated zonoids.*

For our second question we assume now, in addition, that X is a Poisson process. The question concerns the processes of lower dimensional flats that are generated by intersecting hyperplanes of X. For $k \in \{2, \ldots, d\}$ we take, in every realization of X, all intersections of any k hyperplanes in X which have linearly independent normal vectors. This defines a simple process of $(d-k)$-flats. We denote it by X_k, and its intensity measure by Θ_k. Using the strong independence properties of Poisson processes, it can be shown that

$$\Theta_k(A) = \frac{1}{k!} \int_{\mathcal{E}_{d-1}^d} \cdots \int_{\mathcal{E}_{d-1}^d} \mathbf{1}_A(H_1 \cap \cdots \cap H_k) \Theta(dH_1) \cdots \Theta(dH_k) \qquad (97)$$

for $A \in \mathcal{B}(\mathcal{E}_{d-k}^d)$. This formula is similar to (90). From (97) and the regularity of X it can be deduced that X_k is also regular. In particular, the intensity function of X_k is defined; we denote it by γ_k. A computation gives

$$\gamma_k(x) = \frac{1}{k!} \int_{S^{d-1}} \cdots \int_{S^{d-1}} [u_1, \ldots, u_k] \rho_x(du_1) \cdots \rho_x(du_k). \qquad (98)$$

The geometric question we want to answer is whether one can find sharp bounds for the intensity function of the intersection process of order k in terms of the intensity function of the process X itself. The answer comes from a beautiful interpretation of the integral (98). We have mentioned in Subsection 4.1 that the integral representation (84) implies the representation (85) for the k-dimensional projection volume. Further, we know from (17) that averaging the k-dimensional projection volumes over all directions gives the kth intrinsic volume. Applying this to the convex body Π_x, which has the representation (96), we must get a formula for the intrinsic volume $V_k(\Pi_x)$. The result is

$$V_k(\Pi_x) = \frac{2^k}{k!} \int_{S^{d-1}} \cdots \int_{S^{d-1}} [u_1, \ldots, u_k] \, \rho_x(\mathrm{d}u_1) \cdots \rho_x(\mathrm{d}u_k).$$

Comparison with (98) now shows that

$$\gamma_k(x) = V_k(2^{-1}\Pi_x).$$

The intrinsic volumes of convex bodies satisfy well-known inequalities. As a consequence, and with some additional arguments concerning the equality case, the following sharp estimate can be obtained.

Theorem 4.7. *Let X be a regular Poisson hyperplane process in $\mathbb{R}^d$ with intensity function γ. Let $k \in \{2, \ldots, d\}$, and let γ_k be the intensity function of the intersection process X_k of order k. Then*

$$\gamma_k(x) \leq \frac{\binom{d}{k}\kappa_{d-1}^k}{d^k \kappa_{d-k}\kappa_d^{k-1}} \gamma(x)^k$$

for $x \in \mathbb{R}^d$. Equality for all $x \in \mathbb{R}^d$ holds if and only if the process X is stationary and isotropic.

Hints to the literature. Processes of k-flats are treated, for example, in [52]. The contents of this subsection are taken from [49], where full proofs can be found.

References

1. Alexander, R.: Zonoid theory and Hilbert's fourth problem. Geom. Dedicata, **28**, 199–211 (1988)
2. Álvarez Paiva, J.C., Fernandes, E.: Crofton formulas in projective Finsler spaces. Electron. Res. Announc. Amer. Math. Soc., **4**, 91–100 (1998)
3. Federer, H.: Curvature measures. Trans. Amer. Math. Soc., **93**, 418–491 (1959)
4. Giger, H., Hadwiger, H.: Über Treffzahlwahrscheinlichkeiten im Eikörperfeld. Z. Wahrsch. Verw. Gebiete, **10**, 329–334 (1968)
5. Glasauer, S.: Integralgeometrie konvexer Körper im sphärischen Raum. Doctoral Thesis, Universität Freiburg i. Br. (1995)

6. Glasauer, S.: A generalization of intersection formulae of integral geometry. Geom. Dedicata, **68**, 101–121 (1997)

7. Glasauer, S.: Kinematic formulae for support measures of convex bodies. Beiträge Algebra Geom., **40**, 113–124 (1999)

8. Goodey, P., Weil, W.: Representations of mixed measures with applications to Boolean models. Rend. Circ. Mat. Palermo (2) Suppl., **70**, vol. I, 325–346 (2002)

9. Goodey, P., Weil, W.: Translative and kinematic integral formulae for support functions II. Geom. Dedicata, **99**, 103–125 (2003)

10. Groemer, H.: On the extension of additive functionals on classes of convex sets. Pacific J. Math., **75**, 397–410 (1978)

11. Gruber, P., Schneider, R.: Problems in geometric convexity. In: Tölke, J., Wills, J.M. (eds) Contributions to Geometry. Birkhäuser, Basel (1979)

12. Hadwiger, H: Einige Anwendungen eines Funktionalsatzes für konvexe Körper in der räumlichen Integralgeometrie. Monatsh. Math., **54**, 345–353 (1950)

13. Hadwiger, H: Beweis eines Funktionalsatzes für konvexe Körper. Abh. Math. Sem. Univ. Hamburg, **17**, 69–76 (1951)

14. Hadwiger, H: Additive Funktionale k-dimensionaler Eikörper. I. Arch. Math., **3**, 470–478 (1952)

15. Hadwiger, H: Integralsätze im Konvexring. Abh. Math. Sem. Univ. Hamburg, **20**, 136–154 (1956)

16. Hadwiger, H: Vorlesungen über Inhalt, Oberfläche und Isoperimetrie. Springer, Berlin (1957)

17. Hug, D.: Measures, curvatures and currents in convex geometry. Habilitation Thesis, Universität Freiburg i. Br. (1999)

18. Hug, D., Last, G.: On support measures in Minkowski spaces and contact distributions in stochastic geometry. Ann. Probab., **28**, 796–850 (2000)

19. Hug, D., Last, G., Weil, W.: A local Steiner-type formula for general closed sets and applications. Math. Z., **246**, 237–272 (2004)

20. Hug, D., Mani-Levitska, P., Schätzle, R.: Almost transversal intersections of convex surfaces and translative integral formulae. Math. Nachr., **246-247**, 121–155 (2002)

21. Hug, D., Schätzle, R.: Intersections and translative integral formulas for boundaries of convex bodies. Math. Nachr., **226**, 99–128 (2001)

22. Hug, D., Schneider, R.: Kinematic and Crofton formulae of integral geometry: recent variants and extensions. In: Barceló i Vidal, C. (ed) Homenatge al Professor Lluís Santaló i Sors. Universitat de Girona (2002)

23. Kiderlen, M.: Schnittmittelungen und äquivariante Endomorphismen konvexer Körper. Doctoral Thesis, Universität Karlsruhe (1999)

24. Klain, D.A.: A short proof of Hadwiger's characterization theorem. Mathematika **42**, 329–339 (1995)

25. Klain, D.A., Rota, G.-C.: Introduction to Geometric Probability. Cambridge University Press, Cambridge (1997)

26. McMullen, P.: Mixed fibre polytopes. Discrete Comput. Geom., **32**, 521–532 (2004)

27. Pogorelov, A.V.: Hilbert's Fourth Problem. Scripta Series in Mathematics. V.H. Winston & Sons, Washington, D.C.; A Halstead Press Book, John Wiley & Sons, New York (1979) (Russian original: Izdat. "Nauka", Moscow (1974))

28. Rataj, J.: The iterated version of a translative integral formula for sets of positive reach. Rend. Circ. Mat. Palermo (2) Suppl., **46**, 129–138 (1997)

29. Rataj, J.: Remarks on a translative formula for sets of positive reach. Geom. Dedicata, **65**, 59–62 (1997)
30. Rataj, J.: Translative and kinematic formulae for curvature measures of flat sections. Math. Nachr., **197**, 89–101 (1999)
31. Rataj, J.: Absolute curvature measures for unions of sets with positive reach. Preprint 00/22, Universität Karlsruhe (2000)
32. Rataj, J.: A translative integral formula for absolute curvature measures. Geom. Dedicata, **84**, 245–252 (2001)
33. Rataj, J., Zähle, M.: Mixed curvature measures for sets of positive reach and a translative integral formula. Geom. Dedicata, **57**, 259–283 (1995)
34. Rataj, J., Zähle, M.: Curvatures and currents for unions of sets with positive reach, II. Ann. Global Anal. Geom., **20**, 1–21 (2001)
35. Rataj, J., Zähle, M.: A remark on mixed curvature measures of sets with positive reach. Beiträge Algebra Geom., **43**, 171–179 (2002)
36. Rother, W., Zähle, M.: A short proof of a principal kinematic formula and extensions. Trans. Amer. Math. Soc., **321**, 547–558 (1990)
37. Santaló, L.A.: Integral Geometry and Geometric Probability. Encyclopedia of Mathematics and Its Applications, **1**, Addison-Wesley, Reading, MA (1976)
38. Schmidt, V., Spodarev, E.: Joint estimators for the specific intrinsic volumes of stationary random sets. Stochastic Process. Appl., **115**, 959–981 (2005)
39. Schneider, R.: Convex Bodies: the Brunn–Minkowski Theory. Encyclopedia of Mathematics and Its Applications, **44**, Cambridge University Press, Cambridge (1993)
40. Schneider, R.: An extension of the principal kinematic formula of integral geometry. Rend. Circ. Mat. Palermo (2) Suppl., **35**, 275–290 (1994)
41. Schneider, R: On areas and integral geometry in Minkowski spaces. Beiträge Algebra Geom., **38**, 73–86 (1997)
42. Schneider, R.: Convex bodies in exceptional relative positions. J. London Math. Soc., **60**, 617–629 (1999)
43. Schneider, R.: Mixed functionals of convex bodies. Discrete Comput. Geom., **24**, 527–538 (2000)
44. Schneider, R.: On the Busemann area in Minkowski spaces. Beiträge Algebra Geom., **42**, 263–273 (2001)
45. Schneider, R.: Crofton formulas in hypermetric projective Finsler spaces. Arch. Math., **77**, 85–97 (2001)
46. Schneider, R.: On integral geometry in projective Finsler spaces. Izv. Nats. Akad. Armenii Mat., **37**, 34–51 (2002)
47. Schneider, R.: Mixed polytopes. Discrete Comput. Geom., **29**, 575–593 (2003)
48. Schneider, R.: An integral geometric theorem for simple valuations. Beiträge Algebra Geom., **44**, 487–492 (2003)
49. Schneider, R.: Nonstationary Poisson hyperplanes and their induced tessellations. Adv. in Appl. Probab., **35**, 139–158 (2003)
50. Schneider, R., Weil, W.: Translative and kinematic integral formulas for curvature measures. Math. Nachr., **129**, 67–80 (1986)
51. Schneider, R., Weil, W.: Integralgeometrie. Teubner, Stuttgart (1992)
52. Schneider, R., Weil, W.: Stochastische Geometrie. Teubner, Stuttgart (2000)
53. Schneider, R., Wieacker, J.A.: Integral geometry in Minkowski spaces. Adv. Math., **129**, 222–260 (1997)
54. Szabó, Z.I.: Hilbert's Fourth Problem, I. Adv. Math., **59**, 185–301 (1986)

55. Thompson, A.C.: Minkowski Geometry. Encyclopedia of Mathematics and Its Applications, **63**, Cambridge University Press, Cambridge (1996)
56. Weil, W.: Kontinuierliche Linearkombination von Strecken. Math. Z., **148**, 71–84 (1976)
57. Weil, W.: Centrally symmetric convex bodies and distributions, II. Israel J. Math., **32**, 173–182 (1979)
58. Weil, W.: Iterations of translative integral formulae and non-isotropic Poisson processes of particles. Math. Z., **205**, 531–549 (1990)
59. Weil, W.: Translative and kinematic integral formulae for support functions. Geom. Dedicata, **57**, 91–103 (1995)
60. Weil, W.: Mixed measures and functionals of translative integral geometry. Math. Nachr., **223**, 161–184 (2001)
61. Zähle, M.: Integral and current representations of Federer's curvature measures. Arch. Math., **46**, 557–567 (1986)
62. Zähle, M.: Curvatures and currents of sets with positive reach. Geom. Dedicata, **23**, 155–171 (1987)
63. Zähle, M.: Nonosculating sets of positive reach. Geom. Dedicata, **76**, 183–187 (1999)

Random Sets (in Particular Boolean Models)

Wolfgang Weil

Mathematisches Institut II, Universität Karlsruhe
D-76128 Karlsruhe, Germany
e-mail: weil@math.uni-karlsruhe.de

Introduction

Random sets are mathematical models to describe complex spatial data as they arise in modern applications in numerous form, as pictures, maps, digital images, etc. Whenever the geometric structure of an image is essential, a description by a set-valued random variable seems to be appropriate. Mathematically, a random set can simply be defined to be a measurable mapping Z from some abstract probability space $(\Omega, \mathbf{A}, \mathbb{P})$ into a class $\mathcal{F}$ of sets, where the latter is supplied with a σ-algebra $\mathbb{F}$. Here, on the one hand, $\mathcal{F}$ and $\mathbb{F}$ have to be reasonably large to represent structures occurring in practice at least approximately and to allow the basic geometric operations like intersection and union to be measurable operations within the class. On the other hand, $\mathbb{F}$ has to be reasonably small in order to have enough nontrivial examples of distributions on $(\mathcal{F}, \mathbb{F})$. A pair $(\mathcal{F}, \mathbb{F})$ which fulfills both requirements is given by the class $\mathcal{F}$ of all closed subsets of a topological space T and by the Borel σ-algebra $\mathbb{F} = \mathcal{B}(\mathcal{F})$ with respect to the topology of closed convergence on $\mathcal{F}$. Throughout the following, we will concentrate on this case. Random open sets can be defined and discussed in a similar manner, but random closed sets cover more cases which are of interest in applications, since they include random compact sets and, in particular, random finite or locally finite sets (simple point processes). Furthermore, we only work with random subsets $Z \subset \mathbb{R}^d$, other topological spaces T will only be mentioned occasionally. Some aspects of the theory remain unchanged in more general spaces, others make use of the vector space structure of $\mathbb{R}^d$ (or even the finite dimension).

Having clarified now the basic setting for our considerations, the challenge still remains to find a reasonably large class of distributions on $\mathcal{F}$ (we frequently suppress the σ-algebra, in the following), for example to allow some statistical analysis for random sets. In the classical situation of real random variables, various distributions can be constructed using the distribution function as a tool, but there is also the family of normal distributions which plays a central role and from which further distributions can be obtained by variation. In stochastic geometry, there is an analog of the distribution function (the capacity functional of a random closed set), but it cannot be used for the explicit construction of distributions, in general. However, there is a basic random set model, the Boolean model, which allows to calculate various geometric parameters and therefore can be used to fit a specific random set to given spatial data. Actually, Boolean models build a whole family of random sets which is still very rich in structure. Their definition is based on Poisson processes, which play a role in stochastic geometry comparable to the role of the normal distribution in classical statistics.

Boolean models can be classified according to their invariance properties. Those which are stationary and isotropic (thus invariant in distribution with respect to rigid motions) are the best studied ones and there is a large variety of formulas for them. Boolean models which are only stationary (invariant in distribution with respect to translations) have been the object of research for the last years. Boolean models without any invariance properties (or with only partial invariances) are the most complex ones and were studied only recently. In the following, we frequently concentrate on Boolean models, as the basic random sets in stochastic geometry and as ingredients for more general classes of random sets. We present the results for Boolean models with increasing generality, first for stationary and isotropic models, then without the isotropy condition and finally without any invariance assumptions. The section headings are:

1. Random sets, particle processes and Boolean models
2. Mean values of additive functionals
3. Directional data, local densities, nonstationary Boolean models
4. Contact distributions

Since Boolean models arise as union sets of (Poisson) particle processes, geometric functionals compatible with unions are of particular interest (additive functionals) and expectation formulas for such functionals immediately lead to formulas of integral geometric type. For stationary and isotropic models, we will thus make use of kinematic formulas, the general case requires formulas from translative integral geometry. This part will therefore frequently refer to results explained in the contribution [34] by R. Schneider but will also be based on results for point processes (see the chapter [1] by A. Baddeley). In particular, we will use some notations from [34] without further explanation.

The use of integral geometric results has also some influence on the choice of the set class to start with. We will therefore often work with Boolean models having convex grains, but more general results will be mentioned, too.

1 Random Sets, Particle Processes and Boolean Models

In this section, we define Boolean models and explain their role in stochastic geometry. We begin however with introducing the two basic notions in stochastic geometry, random closed sets (RACS) and point processes of compact sets (particle processes).

1.1 Random Closed Sets

Since we already described the ideas behind the notion of a random closed set, we will be quite brief in this subsection and mainly collect the technical setup which we use in the following. From now on, $\mathcal{F}$ denotes the class of closed subsets of $\mathbb{R}^d$ (including the empty set $\emptyset$). Subclasses which we frequently use are the compact sets $\mathcal{C}$ and the convex bodies $\mathcal{K}$ (again both including $\emptyset$). We supply $\mathcal{F}$ with the σ-algebra $\mathbb{F}$ generated by the sets

$$\mathcal{F}_C, \quad C \in \mathcal{C}.$$

Here and in the following we use the notation

$$\mathcal{F}_A := \{F \in \mathcal{F} : F \cap A \neq \emptyset\}$$

for any subset $A \subset \mathbb{R}^d$; and we similarly define

$$\mathcal{F}^A := \{F \in \mathcal{F} : F \cap A = \emptyset\}.$$

There are various other classes which generate the same σ-algebra $\mathbb{F}$, for example

$$\{\mathcal{F}^C : C \in \mathcal{C}\}, \quad \{\mathcal{F}_G : G \subset \mathbb{R}^d \text{ open}\}, \quad \{\mathcal{F}^G : G \subset \mathbb{R}^d \text{ open}\}.$$

Consequently, also $\{\mathcal{F}^C : C \in \mathcal{C}\} \cup \{\mathcal{F}_G : G \subset \mathbb{R}^d \text{ open}\}$ generates $\mathbb{F}$. This set class is of interest since it can serve as a sub-basis of a topology on $\mathcal{F}$, the **topology of closed convergence**. For a sequence $F_j \to F$, convergence in this topology means that each $x \in F$ is limit of a sequence $x_j \to x$, $x_j \in F_j$ (for almost all j), and that each limit point $x = \lim x_{j_k}$ of a (converging) subsequence $x_{j_k} \in F_{j_k}$ lies in F. As it turns out, $\mathbb{F}$ is the Borel σ-algebra with respect to this topology.

The subclasses $\mathcal{C}$ and $\mathcal{K}$, and others which arise later, are supplied with the induced σ-algebras. On $\mathcal{C}$ (and similarly on $\mathcal{K}$), the induced σ-algebra coincides with the Borel σ-algebra generated by the **Hausdorff metric**. Note however that the topology of closed convergence on $\mathcal{C}$ is weaker than the Hausdorff metric topology. For example, the sequence $B^d + kx, k \in \mathbb{N}, x \neq 0$, of balls does not converge in the Hausdorff metric, but it converges in $\mathcal{F}$ (namely to $\emptyset$). Thus, $\mathcal{C}$ and $\mathcal{K}$ are neither closed nor open in $\mathcal{F}$, but they are Borel subsets. The latter is also true for all the subsets of $\mathcal{F}$ which come up later, without that we will mention it in all cases.

The choice of the σ-algebra (respectively the topology) on $\mathcal{F}$ was motivated by the desire to make the standard set transformations, which map closed sets into closed sets, continuous and therefore measurable. As it turns out, all geometric transformations which will occur are measurable, but only a few are continuous, the others have a semi-continuity property. We mention only $(F, F') \mapsto F \cup F'$ (which is continuous), $(F, F') \mapsto F \cap F'$ (which is upper semi-continuous) and $F \mapsto \partial F$ (which is lower semi-continuous). The real- and measure-valued functionals on convex bodies which arise in integral geometry are mostly continuous and therefore measurable. This refers in particular to the continuous valuations like the intrinsic volumes, the curvature measures, the mixed measures and the mixed functionals. Their additive extensions to the convex ring $\mathcal{R}$ are no longer continuous but measurable (this is a deeper result using the existence of measurable selections). We will also use the set class

$$\mathcal{S} := \{F \in \mathcal{F} : F \cap K \in \mathcal{R}, \text{ for all } K \in \mathcal{K}\},$$

in the following. $\mathcal{S}$ consists of all locally finite unions of convex bodies (extended polyconvex sets). As it is explained in more detail in [34], the curvature measures as well as the mixed measures are locally defined and therefore have a (unique additive) extension to $\mathcal{S}$. The extended measures are measurable (as functions on $\mathcal{S}$).

To finish this list of technical pre-requisites, we consider rigid motions (rotations, translations) of closed sets. The action of the group G_d on $\mathcal{F}$ is continuous and thus $(g, F) \mapsto gF$ is measurable, as are $(\vartheta, F) \mapsto \vartheta F, \vartheta \in SO_d$, and $(x, F) \mapsto F + x, x \in \mathbb{R}^d$.

We now come to our basic definitions. A **random closed set (RACS)** Z (in $\mathbb{R}^d$) is a measurable mapping

$$Z : (\Omega, \mathbf{A}, \mathbb{P}) \to (\mathcal{F}, \mathcal{B}(\mathcal{F})).$$

As usual, the image measure

$$\mathbb{P}_Z := Z(\mathbb{P})$$

is the **distribution** of Z. We write $Z \overset{\mathrm{d}}{=} Z'$, if $\mathbb{P}_Z = \mathbb{P}_{Z'}$ (equality in distribution).

Further probabilistic notions will be used without detailed explanation, as long as they are standard. For example, RACS $Z_1, ..., Z_k$ or $Z_1, Z_2, ...$ are (stochastically) **independent** if

$$\mathbb{P}_{(Z_1,...,Z_k)} = \mathbb{P}_{Z_1} \otimes \cdots \otimes \mathbb{P}_{Z_k},$$

respectively

$$\mathbb{P}_{(Z_1, Z_k,...)} = \bigotimes_{i=1}^{\infty} \mathbb{P}_{Z_i}.$$

The measurable transforms mentioned above produce random sets. Thus if Z, Z' are RACS, the following are also RACS: $Z \cup Z', Z \cap Z', gZ$ (for $g \in G_d$).

A RACS Z is **stationary** if $Z \overset{\mathrm{d}}{=} Z + x$, for all $x \in \mathbb{R}^d$, and **isotropic** if $Z \overset{\mathrm{d}}{=} \vartheta Z$, for all $\vartheta \in SO_d$. At the beginning, we will concentrate on RACS which are stationary or even stationary and isotropic. Here is a first result on stationary RACS.

Theorem 1.1. *A stationary RACS Z is almost surely either empty or unbounded.*

If we replace Z by its closed convex hull Z' (which is a stationary convex RACS), the theorem follows from the fact, that Z' (almost surely) only takes values in $\{\emptyset, \mathbb{R}^d\}$. The reader is invited to think about a proof of this simple result.

Although we will not use it in full detail, we want to mention a fundamental result of Choquet. It concerns the **capacity functional** T_Z of a RACS Z,

$$T_Z : \mathcal{C} \to [0,1], \qquad T_Z(C) := \mathbb{P}(Z \cap C \neq \emptyset) = \mathbb{P}_Z(\mathcal{F}_C).$$

Generally, a real functional T on $\mathcal{C}$ is called a Choquet capacity, if it fulfills $0 \leq T \leq 1, T(\emptyset) = 0$ and $T(C_i) \to T(C)$, for every decreasing sequence $C_i \searrow C$. The mapping T is alternating of infinite order, if

$$S_k(C_0; C_1, ..., C_k) \geq 0, \qquad \text{for all } C_0, C_1, ..., C_k \in \mathcal{C}, k \in \mathbb{N}_0.$$

Here, $S_0(C_0) := 1 - T(C_0)$ and

$$S_k(C_0; C_1, ..., C_k) := S_{k-1}(C_0; C_1, ..., C_{k-1}) - S_{k-1}(C_0 \cup C_k; C_1, ..., C_{k-1}),$$

for $k \in \mathbb{N}$.

Theorem 1.2 (Choquet). (a) *The capacity functional T_Z of a RACS Z is an alternating Choquet capacity of infinite order.*

(b) *If T is an alternating Choquet capacity of infinite order, then there is a RACS Z with $T = T_Z$.*

(c) *If $T_Z = T_{Z'}$, then $Z \overset{d}{=} Z'$.*

(a) follows directly from the definition of T_Z. The uniqueness result (c) will be useful for us. It is a consequence of the fact that the class of complements $\mathcal{F}^C, C \in \mathcal{C}$, is $\cap$-stable and generates $\mathcal{B}(\mathcal{F})$. (b) is the genuine result of Choquet and has a longer and more complicated proof, but following the usual lines of extension theorems in measure theory.

The capacity functional T_Z of a RACS Z can be viewed as the analog of the distribution function of a real random variable. Theorem 1.2 thus parallels the continuity and monotonicity properties of distribution functions as well as the corresponding characterization and uniqueness results.

For a stationary RACS Z, the value $p := T_Z(\{x\})$ is independent of $x \in \mathbb{R}^d$, since

$$T_Z(\{x\}) = \mathbb{P}(x \in Z) = \mathbb{P}(0 \in Z - x) = \mathbb{P}(0 \in Z) = T_Z(\{0\}).$$

Moreover,

$$\mathbb{E}\,\lambda_d(Z \cap A) = \mathbb{E} \int_{\mathbb{R}^d} \mathbf{1}_A(x)\mathbf{1}_Z(x)\,\lambda_d(\mathrm{d}x)$$

$$= \int_{\mathbb{R}^d} \mathbf{1}_A(x)\mathbb{E}\,\mathbf{1}_{Z-x}(0)\,\lambda_d(\mathrm{d}x)$$

$$= \mathbb{E}\,\mathbf{1}_Z(0) \int_{\mathbb{R}^d} \mathbf{1}_A(x)\,\lambda_d(\mathrm{d}x)$$

$$= p\,\lambda_d(A),$$

for $A \in \mathcal{B}(\mathbb{R}^d)$, due to the stationarity and Fubini's theorem. The constant p is therefore called the **volume fraction** of Z (later we will frequently denote this quantity by $\overline{V}_d(Z)$).

We also define the **covariance** $C(x,y) := \mathbb{P}(x,y \in Z), x, y \in \mathbb{R}^d$, of Z and note that, for stationary Z,

$$C(x,y) = C(0, y-x) = V_d(Z \cap (Z + x - y)).$$

Hints to the literature. The theory of random closed sets was developed independently by Kendall [19] and Matheron [24]. A first detailed exposition appeared in [25]; for more recent presentations, see [36] and [28].

1.2 Particle Processes

The space $\mathcal{F}' := \mathcal{F} \setminus \{\emptyset\}$ of nonempty closed sets is a locally compact space with countable base. Therefore, one can define and consider point processes X on $\mathcal{F}'$ (see [1]). Formally, a (simple) point process X is a measurable mapping $X : (\Omega, \mathbf{A}, \mathbb{P}) \to (\mathsf{N}, \mathcal{N})$, where N denotes the collection of all locally finite subsets of $\mathcal{F}'$ and $\mathcal{N}$ is the σ-algebra generated by the counting functions

$$N \mapsto \mathrm{card}\,(N \cap \mathcal{A}),$$

for $N \in \mathsf{N}$ and $\mathcal{A} \in \mathcal{B}(\mathcal{F}')$. Alternatively, N can be described as the collection of simple counting measures on $\mathcal{F}'$, the counting functions then have the form $N \mapsto N(\mathcal{A})$. Here, a Borel measure N on $\mathcal{F}'$ is a **counting measure** if it is integer-valued and locally finite, that is finite on all compact subsets of $\mathcal{F}'$. For the latter it is sufficient that

$$N(\mathcal{F}_C) < \infty, \qquad \text{for all } C \in \mathcal{C}. \tag{1}$$

The counting measure N is **simple**, if $N(\{F\}) \leq 1$, for all $F \in \mathcal{F}'$, that means there are no multiple points occurring in N. For the following, it is convenient to use both interpretations of (simple) point processes X on $\mathcal{F}'$ simultaneously. Thus, we will interpret X as a random countable collection of closed sets, but will also view it as a random measure on $\mathcal{F}'$, such that expressions like $X(\mathcal{A})$, $\mathcal{A} \in \mathcal{B}(\mathcal{F}')$, make sense (and describe the number of 'points' in X which lie in $\mathcal{A}$).

For a point process X on $\mathcal{F}'$, let Θ be the **intensity measure**. For a Borel set $\mathcal{A} \in \mathcal{B}(\mathcal{F}')$, $\Theta(\mathcal{A})$ gives the mean number of points in X which lie in $\mathcal{A}$. In the language of counting measures,

$$\Theta(\mathcal{A}) = \mathbb{E}X(\mathcal{A}).$$

Whereas X is, by definition, locally finite (at least almost surely), the intensity measure Θ need not be. However, we will make the corresponding assumption throughout the following, that is, we generally assume

$$\Theta(\mathcal{F}_C) < \infty, \qquad \text{for all } C \in \mathcal{C}. \tag{2}$$

We also assume that Θ is not the zero measure since then the point process X would be empty (with probability 1), a case which is not very interesting, but also has to be excluded in some of the later results.

Although there are a number of interesting results for point processes on $\mathcal{F}'$, we now restrict our attention to point processes on $\mathcal{C}' := \mathcal{C} \setminus \{\emptyset\}$, that is, to point processes X on $\mathcal{F}'$ which are concentrated on $\mathcal{C}'$. The latter is the case, if and only if Θ is concentrated on $\mathcal{C}'$. We call such point processes **particle processes**. Why did we make this detour via point processes on $\mathcal{F}'$? Since $\mathcal{C}'$ with the Hausdorff metric is also a locally compact space with countable base, we could have defined a particle process directly as a point process on the metric space $\mathcal{C}'$. One reason is that the sets $\mathcal{F}_C, C \in \mathcal{C}$, are compact in $\mathcal{F}'$, but the corresponding sets $\mathcal{C}_C := \mathcal{F}_C \cap \mathcal{C}', C \in \mathcal{C}$, are not compact in the Hausdorff metric. Thus the condition of local finiteness for measures on $\mathcal{C}'$ would be weaker than (1) and not sufficient for our later purposes. A second aspect is that there is another important family of point processes in stochastic geometry which should be at least mentioned here, the q-**flat processes**. These are point processes of closed sets which are concentrated on the space $\mathcal{E}_q^d$ of q-flats (q-dimensional affine subspaces). $\mathcal{E}_q^d$ is also a measurable subset of $\mathcal{F}'$. Processes of flats are very interesting objects and show also close connections to convex geometry. Some results of this kind are discussed in [34]. Due to lack of time, we will not consider them further. Another very interesting class, which will not be treated here, are the **random mosaics**. A random mosaic can be defined as a particle process X where the particles are convex polytopes which tile the space. Alternatively, but mathematically less informative, one can consider the union of the boundaries of the tiles and call the RACS Z made up by these boundaries a random mosaic. Because there is a strong dependence between the cells of a random mosaic, random mosaics and Boolean models are far from each other. We refer to the contribution [13], for some results on random mosaics and further references.

The definition of invariance properties of a particle process X is now straight-forward. Rigid motions g act in a natural way on collections of sets and on (random) measures η (on $\mathcal{C}'$). Namely,

$$g\mathcal{A} := \{gK : K \in \mathcal{A}\} \quad \text{and} \quad g\eta(\mathcal{A}) := \eta(g^{-1}\mathcal{A}), \quad \mathcal{A} \in \mathcal{B}(\mathcal{C}').$$

Therefore, X is called **stationary** (respectively **isotropic**) if $X \overset{\mathrm{d}}{=} X + x$, for all translations x, (respectively $X \overset{\mathrm{d}}{=} \vartheta X$, for all rotations ϑ). Here, we use distributions of particle processes and equality in distribution in the obvious way, without copying the definitions which we have described in more detail for RACS.

Particle processes can also be interpreted as marked point processes in $\mathbb{R}^d$ with mark space $\mathcal{C}'$, if we associate with each particle K a pair (x, K') such that $K = x + K'$. The idea is that x represents the 'location' of K, whereas K' represents the 'form'. Such a representation is especially helpful in the stationary case and we will use it directly or indirectly throughout the following. Apparently, there is no natural decomposition of this kind, any suitable center map $c : \mathcal{C}' \to \mathbb{R}^d$ will produce a corresponding pair $(c(K), K - c(K))$. In the following we work with one specific center map, which is compatible with rigid motions, namely we choose $c(K)$ to be the **midpoint of the circumsphere** of K. The marks are then concentrated on $\mathcal{C}_0 := \{K \in \mathcal{C}' : c(K) = 0\}$. For some applications, different center maps have been used (for example lower tangent points of the particles). In the case of convex particles, the Steiner point is also a natural choice.

For stationary particle processes, the representation as marked point process leads to a decomposition of the intensity measure (we remind the reader that we always assume $\Theta \not\equiv 0$).

Theorem 1.3. *For a stationary particle process X, the intensity measure Θ is translation invariant and has a decomposition*

$$\Theta(\mathcal{A}) = \gamma \int_{\mathcal{C}_0} \int_{\mathbb{R}^d} \mathbf{1}_{\mathcal{A}}(x + K) \, \lambda_d(\mathrm{d}x) \mathbb{Q}(\mathrm{d}K), \qquad \mathcal{A} \in \mathcal{B}(\mathcal{C}'), \qquad (3)$$

with a constant $\gamma > 0$ and a probability measure $\mathbb{Q}$ on $\mathcal{C}_0$.

If X is isotropic, then $\mathbb{Q}$ is rotation invariant.

We call γ the **intensity** of X and $\mathbb{Q}$ the **grain distribution**. The marked point process $\Psi := \{(c(K), K - c(K)) : K \in X\}$ is sometimes called a **germ-grain process** since we can think of the particles K as grains grown around a germ.

We shortly indicate the proof of Theorem 1.3. The translation invariance of Θ is obvious. The image measure Θ' of Θ under $K \mapsto (c(K), K - c(K))$ is a measure on $\mathbb{R}^d \times \mathcal{C}_0$. The translation invariance of Θ implies that $\Theta' = \lambda_d \otimes \rho$ with some measure ρ. The local finiteness of Θ yields that ρ is finite. (3) thus follows with $\gamma := \rho(\mathcal{C}_0)$ and $\mathbb{Q} := \frac{1}{\gamma}\rho$. If X is isotropic, Θ is rotation invariant, and therefore Θ' is rotation invariant in the second component. Thus, $\mathbb{Q}$ is rotation invariant.

The representation (3) is unique with respect to the center map c which we used. A different center map c' can produce a different representation of X as a marked point process, and therefore a different decomposition of Θ. As we shall show below, this does not affect the intensity γ which will be the same for each representation, but it will affect the grain distribution $\mathbb{Q}'$ which will then live on a different space $\mathcal{C}_1 := \{C \in \mathcal{C}' : c'(C) = 0\}$. However, on can express $\mathbb{Q}$ and $\mathbb{Q}'$ as image measures of each other under a certain transformation which is connected with the center maps c, c'. If the center map c' is translation covariant but not compatible with rotations, the statement about isotropic X in Theorem 1.3 is no longer true, in general. This is another reason why we work with the circumcenter c, in the following.

The fact that γ depends only on X and not on c follows from the representation

$$\gamma = \lim_{r \to \infty} \frac{1}{\lambda_d(rB^d)} \Theta(\mathcal{F}_{rB^d}). \tag{4}$$

(4) is a consequence of (3), since

$$\frac{1}{\lambda_d(rB^d)} \Theta(\mathcal{F}_{rB^d}) = \frac{\gamma}{\kappa_d r^d} \int_{\mathcal{C}_0} \int_{\mathbb{R}^d} 1_{\mathcal{F}_{rB^d}}(x + K) \lambda_d(dx) \mathbb{Q}(dK)$$

$$= \frac{\gamma}{\kappa_d r^d} \int_{\mathcal{C}_0} \lambda_d(K + rB^d) \mathbb{Q}(dK)$$

$$= \frac{\gamma}{\kappa_d} \int_{\mathcal{C}_0} \lambda_d(\frac{1}{r}K + B^d) \mathbb{Q}(dK).$$

For $r \to \infty$, we have $\lambda_d(\frac{1}{r}K + B^d) \to \lambda_d(B^d) = \kappa_d$ und thus the result follows from Lebesgue's dominated convergence theorem.

In view of (4) we may interpret γ as the mean number of particles in X per unit volume of $\mathbb{R}^d$; we also speak of the **particle density**.

So far, we have worked with particle processes in general, now we want to mention a special class of them, the Poisson processes. A particle process X (with intensity measure Θ) is a **Poisson (particle) process**, if

$$\mathbb{P}(X(\mathcal{A}) = k) = e^{-\Theta(\mathcal{A})} \frac{\Theta(\mathcal{A})^k}{k!}, \qquad \text{for } k \in \mathbb{N}_0, \ \mathcal{A} \in \mathcal{B}(\mathcal{C}'), \tag{5}$$

and, for mutually disjoint $\mathcal{A}_1, \mathcal{A}_2, \ldots \in \mathcal{B}(\mathcal{C}')$,

$$X(\mathcal{A}_1), X(\mathcal{A}_2), \ldots \text{ are independent.} \tag{6}$$

Conditions (5) and (6) are not independent. In fact, for intensity measures Θ without atoms, (5) and (6) are equivalent (see [36], for more details). Since we only consider simple particle processes throughout these lectures, the intensity measure Θ of a Poisson process is atom free, automatically.

Poisson processes actually can be defined on quite arbitrary (measurable) spaces and each measure Θ (which is atom free and fulfills a suitable finiteness condition) gives rise to a (simple) Poisson process which is uniquely determined in distribution and which has Θ as intensity measure. Hence, knowing the intensity measure Θ already determines the whole (Poisson) process. This uniqueness property makes the class of Poisson processes so important for results in stochastic geometry, but also for the statistical analysis of random point fields. In particular, the uniqueness implies that a Poisson process is stationary (isotropic), if and only if Θ is translation invariant (rotation invariant). We refer to [1], for further properties of Poisson processes (in $\mathbb{R}^d$).

If we use the representation of a Poisson particle process X as a marked point process Ψ (based on the center map c), the underlying point process $X_0 := \{c(K) : K \in X\}$ will be a Poisson process. Vice versa, we can start with a Poisson process X_0 in $\mathbb{R}^d$ (with intensity measure Ξ) and can add to each point $\xi_i \in X_0$ a random set Z_i independently (from each other and from X) with a given distribution $\mathbb{Q}$ on $\mathcal{C}_0$, say. Then, $X := \{\xi_1 + Z_1, \xi_2 + Z_2, ...\}$ is a Poisson particle process and the intensity measure Θ of X is the image of $\Xi \otimes \mathbb{Q}$ under $(x, K) \mapsto x + K$. In general, however, not every Poisson particle process X is obtained from a Poisson process X_0 on $\mathbb{R}^d$ by independent marking, since for the intensity measure Θ the image under $K \mapsto (c(K), K - c(K))$ need not be a product measure.

For a stationary Poisson particle process X the situation is simpler, since then we can apply Theorem 1.3.

Theorem 1.4. *For a stationary Poisson process X on $\mathcal{C}$, let γ be the intensity and $\mathbb{Q}$ the grain distribution. If X_0 denotes the stationary Poisson process on $\mathbb{R}^d$ with intensity measure $\gamma \lambda_d$, then (up to equivalence in distribution) X is obtained from X_0 by independent marking and $\mathbb{Q}$ is the corresponding mark distribution.*

X is isotropic, if and only if $\mathbb{Q}$ is rotation invariant.

The above considerations make also clear how to simulate a (stationary) Poisson particle process X, given γ and $\mathbb{Q}$, namely by generating realizations $X_0(\omega) = \{\xi_1(\omega), \xi_2(\omega), ...\}$, $Z_1(\omega), Z_2(\omega), ...$ and finally $X(\omega) = \{\xi_1(\omega) + Z_1(\omega), \xi_2(\omega) + Z_2(\omega), ...\}$. If the simulation is performed in a bounded window W, edge effects have to be taken into account. Depending on the size of the random sets Z_i, the realizations of X_0 have to be generated in a larger neighborhood of W in order to allow for particles $K_i = \xi_i(\omega) + Z_i(\omega)$ which intersect W although the center point $c(K_i) = \xi_i(\omega)$ lies outside W.

Hints to the literature. There are numerous books on point processes and many of them work in general spaces (e.g. [4]). Point processes of geometric objects are treated in [25], [23] and [36].

1.3 Boolean Models

Having now defined the two basic notions in stochastic geometry, the RACS and the particle processes, we can start looking for examples. What are interesting random sets which can serve as models for random structures as they appear in practical applications? At this stage we notice that Theorem 1.2 is not as helpful as its real-valued counterpart. Whereas distribution functions on the real line are easy to construct and lead to a large variety of explicit (families of) distributions, the conditions for an alternating Choquet capacity of infinite order are more complex and the procedure to define a corresponding distribution on $\mathcal{F}$ is far from being constructive. However, particle processes are easier to construct and then we can use the following simple fact.

Theorem 1.5. *If X is a particle process, then*

$$Z := \bigcup_{K \in X} K$$

is a RACS. Moreover, if X is stationary (isotropic), then Z is stationary (isotropic).

The proof is simple and left to the reader. A bit more challenging (but still simple) is a reverse statement: Each RACS Z is the union set of a particle process X, and if Z is stationary (isotropic), X can be chosen to be stationary (isotropic).

If the particle process X is concentrated on $\mathcal{K}$ (we speak of convex particles then), the union set Z takes its values in $\mathcal{S}$. We call Z a **random $\mathcal{S}$-set**. For random $\mathcal{S}$-sets Z there is also a reverse statement, which is not as obvious anymore.

Theorem 1.6. *Each random $\mathcal{S}$-set Z is the union set of a process X of convex particles, and if Z is stationary (isotropic), X can be chosen to be stationary (isotropic).*

It is now easy to construct some examples of random $\mathcal{S}$-sets. For instance, let ξ be a nonnegative real random variable and K a fixed convex body, then $Z_0 := \xi K$ is a random convex body. The collection $X := \{z + \xi K : z \in \mathbb{Z}^d\}$ is then a particle process and its union set Z a random $\mathcal{S}$-set. In order to make X and Z stationary, we can add a uniform random translation $\tau \in [0,1]^d$ (independently of ξ), the distribution of τ thus being the Lebesgue measure on the unit cube $[0,1]^d$. In addition, we can make X and Z isotropic by applying a subsequent random rotation $\vartheta \in SO_d$ (again independently), the distribution of which is given by the Haar probability measure on SO_d. Although the resulting random set Z is now stationary and isotropic, it looks pretty regular. There are some obvious modifications which would add some more randomness to this construction. Using an enumeration $z_1, z_2, \ldots$ of $\mathbb{Z}^d$, we could replace $z_i + \xi K$ by $z_i + \xi_i K$, where $\xi_1, \xi_2, \ldots$ are independent copies of ξ, or even by $z_i + \xi_i K_i$, where we use a sequence $K_1, K_2, \ldots$ of convex bodies. Of course, we could even start with $X := \{z_i + Z_i : i = 1, 2, \ldots\}$, where $Z_1, Z_2, \ldots$ is a sequence of (independent or dependent) random sets with values in $\mathcal{K}, \mathcal{R}$ or $\mathcal{C}$ (in the latter case, the union set Z will be a RACS, but in general not a random $\mathcal{S}$-set).

Even with these modifications, the outcomes will be too regular to be useful for practical applications. But we can use the principle just described also to produce more interesting examples. Namely, we can start with a point process X_0 in $\mathbb{R}^d$, choose a (measurable) enumeration $X_0 = \{\xi_1, \xi_2, ...\}$, and then 'attach' random (compact or convex) sets $Z_1, Z_2, ...$ to the points and consider

$$Z := \bigcup_{i=1}^{\infty} (\xi_i + Z_i). \tag{7}$$

We will only consider the case where the Z_i are i.i.d. random compact sets, Z is then called a **germ-grain model**, the Z_i are called the **grains** of Z and their common distribution $\mathbb{Q}$ is called the **distribution of the typical grain** (or grain distribution). If $\mathbb{Q}(\mathcal{K}) = 1$, the germ-grain model Z has **convex grains**. If X_0 is stationary, then Z is stationary, and if X_0 is in addition isotropic and $\mathbb{P}_{Z_1}$ is rotation invariant, then Z is isotropic. Since there are many well-studied classes of point processes in $\mathbb{R}^d$, we can thus produce a large variety of random sets Z. However, since the particles of the process $X := \{\xi_1 + Z_1, \xi_2 + Z_2, ...\}$ may overlap, it is in general difficult to calculate geometric functionals of Z, even for well-established point processes X_0. The exception is the class of Poisson processes, for which a rich variety of formulas for the union sets Z are known. This is the reason why we will concentrate on Poisson processes in the following.

A RACS Z is a **Boolean model** if it is the union set of a Poisson particle process X. In particular, if we start with a Poisson process X_0 in $\mathbb{R}^d$, the corresponding germ-grain model with grain distribution $\mathbb{Q}$ is a Boolean model. Not every Boolean model arises in this way since not every Poisson particle process X is obtained from a Poisson process in $\mathbb{R}^d$ by independent marking (as we mentioned already). Vice versa, for a stationary Boolean model, the center map c produces a representation as germ-grain model, but of a special kind, namely with grain distribution $\mathbb{Q}$ concentrated on $\mathcal{C}_0$. The correspondence between Boolean models Z and Poisson particle processes X is one-to-one, as we will show now. Our argument is based on the fact that the capacity functional T_Z of a Boolean model Z can be expressed in terms of the intensity measure Θ of X. Namely

$$\begin{aligned} T_Z(C) = \mathbb{P}(Z \cap C \neq \emptyset) &= \mathbb{P}(X(\mathcal{F}_C) > 0) \\ &= 1 - \mathbb{P}(X(\mathcal{F}_C) = 0) \\ &= 1 - e^{-\Theta(\mathcal{F}_C)}, \end{aligned}$$

for $C \in \mathcal{C}$. Hence, if Z is the union set of another Poisson particle process X' as well (with intensity measure Θ'), we obtain

$$\Theta(\mathcal{F}_C) = \Theta'(\mathcal{F}_C), \qquad C \in \mathcal{C},$$

and therefore $\Theta = \Theta'$ (this implication is not immediate, but needs a bit of work; see [36], for details). From $\Theta = \Theta'$, we get $X \stackrel{d}{=} X'$ and hence the following result.

Theorem 1.7. *Let X, X' be Poisson particle processes with the same union set,*

$$\bigcup_{K \in X} K \stackrel{d}{=} \bigcup_{K' \in X'} K'.$$

Then,

$$X \stackrel{d}{=} X'.$$

The fact that $\Theta = \Theta'$ implies $X \stackrel{d}{=} X'$ can be deduced from general results in point process theory. It follows however also from Theorem 1.2 (in its general version, for RACS in a topological space T), since $\Theta = \Theta'$ implies $\mathbb{P}(X(A) = 0) = \mathbb{P}(X'(A) = 0)$ (from (5)). Therefore the (locally finite) RACS X and X' in $T := \mathcal{C}'$ have the same capacity functional.

For the remainder of this section, we concentrate on stationary Boolean models Z and their representation (7), yielding the intensity γ and the grain distribution $\mathbb{Q}$. Such stationary Boolean models can easily be simulated and produce interesting RACS even for simple distributions $\mathbb{Q}$ (for example, in the plane, for circles with random radii). In order to fit such a model to given (spatial) data, it is important to express geometric quantities of Z in terms of γ and $\mathbb{Q}$. For the capacity functional T_Z, such a result follows now from Theorem 1.3 in conjunction with Theorem 1.7. For a set $A \subset \mathbb{R}^d$, we use A^* to denote the reflection of A in the origin.

Theorem 1.8. *Let Z be a stationary Boolean model with intensity γ and grain distribution $\mathbb{Q}$. Then, the capacity functional T_Z of Z has the form*

$$T_Z(C) = 1 - \exp\left(-\gamma \int_{\mathcal{C}_0} \lambda_d(K + C^*)\mathbb{Q}(dK)\right), \qquad C \in \mathcal{C}, \qquad (8)$$

and the volume fraction p of Z fulfills

$$p = 1 - \exp\left(-\gamma \int_{\mathcal{C}_0} \lambda_d(K)\mathbb{Q}(dK)\right). \qquad (9)$$

Proof. Equation (8) follows from (3) since $(K + x) \cap C \neq \emptyset$ is equivalent to $x \in K^* + C$, and thus

$$\int_{\mathbb{R}^d} \mathbf{1}_{\mathcal{F}_C}(K + x)\lambda_d(dx) = \int_{\mathbb{R}^d} \mathbf{1}_{K^* + C}(x)\lambda_d(dx)$$
$$= \lambda_d(K + C^*).$$

Putting $C = \{0\}$ in (8) yields (9). $\qquad \square$

The capacity functional is of interest for statistical purposes, since it is easily estimated, using modern image analysing equipment. To be more precise, consider

$$T_Z(C) = \mathbb{P}(Z \cap C \neq \emptyset)$$

for a random set Z and a given 'test set' C. Since $\mathbb{P}(Z \cap C \neq \emptyset)$ equals the volume fraction of $Z + C^*$, a simple estimator of $T_Z(C)$ arises from counting pixels of a digitized image of $(Z + C^*) \cap [0,1]^d$.

Which information on γ and $\mathbb{Q}$ is contained in $T_Z(C)$? We can get a more precise answer if we choose C to be convex and assume that the Boolean model has convex grains. Then $\mathbb{Q}$ is concentrated on $\mathcal{K}_0 := \{K \in \mathcal{K} : c(K) = 0\}$ and we can use the mixed volume expansion for convex bodies. For example, if C is a ball $rB^d, r > 0$, the Steiner formula (see [34]) gives us

$$T_Z(rB^d) = 1 - \exp\left(-\gamma \int_{\mathcal{K}_0} \lambda_d(K + rB^d)\mathbb{Q}(dK)\right)$$

$$= 1 - \exp\left(-\gamma \sum_{i=0}^{d} r^{d-i} \kappa_{d-i} \int_{\mathcal{K}_0} V_i(K)\mathbb{Q}(dK)\right)$$

$$= 1 - \exp\left(-\sum_{i=0}^{d} r^{d-i} \kappa_{d-i} \overline{V}_i(X)\right).$$

Here, $V_i(K)$ denotes the ith **intrinsic volume** of K and the mean values $\overline{V}_i(X)$ of X are defined by

$$\overline{V}_i(X) := \gamma \int_{\mathcal{K}_0} V_i(K)\mathbb{Q}(dK), \qquad i = 0, ..., d.$$

We have $\overline{V}_0(X) = \gamma$. For the $\overline{V}_i(X)$, different names are used in the literature. They are called **mean Minkowski functionals**, **specific intrinsic volumes** or **quermass densities**. We will mostly use the latter name, although it is slightly irritating since the quantities are based on the intrinsic volumes, not on the quermassintegrals.

Hence, if we estimate $T_Z(rB^d)$ from realizations of Z, for different values of r, we obtain an empirical function $\hat{f} : r \mapsto \hat{T}_Z(rB^d)$. Fitting a polynomial (of order d) to

$$-\ln(1 - \hat{f})$$

gives us estimators for $\overline{V}_0(X), ..., \overline{V}_d(X)$ and thus for γ and the mean values

$$\int_{\mathcal{K}_0} V_i(K)\mathbb{Q}(dK), \qquad i = 1, ..., d.$$

Hints to the literature. In addition to the books [25] and [36], which we already mentioned, we refer to [9], [27] and [37] for results on Boolean models. Applications of Boolean models in statistical physics are described by Mecke (see [26] and the references given there).

2 Mean Values of Additive Functionals

The formula for the capacity functional, which we just gave in the case of convex grains, connects certain geometrical mean values of the Boolean model, the volume density of $Z + rB^d$, with mean values of the underlying Poisson process X, the quermass densities. Formulas of this kind can be used for the statistical estimation of particle quantities and in particular, for the estimation of the intensity γ. In this section, we discuss related formulas for other geometric quantities, like the surface area. Since Z is defined as the union set of the particles in X, functionals which are adapted to unions and intersections are of particular interest. Therefore, we concentrate on additive functionals and their mean values. We start with a general result for Boolean models. Then we consider general RACS and particle processes and discuss and compare different possible approaches for mean values of additive functionals. Results from integral geometry will be especially helpful here. In the last subsection, we come back to Boolean models and give explicit results for the quermass densities in the stationary and isotropic case.

2.1 A General Formula for Boolean Models

We consider a Boolean model Z in $\mathbb{R}^d$ with convex grains and an additive (and measurable) functional φ on $\mathcal{R}$. Additivity in the sequel always includes the convention $\varphi(\emptyset) = 0$. How can we define a mean value of φ for Z? Since Z may be unbounded (for example in the stationary case), it does not make much sense to work with $\varphi(Z)$, namely because Z is then not in the convex ring anymore and extensions of φ to $\mathcal{S}$, if they exist, usually yield $\varphi(Z) = \infty$. It seems more promising to consider $\varphi(Z \cap K_0)$ instead, where $K_0 \in \mathcal{K}$ is a suitably chosen bounded set. We can think of K_0 as a **sampling window** in which we observe the realizations of Z. This corresponds to many practical situations, where natural sampling windows arise in the form of boxes (rectangles) or balls (circles), for example as boundaries in photographic or microscopic images. Hence the question arises how to express $\mathbb{E}\,\varphi(Z \cap K_0)$ in terms of the intensity measure Θ.

We first describe a corresponding result in rather loose form before we give a rigorous formulation and proof. By definition of Z and the additivity of φ (used in form of the inclusion-exclusion formula),

$$\varphi(Z \cap K_0) = \varphi\left(\bigcup_{K \in X}(K \cap K_0)\right)$$

$$= \sum_{k=1}^{N}(-1)^{k+1}\sum_{1 \leq i_1 < i_2 < \cdots < i_k \leq N}\varphi\big(K_0 \cap K_{i_1} \cap \cdots \cap K_{i_k}\big).$$

Here N is the (random) number of grains $K \in X$ hitting K_0 and $K_1, \ldots, K_N$ is an enumeration of these grains. Since $\varphi(\emptyset) = 0$, we can simplify this formula by using the product particle process $X_{\neq}^k := \{(K_1, \ldots, K_k) \in X^k : K_i \text{ pairwise different}\}$, and get

$$\varphi(Z \cap K_0) = \sum_{k=1}^{\infty} \frac{(-1)^{k+1}}{k!} \sum_{(K_1,\dots,K_k) \in X_{\neq}^k} \varphi(K_0 \cap K_1 \cap \cdots \cap K_k).$$

Turning now to the expectation, the independence properties of the Poisson process yield

$$\mathbb{E}\,\varphi(Z \cap K_0)$$

$$= \sum_{k=1}^{\infty} \frac{(-1)^{k+1}}{k!} \mathbb{E} \sum_{(K_1,\dots,K_k) \in X_{\neq}^k} \varphi(K_0 \cap K_1 \cap \cdots \cap K_k)$$

$$= \sum_{k=1}^{\infty} \frac{(-1)^{k+1}}{k!} \int_{\mathcal{K}'} \cdots \int_{\mathcal{K}'} \varphi(K_0 \cap K_1 \cap \cdots \cap K_k) \Theta(\mathrm{d}K_1) \cdots \Theta(\mathrm{d}K_k),$$

which is our desired result.

Our derivation was not totally correct, since we did not pay attention to integrability requirements when we exchanged expectation and summation. In view of the alternating sign $(-1)^{k+1}$ in the sum, this may be problematic and requires us to impose a further restriction on φ. We call a functional $\varphi : \mathcal{R} \to \mathbb{R}$ **conditionally bounded**, if φ is bounded on each set $\{K \in \mathcal{K} : K \subset K'\}, K' \in \mathcal{K}$. The intrinsic volumes V_j, $j = 0, \dots, d$, are examples of additive (and measurable) functionals on $\mathcal{R}$, which are monotonic (and continuous) on $\mathcal{K}$ and therefore conditionally bounded.

Now we can formulate a precise result. We remark that a corresponding theorem holds for Boolean models with grains in $\mathcal{R}$ and $K_0 \in \mathcal{R}$, but this requires additional integrability conditions which we want to avoid here.

Theorem 2.1. *Let Z be a Boolean model with convex grains and let Θ be the intensity measure of the underlying Poisson particle process X on $\mathcal{K}'$. Let $\varphi : \mathcal{R} \to \mathbb{R}$ be additive, measurable and conditionally bounded. Then, for each $K_0 \in \mathcal{K}$, the random variable $\varphi(Z \cap K_0)$ is integrable and*

$$\mathbb{E}\,\varphi(Z \cap K_0)$$

$$= \sum_{k=1}^{\infty} \frac{(-1)^{k+1}}{k!} \int_{\mathcal{K}'} \cdots \int_{\mathcal{K}'} \varphi(K_0 \cap K_1 \cap \cdots \cap K_k) \Theta(\mathrm{d}K_1) \cdots \Theta(\mathrm{d}K_k). \quad (10)$$

Proof. Let $c = c_{K_0}$ be an upper bound for $|\varphi|$ on $\{M \in \mathcal{K} : M \subset K_0\}$. Then

$$|\varphi(Z \cap K_0)| \leq \sum_{k=1}^{N} \frac{1}{k!} \sum_{(K_1,\dots,K_k) \in X_{\neq}^k} |\varphi(K_0 \cap K_1 \cap \cdots \cap K_k)|$$

$$\leq \sum_{k=1}^{N} \binom{N}{k} c \leq c\, 2^N = c\, 2^{X(\mathcal{F}_{K_0})}.$$

Here, N is again the (random) number of particles in X which intersect K_0. The right-hand side is integrable since

$$\mathbb{E}\, 2^{X(\mathcal{F}_{K_0})} = \sum_{k=0}^{\infty} 2^k \mathbb{P}(X(\mathcal{F}_{K_0}) = k)$$

$$= e^{-\Theta(\mathcal{F}_{K_0})} \sum_{k=0}^{\infty} 2^k \frac{\Theta(\mathcal{F}_{K_0})^k}{k!}$$

$$= e^{-\Theta(\mathcal{F}_{K_0})} e^{2\Theta(\mathcal{F}_{K_0})} = e^{\Theta(\mathcal{F}_{K_0})} < \infty,$$

by (2). This yields the integrability of $\varphi(Z \cap K_0)$, but also justifies the interchange of expectation and summation, which we performed in the derivation above.

The equation

$$\mathbb{E} \sum_{(K_1,\ldots,K_k) \in X_{\neq}^k} \psi(K_1,\ldots,K_k)$$

$$= \int_{\mathcal{K}'} \cdots \int_{\mathcal{K}'} \psi(K_1,\ldots,K_k) \Theta(\mathrm{d}K_1) \cdots \Theta(\mathrm{d}K_k),$$

which we used for $\psi(K_1,\ldots,K_k) := \varphi(K_0 \cap K_1 \cap \cdots \cap K_k)$ and which we explained with the independence properties of X actually holds for integrable ψ and is formally a consequence of Campbell's theorem (applied to $X_{\neq}^k$) together with the fact that the intensity measure of $X_{\neq}^k$ (usually called the factorial moment measure of X) is the product measure Θ^k. $\square$

For a continuous and additive functional $\varphi : \mathcal{K} \to \mathbb{R}$, the conditions of Theorem 2.1 are fulfilled automatically. It is conditionally bounded and measurable on $\mathcal{K}$ and it has a (unique) additive extension to $\mathcal{R}$ which is measurable (and conditionally bounded). Examples for such functionals are the intrinsic volumes, but also other geometric quantities which we will consider later (mixed volumes, mixed functionals, surface area measures, support functions).

If Z is stationary, (3) yields

$$\mathbb{E}\, \varphi(Z \cap K_0)$$

$$= \sum_{k=1}^{\infty} \frac{(-1)^{k+1}}{k!} \gamma^k \int_{\mathcal{K}_0} \cdots \int_{\mathcal{K}_0} \Phi(K_0, K_1, \ldots, K_k) \mathbb{Q}(\mathrm{d}K_1) \cdots \mathbb{Q}(\mathrm{d}K_k) \quad (11)$$

with

$$\Phi(K_0, K_1, \ldots, K_k)$$

$$:= \int_{\mathbb{R}^d} \cdots \int_{\mathbb{R}^d} \varphi(K_0 \cap x_1 K_1 \cap \cdots \cap x_k K_k) \lambda_d(\mathrm{d}x_1) \cdots \lambda_d(\mathrm{d}x_k).$$

(Here and in the following, it is convenient to use the operational notation $xK := K + x$.)

For example, we can put $\varphi = V_d$ (the volume or Lebesgue measure), where

$$\Phi(K_0, K_1, \ldots, K_k) = V_d(K_0)V_d(K_1)\cdots V_d(K_k),$$

hence we obtain

$$
\begin{aligned}
\mathbb{E}\,V_d(Z \cap K_0) &= V_d(K_0)\sum_{k=1}^{\infty}\frac{(-1)^{k+1}}{k!}\left(\gamma\int_{\mathcal{K}_0}V_d(K)\mathbb{Q}(\mathrm{d}K)\right)^k \\
&= V_d(K_0)\left(1 - \exp\left(-\gamma\int_{\mathcal{K}_0}V_d(K)\mathbb{Q}(\mathrm{d}K)\right)\right) \\
&= V_d(K_0)\left(1 - e^{-\overline{V}_d(X)}\right).
\end{aligned}
$$

For $V_d(K_0) > 0$, we have $\mathbb{E}\,V_d(Z \cap K_0)/V_d(K_0) = p$ and thus we get the formula which we derived already in a more direct way (and in a slightly more general situation, namely for compact grains) in Theorem 1.8.

As another example, we choose $\varphi = V_{d-1}$ (which is half the surface area). Then we get from the translative integral formula for V_{d-1}, which is explained in [34],

$$
\begin{aligned}
\Phi(K_0, K_1, \ldots, K_k) &\\
= \int_{\mathbb{R}^d}\cdots\int_{\mathbb{R}^d} & V_{d-1}\left(K_0 \cap x_1 K_1 \cap \cdots \cap x_k K_k\right)\lambda_d(\mathrm{d}x_1)\cdots\lambda_d(\mathrm{d}x_k) \\
= \sum_{i=0}^{k} & V_d(K_0)\cdots V_d(K_{i-1})V_{d-1}(K_i)V_d(K_{i+1})\cdots V_d(K_d),
\end{aligned}
$$

and so

$$
\begin{aligned}
\mathbb{E}\,V_{d-1}(Z \cap K_0) &= \sum_{k=1}^{\infty}\frac{(-1)^{k+1}}{k!}\Big(V_{d-1}(K_0)\overline{V}_d(X)^k \\
&\qquad\qquad + kV_d(K_0)\overline{V}_{d-1}(X)\overline{V}_d(X)^{k-1}\Big) \\
&= V_d(K_0)\overline{V}_{d-1}(X)e^{-\overline{V}_d(X)} + V_{d-1}(K_0)\left(1 - e^{-\overline{V}_d(X)}\right).
\end{aligned}
$$

If we consider here the normalized value $\mathbb{E}\,V_{d-1}(Z \cap K_0)/V_d(K_0)$ (for $V_d(K_0) > 0$), this is no longer independent of K_0 (as it was the case with volume), but is influenced by the shape of the boundary ∂K_0. In order to eliminate these boundary effects, we may replace our sampling window K_0 (with inner points) by $rK_0, r > 0$, and let $r \to \infty$. Then $V_{d-1}(rK_0)/V_d(rK_0) = c/r \to 0$, and we see that $\mathbb{E}\,V_{d-1}(Z \cap rK_0)/V_d(rK_0)$ has a limit which we denote by $\overline{V}_{d-1}(Z)$,

$$\overline{V}_{d-1}(Z) := \lim_{r\to\infty}\frac{\mathbb{E}\,V_{d-1}(Z \cap rK_0)}{V_d(rK_0)},$$

and which satisfies

$$\overline{V}_{d-1}(Z) = \overline{V}_{d-1}(X)e^{-\overline{V}_d(X)}. \tag{12}$$

We call $\overline{V}_{d-1}(Z)$ the **surface area density** of Z.

For the other intrinsic volumes V_j, $0 \le j \le d-2$, the situation is not as simple anymore since for them a translative integral formula looks more complicated and the iterated version is even more technical. We will come back to this problem later.

Now we assume that Z is stationary and isotropic. Since then $\mathbb{Q}$ is rotation invariant, we may replace the translation integrals by integrals over rigid motions and obtain (11) with

$$\Phi(K_0, K_1, \ldots, K_k)$$
$$= \int_{G_d} \cdots \int_{G_d} \varphi(K_0 \cap g_1 K_1 \cap \cdots \cap g_k K_k) \mu(dg_1) \cdots \mu(dg_k).$$

If φ is continuous on $\mathcal{K}$, we can apply Hadwiger's (iterated) kinematic formula (Theorem 1.6 in [34]),

$$\int_{G_d} \cdots \int_{G_d} \varphi(K_0 \cap g_1 K_1 \cap \cdots \cap g_k K_k) \mu(dg_1) \cdots \mu(dg_k)$$
$$= \sum_{\substack{m_0, \ldots, m_k = 0 \\ m_0 + \cdots + m_k = kd}}^{d} c_{d-m_0, d, \ldots, d}^{d, m_1, \ldots, m_k} \varphi_{m_0}(K_0) V_{m_1}(K_1) \cdots V_{m_k}(K_k),$$

with

$$c_{d-m_0, d, \ldots, d}^{d, m_1, \ldots, m_k} = \frac{d! \kappa_d}{(d - m_0)! \kappa_{d-m_0}} \frac{m_1! \kappa_{m_1}}{d! \kappa_d} \cdots \frac{m_k! \kappa_{m_k}}{d! \kappa_d} = c_{d-m_0}^d c_d^{m_1} \cdots c_d^{m_k}$$

and with the Crofton integrals $\varphi_{m_0}(K_0)$, $m_0 = 0, \ldots, d$. We obtain

$$\mathbb{E}\,\varphi(Z \cap K_0)$$
$$= \sum_{k=1}^{\infty} \frac{(-1)^{k+1}}{k!} \sum_{\substack{m_0, \ldots, m_k = 0 \\ m_0 + \cdots + m_k = kd}}^{d} c_{d-m_0, d, \ldots, d}^{d, m_1, \ldots, m_k} \varphi_{m_0}(K_0) \overline{V}_{m_1}(X) \cdots \overline{V}_{m_k}(X)$$
$$= \sum_{k=1}^{\infty} \frac{(-1)^{k+1}}{k!} \sum_{m=0}^{d} c_{d-m}^d \varphi_m(K_0) \sum_{\substack{m_1, \ldots, m_k = 0 \\ m_1 + \cdots + m_k = kd-m}}^{d} \prod_{i=1}^{k} c_d^{m_i} \overline{V}_{m_i}(X)$$
$$= \varphi(K_0)\left(1 - e^{-\overline{V}_d(X)}\right)$$
$$+ \sum_{m=1}^{d} c_{d-m}^d \varphi_m(K_0) \sum_{k=1}^{\infty} \frac{(-1)^{k+1}}{k!} \sum_{\substack{m_1, \ldots, m_k = 0 \\ m_1 + \cdots + m_k = kd-m}}^{d} \prod_{i=1}^{k} c_d^{m_i} \overline{V}_{m_i}(X).$$

We notice that, in the last sum, the number s of the indices m_i which are smaller than d ranges between 1 and m. Therefore, we can re-arrange the last two sums and get

$$\sum_{k=1}^{\infty} \frac{(-1)^{k+1}}{k!} \sum_{\substack{m_1,\dots,m_k=0 \\ m_1+\cdots+m_k=kd-m}}^{d} \prod_{i=1}^{k} c_d^{m_i} \overline{V}_{m_i}(X)$$

$$= \sum_{s=1}^{m} \sum_{r=0}^{\infty} \binom{r+s}{r} \frac{(-1)^{r+s+1}}{(r+s)!} \overline{V}_d(X)^r \sum_{\substack{m_1,\dots,m_s=0 \\ m_1+\cdots+m_s=sd-m}}^{d-1} \prod_{i=1}^{s} c_d^{m_i} \overline{V}_{m_i}(X)$$

$$= -\mathrm{e}^{-\overline{V}_d(X)} \sum_{s=1}^{m} \frac{(-1)^s}{s!} \sum_{\substack{m_1,\dots,m_s=0 \\ m_1+\cdots+m_s=sd-m}}^{d-1} \prod_{i=1}^{s} c_d^{m_i} \overline{V}_{m_i}(X).$$

Altogether we obtain the following result.

Theorem 2.2. *Let Z be a stationary and isotropic Boolean model with convex grains, $K_0 \in \mathcal{K}$, and let $\varphi : \mathcal{K} \to \mathbb{R}$ be additive and continuous. Then,*

$$\mathbb{E}\,\varphi(Z \cap K_0) = \varphi(K_0)\left(1 - \mathrm{e}^{-\overline{V}_d(X)}\right) - \mathrm{e}^{-\overline{V}_d(X)} \sum_{m=1}^{d} c_{d-m}^d \varphi_m(K_0)$$

$$\times \sum_{s=1}^{m} \frac{(-1)^s}{s!} \sum_{\substack{m_1,\dots,m_s=0 \\ m_1+\cdots+m_s=sd-m}}^{d-1} \prod_{i=1}^{s} c_d^{m_i} \overline{V}_{m_i}(X). \tag{13}$$

In general, the expectation $\mathbb{E}\,\varphi(Z \cap K_0)$ will depend on the shape and even the location of the sampling window K_0. To delete the dependence on the location, we may concentrate on translation invariant φ. But still the dependence on the shape of K_0 remains and is expressed explicitly by (13). In order to obtain a mean value of φ for Z which does not depend on a specific sampling window, it is tempting to proceed as in the case of the surface area density $\overline{V}_{d-1}(Z)$ and consider

$$\lim_{r \to \infty} \frac{\mathbb{E}\,\varphi(Z \cap rK_0)}{V_d(rK_0)},$$

provided this limit exists. As we shall show next, for a translation invariant additive functional φ, this is indeed the case. In fact, a corresponding limit result holds more generally for stationary RACS, which need not be isotropic. We will discuss this and similar results for particle processes in the next two subsections. In Section 2.4, we come back to Boolean models and apply the above result to $\varphi = V_j$, $j = 0, \dots, d-1$.

Hints to the literature. The basic formula (10) appears in [36].

2.2 Mean Values for RACS

The following considerations on mean values of additive functionals for stationary RACS are based on a result for valuations which we explain first. We denote by $C^d := [0,1]^d$ the **unit cube** and by $\partial^+ C^d := \{x = (x_1, \dots, x_d) \in C^d : \max_{1 \le i \le d} x_i = 1\}$ the '**upper right boundary**' of C^d. Note that $\partial^+ C^d \in \mathcal{R}$.

Lemma 2.1. *Let $\varphi : \mathcal{R} \to \mathbb{R}$ be additive, translation invariant and condition-ally bounded. Then,*

$$\lim_{r \to \infty} \frac{\varphi(rK)}{V_d(rK)} = \varphi(C^d) - \varphi(\partial^+ C^d),$$

for each $K \in \mathcal{K}$ with $V_d(K) > 0$.

Proof. The additivity can be used to show

$$\varphi(M) = \sum_{z \in \mathbb{Z}^d} \left(\varphi(M \cap zC^d) - \varphi(M \cap z\partial^+ C^d) \right),$$

for all $M \in \mathcal{R}$ (we omit the details of this slightly lengthy derivation). In particular,

$$\varphi(rK) = \sum_{z \in \mathbb{Z}^d} \left(\varphi(rK \cap zC^d) - \varphi(rK \cap z\partial^+ C^d) \right),$$

for $r > 0$ and our given K, were we may assume $0 \in$ int K.

We define two sets of lattice points,

$$Z_r^1 := \{ z \in \mathbb{Z}^d : rK \cap zC^d \neq \emptyset, zC^d \not\subset rK \}$$

and

$$Z_r^2 := \{ z \in \mathbb{Z}^d : zC^d \subset rK \}.$$

Then,

$$\lim_{r \to \infty} \frac{|Z_r^1|}{V_d(rK)} = 0, \qquad \lim_{r \to \infty} \frac{|Z_r^2|}{V_d(rK)} = 1.$$

Consequently,

$$\frac{1}{V_d(rK)} \left| \sum_{z \in Z_r^1} \left(\varphi(rK \cap zC^d) - \varphi(rK \cap z\partial^+ C^d) \right) \right|$$

$$\leq c(C^d) \frac{|Z_r^1|}{V_d(rK)} \to 0 \qquad (r \to \infty)$$

(were the constant $c(C^d)$ arises since φ is conditionally bounded) and therefore

$$\lim_{r \to \infty} \frac{\varphi(rK)}{V_d(rK)} = \lim_{r \to \infty} \frac{1}{V_d(rK)} \sum_{z \in Z_r^2} \left(\varphi(rK \cap zC^d) - \varphi(rK \cap z\partial^+ C^d) \right)$$

$$= (\varphi(C^d) - \varphi(\partial^+ C^d)) \lim_{r \to \infty} \frac{|Z_r^2|}{V_d(rK)}$$

$$= \varphi(C^d) - \varphi(\partial^+ C^d).$$

$\square$

Now we turn to a stationary RACS Z with values in $\mathcal{S}$. In contrast to the case of Boolean models, we need an additional integrability condition here, and we choose

$$\mathbb{E}\, 2^{N(Z \cap C^d)} < \infty, \tag{14}$$

which, although not optimal, is simple enough and works for all valuations φ. Here, for $K \in \mathcal{R}$, $N(K)$ is the minimal number m of convex bodies $K_1, ..., K_m$ with $K = \bigcup_{i=1}^{m} K_i$. Condition (14) guarantees that the realizations of Z do not become too complex in structure.

Theorem 2.3. *Let Z be a stationary random $\mathcal{S}$-set fulfilling (14) and let $\varphi : \mathcal{R} \to \mathbb{R}$ be additive, translation invariant, measurable and conditionally bounded. Then, for every $K \in \mathcal{K}$ with $V_d(K) > 0$, the limit*

$$\overline{\varphi}(Z) := \lim_{r \to \infty} \frac{\mathbb{E}\, \varphi(Z \cap rK)}{V_d(rK)}$$

exists and satisfies

$$\overline{\varphi}(Z) = \mathbb{E}\left(\varphi(Z \cap C^d) - \varphi(Z \cap \partial^+ C^d)\right).$$

Hence, $\overline{\varphi}(Z)$ is independent of K.

Proof. Consider $M \in \mathcal{K}$ with $M \subset C^d$. For each realization $Z(\omega)$ of Z, we use a representation

$$Z(\omega) \cap M = \bigcup_{i=1}^{N_M(\omega)} K_i(\omega)$$

with $K_i(\omega) \in \mathcal{K}$ and $N_M(\omega) := N(Z(\omega) \cap M)$. The inclusion-exclusion formula yields

$$\varphi(Z(\omega) \cap M) = \sum_{k=1}^{N_M(\omega)} (-1)^{k+1} \sum_{1 \leq i_1 < i_2 < \cdots < i_k \leq N_M(\omega)} \varphi\left(K_{i_1}(\omega) \cap \cdots \cap K_{i_k}(\omega)\right).$$

Therefore,

$$\mathbb{E}|\varphi(Z \cap M)| \leq c\, \mathbb{E} \sum_{k=1}^{N_M} \binom{N_M}{k}$$

$$\leq c\, \mathbb{E}\, 2^{N_M} < \infty, \tag{15}$$

by (14). Here, c is an upper bound for $|\varphi|$ on $\{M' \in \mathcal{K} : M' \subset C^d\}$ (which exists, since φ is conditionally bounded).

Hence, $\varphi(Z \cap M)$ is integrable for every $M \in \mathcal{K}$, $M \subset C^d$, but then, by the inclusion-exclusion formula, also for all $M \in \mathcal{R}$. Consequently, all expectations, which appear in the theorem, exist. We use this to define a functional

$$\phi : \mathcal{R} \to \mathbb{R}, \qquad \phi(M) := \mathbb{E}\varphi(Z \cap M).$$

ϕ is additive, translation invariant (here we use the stationarity), and conditionally bounded (this follows from (15)). Lemma 2.1 now yields the asserted result. □

Theorem 2.3 shows that $\varphi(Z \cap C^d) - \varphi(Z \cap \partial^+ C^d)$ is an unbiased estimator for $\overline{\varphi}(Z)$.

The results, obtained so far, hold in particular for the case $\varphi = V_j$, $j \in \{0, ..., d\}$. Here we write $\overline{V}_j(Z)$ for the corresponding density (for $j = d$ we get the volume fraction $p = \overline{V}_d(Z)$ again). The following result gives a formula of kinematic type. We recall from [34] that ν is the invariant probability measure on SO_d and we denote by $\mathbb{E}_\nu$ the expectation with respect to ν. In the proof, we also use the invariant measure μ on the group G_d of rigid motions.

Theorem 2.4. *Let Z be a stationary random $\mathcal{S}$-set fulfilling (14). Let $K \in \mathcal{K}$ and let ϑ be a random rotation with distribution ν and independent of Z. Then*

$$\mathbb{E}_\nu \mathbb{E} V_j(Z \cap \vartheta K) = \sum_{k=j}^{d} c_{j,d}^{k,d+j-k} V_k(K) \overline{V}_{d+j-k}(Z),$$

for $j = 0, ..., d$.

Proof. As in the proof of Theorem 2.3, one shows that

$$(x, \vartheta, \omega) \mapsto V_j(Z(\omega) \cap \vartheta K \cap x B^d)$$

is $\lambda_d \otimes \nu \otimes \mathbb{P}$-integrable. The invariance properties of V_j and Z then show that

$$\mathbb{E}_\nu \mathbb{E} V_j(Z \cap \vartheta K \cap x r B^d) = \mathbb{E}_\nu \mathbb{E} V_j(Z \cap (-x)\vartheta K \cap r B^d),$$

for all $x \in \mathbb{R}^d$. Integration over x and Fubini's theorem yield

$$\mathbb{E}_\nu \mathbb{E} \int_{\mathbb{R}^d} V_j(Z \cap \vartheta K \cap x r B^d) \, \lambda_d(dx) = \mathbb{E} \int_{G_d} V_j(Z \cap g K \cap r B^d) \mu(dg).$$

Since B^d is rotation invariant, we can replace the integral over $\mathbb{R}^d$ on the left-hand side also by an integration over G_d. The principal kinematic formula thus gives

$$\sum_{k=j}^{d} c_{j,d}^{k,d+j-k} \mathbb{E}_\nu \mathbb{E} V_k(Z \cap \vartheta K) V_{d+j-k}(r B^d)$$

$$= \sum_{k=j}^{d} c_{j,d}^{k,d+j-k} V_k(K) \mathbb{E} V_{d+j-k}(Z \cap r B^d).$$

We divide both sides by $V_d(r B^d)$ and let $r \to \infty$. The left-hand side then converges to

$$\mathbb{E}_\nu \, \mathbb{E} \, V_j(Z \cap \vartheta K),$$

and the right-hand side converges to

$$\sum_{k=j}^{d} c_{j,d}^{k,d+j-k} V_k(K) \overline{V}_{d+j-k}(Z).$$

$\square$

The expectation $\mathbb{E}_\nu$ can be omitted if $j = d$ or $j = d - 1$ or $K = B^d$ or if Z is isotropic.

Theorem 2.4 is useful for two reasons. First, it describes the bias, if the value $V_j(Z \cap \vartheta K)$, for a randomly rotated sampling window K, is used as an estimator for $\overline{V}_j(Z)$. Second, it provides us with a further unbiased estimator, if we solve the corresponding (triangular) system of linear equations with unknowns $\overline{V}_j(Z), j = 0, ..., n$. For example, if $K = B^d$, we get

$$\overline{V}_j(Z) = \sum_{i=j}^{d} a_{ij} \mathbb{E} \, V_i(Z \cap B^d), \qquad j = 0, ..., d,$$

with constants a_{ij}, which can be given explicitly. Hence,

$$\sum_{i=j}^{d} a_{ij} V_i(Z \cap B^d)$$

is an unbiased estimator for $\overline{V}_j(Z)$.

Hints to the literature. Also for this section, [36] is the main reference. The slightly more general version of Theorem 2.4 was taken from [45]. Joint estimators of the quermass densities, based on the above-mentioned linear equations, are studied in [32].

2.3 Mean Values for Particle Processes

For a stationary particle process X, Theorem 1.3 immediately allows us to define a mean value $\overline{\varphi}(X)$, for any translation invariant, measurable function $\varphi : \mathcal{C}' \to \mathbb{R}$ (which is either nonnegative or $\mathbb{Q}$-integrable), namely by

$$\overline{\varphi}(X) := \gamma \int_{\mathcal{C}_0} \varphi(K) \mathbb{Q}(\mathrm{d}K).$$

The following alternative representations of $\overline{\varphi}(X)$ follow with standard techniques (Campbell's theorem, majorized convergence).

Theorem 2.5. *Let X be a stationary particle process and $\varphi : \mathcal{C}' \to \mathbb{R}$ translation invariant, measurable and $\mathbb{Q}$-integrable (or nonnegative). Then,*
(a) *for all Borel sets $A \subset \mathbb{R}^d$ with $0 < \lambda_d(A) < \infty$,*

$$\overline{\varphi}(X) = \frac{1}{\lambda_d(A)} \mathbb{E} \sum_{K \in X, c(K) \in A} \varphi(K),$$

(b) *for all $K_0 \in \mathcal{K}$ with $V_d(K_0) > 0$,*

$$\overline{\varphi}(X) = \lim_{r \to \infty} \frac{1}{V_d(rK_0)} \mathbb{E} \sum_{K \in X, K \subset rK_0} \varphi(K),$$

(c) *for all $K_0 \in \mathcal{K}$ with $V_d(K_0) > 0$,*

$$\overline{\varphi}(X) = \lim_{r \to \infty} \frac{1}{V_d(rK_0)} \mathbb{E} \sum_{K \in X, K \cap rK_0 \neq \emptyset} \varphi(K)$$

(if we assume, in addition that $\int_{\mathcal{C}_0} |\varphi(K)| V_d(K + B^d) \mathbb{Q}(dK) < \infty$).

Notice that this theorem yields some natural, unbiased or asymptotically unbiased estimators of $\overline{\varphi}(X)$, which have however some disadvantages, since they either require knowledge of particles outside the window A (resp. rK_0) or they do not use the full information observable in rK_0 (this is the case in (b)). We can find better estimators for additive and conditionally bounded functionals φ on $\mathcal{R}$ and processes X with particles in $\mathcal{R}'$, satisfying a certain integrability condition. We assume

$$\int_{\mathcal{R}_0} 2^{N(K)} V_d(K + B^d) \mathbb{Q}(dK) < \infty. \tag{16}$$

The following result is proved similarly to Lemma 2.1 and Theorem 2.3.

Theorem 2.6. *Let X be a stationary process of particles in $\mathcal{R}'$ satisfying (16). Let $\varphi : \mathcal{R}' \to \mathbb{R}$ be translation invariant, additive, measurable and conditionally bounded. Then φ is $\mathbb{Q}$-integrable and*

$$\overline{\varphi}(X) = \lim_{r \to \infty} \frac{1}{V_d(rK_0)} \mathbb{E} \sum_{K \in X} \varphi(K \cap rK_0),$$

for all $K_0 \in \mathcal{K}$ with $V_d(K_0) > 0$. Moreover,

$$\overline{\varphi}(X) = \mathbb{E} \sum_{K \in X} \left(\varphi(K \cap C^d) - \varphi(K \cap \partial^+ C^d) \right).$$

The choice $\varphi = V_j$ provides us therefore with two further representations (and estimators) of the quermass densities $\overline{V}_j(X)$.

We also get an analogue of Theorem 2.4. The proof is even simpler here and only requires Campbell's theorem and the principal kinematic formula. If we consider processes X with convex particles, we can even skip condition (16).

Theorem 2.7. *Let X be a stationary process of particles in $\mathcal{K}'$ and $K \in \mathcal{K}$. Let ϑ be a random rotation with distribution ν and independent of Z. Then*

$$\mathbb{E}_\nu \, \mathbb{E} \sum_{M \in X} V_j(M \cap \vartheta K) = \sum_{k=j}^{d} c_{j,d}^{k,d+j-k} V_k(K) \overline{V}_{d+j-k}(X),$$

for $j = 0, ..., d$.

Again, the expectation $\mathbb{E}_\nu$ can be omitted in any of the cases $j = d$, $j = d - 1$, $K = B^d$ or if X is isotropic.

As in the case of RACS, these results produce various (unbiased) estimators for $\overline{V}_j(X)$.

Hints to the literature. Again, [36] is the main reference and Theorem 2.7 was taken from [45].

2.4 Quermass Densities of Boolean Models

We return now to Boolean models Z and assume first that Z is stationary and isotropic. We apply Theorem 2.2 to $\varphi = V_j$. Since $(V_j)_m = 0$, for $m > d - j$, and $(V_j)_m = c_{j,d}^{d-m,m+j} V_{m+j}$, for $m = 0, ..., d - j$, by the Crofton formula (see [34]), we obtain the following result.

Theorem 2.8. *Let Z be a stationary and isotropic Boolean model with convex grains. Then,*

$$\overline{V}_d(Z) = 1 - e^{-\overline{V}_d(X)}$$

and

$$\overline{V}_j(Z)$$

$$= e^{-\overline{V}_d(X)} \left(\overline{V}_j(X) - \sum_{s=2}^{d-j} \frac{(-1)^s}{s!} \, c_j^d \sum_{\substack{m_1,...,m_s=j+1 \\ m_1+\cdots+m_s=(s-1)d+j}}^{d-1} \prod_{i=1}^{s} c_d^{m_i} \overline{V}_{m_i}(X) \right),$$

for $j = 0, \ldots, d - 1$.

Although these formulas still look very technical, they are quite useful for practical applications. Of course, these applications mostly appear in the planar or spatial situation. Therefore, we discuss these cases shortly. We use A and U, for the area and the boundary length in the plane, V, S and M, for the volume, the surface area and the (additively extended) integral mean curvature in three-dimensional space, and χ for the Euler characteristic (for convex bodies $K \subset \mathbb{R}^3$, the integral mean curvature $M(K)$ is 2π times the mean width of K).

Corollary 2.1. *For a stationary and isotropic Boolean model Z in $\mathbb{R}^2$ with convex grains, we have*

$$\overline{A}(Z) = 1 - e^{-\overline{A}(X)},$$

$$\overline{U}(Z) = e^{-\overline{A}(X)}\overline{U}(X),$$

$$\overline{\chi}(Z) = e^{-\overline{A}(X)}\left(\gamma - \frac{1}{4\pi}\overline{U}(X)^2\right).$$

Corollary 2.2. *For a stationary and isotropic Boolean model Z in $\mathbb{R}^3$ with convex grains, we have*

$$\overline{V}(Z) = 1 - e^{-\overline{V}(X)},$$

$$\overline{S}(Z) = e^{-\overline{V}(X)}\overline{S}(X),$$

$$\overline{M}(Z) = e^{-\overline{V}(X)}\left(\overline{M}(X) - \frac{\pi^2}{32}\overline{S}(X)^2\right),$$

$$\overline{\chi}(Z) = e^{-\overline{V}(X)}\left(\gamma - \frac{1}{4\pi}\overline{M}(X)\overline{S}(X) + \frac{\pi}{384}\overline{S}(X)^3\right).$$

Since we now have several possibilities to estimate the densities of Z on the left-hand side, these equations allow the estimation of the particle means and therefore of the intensity γ. An important aspect is that the formulas hold for Boolean models Z with grains in $\mathcal{R}$ as well (under an additional integrability assumption). If the grains K obey $\chi(K) = 1$, $\mathbb{Q}$-almost surely (in the plane this follows, for example, if the grains are all simply connected), then the results hold true without any change. Otherwise, γ has to be replaced by $\overline{\chi}(X)$ (and then we do not get an estimation of the intensity itself).

What changes if we skip the isotropy assumption? For a stationary Boolean model Z (again we assume convex grains, for simplicity) and $\varphi = V_j$, we can use (11) as a starting point and apply the iterated translative formula (Theorem 3.1 in [34]). We obtain

$$\mathbb{E}\,V_j(Z \cap K_0)$$

$$= \sum_{k=1}^{\infty} \frac{(-1)^{k+1}}{k!}\gamma^k \sum_{\substack{m_0,\dots,m_k=j \\ m_0+\cdots+m_k=kd+j}}^{d} \int_{K_0}\cdots\int_{K_0} V_{m_0,\dots,m_k}^{(j)}(K_0,\dots,K_k)$$

$$\times \mathbb{Q}(\mathrm{d}K_1)\cdots\mathbb{Q}(\mathrm{d}K_k).$$

Again, we replace K_0 by rK_0, normalize by $V_d(rK_0)$ and let $r \to \infty$. Then, due to the homogeneity properties of mixed functionals (see [34]), all summands on the right-hand side with $m_0 < d$ disappear asymptotically. For $m_0 = d$, we can use the decomposition property of mixed functionals and get, with essentially the same arguments as in the isotropic case,

$$\overline{V}_j(Z) = \lim_{r \to \infty} \frac{\mathbb{E}\, V_j(Z \cap rK_0)}{V_d(rK_0)}$$

$$= \sum_{k=1}^{\infty} \frac{(-1)^{k+1}}{k!} \gamma^k \sum_{\substack{m_1,\dots,m_k=j \\ m_1+\cdots+m_k=(k-1)d+j}}^{d} \int_{\mathcal{K}_0} \cdots \int_{\mathcal{K}_0} V_{m_1,\dots,m_k}^{(j)}(K_1,\dots,K_k)$$

$$\times\, \mathbb{Q}(dK_1) \cdots \mathbb{Q}(dK_k)$$

$$= \sum_{s=1}^{d-j} \sum_{r=0}^{\infty} \binom{r+s}{r} \frac{(-1)^{r+s+1}}{(r+s)!} \overline{V}_d(X)^r \gamma^s$$

$$\times \sum_{\substack{m_1,\dots,m_s=j \\ m_1+\cdots+m_s=(s-1)d+j}}^{d-1} \int_{\mathcal{K}_0} \cdots \int_{\mathcal{K}_0} V_{m_1,\dots,m_s}^{(j)}(K_1,\dots,K_s)\mathbb{Q}(dK_1)\cdots\mathbb{Q}(dK_s)$$

$$= -\mathrm{e}^{-\overline{V}_d(X)} \sum_{s=1}^{d-j} \frac{(-1)^s}{s!} \sum_{\substack{m_1,\dots,m_s=j \\ m_1+\cdots+m_s=(s-1)d+j}}^{d-1} \overline{V}_{m_1,\dots,m_s}^{(j)}(X,\dots,X)$$

$$= \mathrm{e}^{-\overline{V}_d(X)} \left(\overline{V}_j(X) - \sum_{s=2}^{d-j} \frac{(-1)^s}{s!} \sum_{\substack{m_1,\dots,m_s=j+1 \\ m_1+\cdots+m_s=(s-1)d+j}}^{d-1} \overline{V}_{m_1,\dots,m_s}^{(j)}(X,\dots,X) \right).$$

Here the mixed densities of X are defined as

$$\overline{V}_{m_1,\dots,m_s}^{(j)}(X,\dots,X) := \gamma^s \int_{\mathcal{K}_0} \cdots \int_{\mathcal{K}_0} V_{m_1,\dots,m_s}^{(j)}(K_1,\dots,K_s)\mathbb{Q}(dK_1)\cdots\mathbb{Q}(dK_s).$$

Hence, we arrive at the following result.

Theorem 2.9. *Let Z be a stationary Boolean model with convex grains. Then,*

$$\overline{V}_d(Z) = 1 - \mathrm{e}^{-\overline{V}_d(X)},$$

$$\overline{V}_{d-1}(Z) = \mathrm{e}^{-\overline{V}_d(X)}\overline{V}_{d-1}(X),$$

and

$$\overline{V}_j(Z)$$

$$= \mathrm{e}^{-\overline{V}_d(X)} \left(\overline{V}_j(X) - \sum_{s=2}^{d-j} \frac{(-1)^s}{s!} \sum_{\substack{m_1,\dots,m_s=j+1 \\ m_1+\cdots+m_s=(s-1)d+j}}^{d-1} \overline{V}_{m_1,\dots,m_s}^{(j)}(X,\dots,X) \right),$$

for $j = 0,\dots,d-2$.

For $d = 2$, only the formula for the Euler characteristic changes and we have

$$\overline{A}(Z) = 1 - \mathrm{e}^{-\overline{A}(X)},$$

$$\overline{U}(Z) = \mathrm{e}^{-\overline{A}(X)}\overline{U}(X),$$

$$\overline{\chi}(Z) = \mathrm{e}^{-\overline{A}(X)} \left(\gamma - \overline{A}(X,X^*) \right),$$

where

$$A(X, X^*) := \gamma^2 \int_{\mathcal{K}_0} \int_{\mathcal{K}_0} A(K, M^*) \mathbb{Q}(dK) \mathbb{Q}(dM).$$

Here, we made use of the fact that the mixed functional $V_{1,1}^{(0)}(K, M)$ in the plane equals the mixed area $A(K, M^*)$ of K and the reflection M^* of M. It is obvious that the formulas can no longer be used directly for the estimation of γ. Hence, we need more (local) information for the statistical analysis of nonisotropic Boolean models and this will be discussed in Section 3.

We emphasize again, that the formulas for the quermass densities hold also for grains in $\mathcal{R}$ having Euler characteristic one (in the plane, this is fulfilled if the grains are simply connected). For general grains in $\mathcal{R}$, in the formula for $\overline{\chi}(Z)$, γ has to be replaced by $\overline{\chi}(X)$.

For convex grains, there are further methods to estimate γ. One possibility is to use the capacity functional, as we have mentioned already. A further one is to use associated points (the so-called tangent count). We describe the basis of this method shortly. We associate a boundary point $z(K)$ with each particle $K \in X$, in a suitable way, and then count the number $\chi^+(Z \cap K_0)$ of associated points which are visible in $Z \cap K_0$ (these are the associated points which lie in K_0 and are not covered by other particles). The associated points $z(K), K \in X$, build a stationary Poisson process $\tilde{X}_0$ with the same intensity as the process X_0 of center points, namely γ. The associated points which are not covered by other particles, build a stationary point process X_0' which is obtained from $\tilde{X}_0$ by a thinning and has intensity $\gamma e^{-\overline{V}_d(X)}$. This follows from Slivnyak's theorem. The probability that a given associated point $x \in \tilde{X}_0$ is not covered by other particles equals the probability that an arbitrary point $x \in \mathbb{R}^d$ is not covered by any particle, hence that $x \notin Z$. Since $\mathbb{P}(x \notin Z) = 1 - p$, the thinning probability is $e^{-\overline{V}_d(X)}$. Therefore,

$$\mathbb{E}\chi^+(Z \cap K_0) = V_d(K_0)\gamma e^{-\overline{V}_d(X)}. \tag{17}$$

In the plane, one can choose the "lower left tangent point" as associated point (the left most boundary point in the lower part of a particle with a horizontal tangent) and then has to count these points in $Z \cap K_0$. Together with the formula for the area density, this yields a simple estimator for γ.

Hints to the literature. The formulas for the quermass densities of stationary and isotropic Boolean models have a long history, beginning with results by Miles and Davy and including contributions by Kellerer, Weil and Wieacker, and Zähle. More details and references can be found in [36]. For $d = 3$, some of the formulas given in the literature contain incorrect constants. In [46], an attempt was made to present the correct formula for $\overline{\chi}(Z)$, however the printer of this paper mixed up π and π^2 such that, in the end, both formulas given there for $\overline{\chi}(Z)$ and $\overline{M}(Z)$ ended up wrongly. Theorem 2.9 was proved in [40]. The formula (17) on the 'tangent count' is classical (see [37]). Molchanov and Stoyan [29] (see also [27]) have shown that, in the case of convex grains with interior points, the point process of visible tangent points (together with the covariance C of Z) determines the grain distribution $\mathbb{Q}$ (and γ) uniquely.

2.5 Ergodicity

We do not want to go deeper into ergodic theory here. Thus, we only mention that stationary Poisson particle processes X and stationary Boolean models Z fulfill a mixing condition and are therefore **ergodic** (w.r.t. the group of lattice translations).

For a stationary RACS Z, the mixing condition says

$$\lim_{\|x\|\to\infty} (1 - T_Z(C_1 \cup xC_2)) = (1 - T_Z(C_1))(1 - T_Z(C_2)),$$

for all $C_1, C_2 \in \mathcal{C}$, and it can be easily checked that this holds for a Boolean model Z. For a stationary Poisson process X of convex particles, the mixing condition is formulated and proved for the locally finite RACS X in the space $\mathcal{F}'$.

If Z is a stationary Boolean model with convex grains, X the corresponding Poisson particle process and $\varphi : \mathcal{R} \to \mathbb{R}$ additive, translation invariant, measurable and conditionally bounded, then a general ergodic theorem by Tempel'man implies that

$$\overline{\varphi}(Z) = \lim_{r\to\infty} \frac{\varphi(Z(\omega) \cap rK)}{V_n(rK)}$$

and

$$\overline{\varphi}(X) = \lim_{r\to\infty} \frac{1}{V_n(rK)} \sum_{M \in X(\omega)} \varphi(M \cap rK)$$

hold for $K \in \mathcal{K}$ with $V_d(K) > 0$ and $\mathbb{P}$-almost all $\omega \in \Omega$ (without the expectation!). In particular, for $\varphi = V_j$, the right-hand sides provide consistent estimators for the quermass densities $\overline{V}_j(Z)$ and $\overline{V}_j(X)$.

Hints to the literature. The above individual ergodic theorems are discussed in more detail in [36]. More general results for spatial processes, including statistical ergodic theorems, go back to Nguyen and Zessin [30]. See also Heinrich [10], for mixing properties of germ-grain models.

3 Directional Data, Local Densities, Nonstationary Boolean Models

In this section, we extend the results on mean values of RACS, particle processes and Boolean models in various directions. First, we consider functionals which reflect the directional behavior of a random structure. Then, we give local interpretations of the quermass densities and other mean values by using curvature measures and their generalizations. Finally, we extend the basic formulas for Boolean models to the nonstationary case.

3.1 Directional Data and Associated Bodies

Motion invariant functionals like the intrinsic volumes may give some information on the shape of a convex body K (e.g. if we consider the isoperimetric quotient of K) but they do not give any information about the orientation. It is therefore reasonable that, for stationary nonisotropic Boolean models Z, the densities $\overline{V}_j(Z), j = 0, ..., d$, are not sufficient to estimate the intensity γ. Therefore, we consider additive functionals φ now, which better reflect the orientation. Two different but related approaches exist here.

First, we observe that the intrinsic volumes are mixed volumes (with the unit ball B^d),

$$V_j(K) = \binom{d}{j} \frac{1}{\kappa_{d-j}} V(K\,[j], B^d\,[d-j]), \qquad j = 0, ..., d.$$

Therefore, an obvious generalization is to consider functionals

$$V(K\,[j], M\,[d-j]),$$

for $j \in \{0, ..., d\}$ and $M \in \mathcal{K}$. For example, if M is a segment s, the mixed volume $V(K\,[d-1], s\,[1])$ is (proportional to) the $(d-1)$-volume of the projection of K orthogonal to s.

We note that we can also work with mixed translative functionals here since

$$\binom{d}{j} V(K\,[j], M\,[d-j]) = V^{(0)}_{j,d-j}(K, M^*).$$

For fixed M, the functional $K \mapsto V^{(0)}_{j,d-j}(K, M)$ is additive, translation invariant and continuous on $\mathcal{K}$ and therefore has a unique additive extension to $\mathcal{R}$ which is measurable and conditionally bounded. By Theorem 2.3, the density

$$\overline{V}^{(0)}_{j,d-j}(Z, M) := \lim_{r \to \infty} \frac{\mathbb{E}V^{(0)}_{j,d-j}(Z \cap rK_0, M)}{V_d(rK_0)}$$

exists. Also, densities of general mixed functionals exist for X, either obtained in a similar way, as limits for increasing sampling windows rK_0, $r \to \infty$, or simply as integrals with respect to $\mathbb{Q}$,

$$V^{(0)}_{m_1,...,m_s,d-j}(X, ..., X, M)$$
$$:= \gamma^s \int_{K_0} \cdots \int_{K_0} V^{(0)}_{m_1,...,m_s,d-j}(K_1, ..., K_s, M)\mathbb{Q}(dK_1) \cdots \mathbb{Q}(dK_k),$$

for $s = 1, ..., d-1$, and $m_i \in \{j, ..., d\}$ with $m_1 + \cdots + m_s = (s-1)d + j$. Since $K \mapsto V^{(0)}_{j,d-j}(K, M)$ satisfies an iterated translative integral formula, similarly to the one for $K \mapsto V_j(K)$, we obtain the following result, similarly to the proof of Theorem 2.9.

Theorem 3.1. *Let Z be a stationary Boolean model with convex grains. Then,*

$$\overline{V}^{(0)}_{j,d-j}(Z,M) = e^{-\overline{V}_d(X)} \left(\overline{V}^{(0)}_{j,d-j}(X,M) \right.$$

$$\left. - \sum_{s=2}^{d-j} \frac{(-1)^s}{s!} \sum_{\substack{m_1,\ldots,m_s=j+1 \\ m_1+\cdots+m_s=(s-1)d+j}}^{d-1} \overline{V}^{(0)}_{m_1,\ldots,m_s,d-j}(X,\ldots,X,M) \right),$$

for $j = 0,\ldots,d-1$ and $M \in \mathcal{K}$.

For $j = 0$, the formula above coincides with the one given in Theorem 2.9.

In dimension $d = 2$, the resulting formulas read (after simple modifications)

$$\overline{A}(Z) = 1 - e^{-\overline{A}(X)},$$

$$\overline{A}(Z,M) = e^{-\overline{A}(X)}\overline{A}(X,M), \qquad M \in \mathcal{K},$$

$$\overline{\chi}(Z) = e^{-\overline{A}(X)} \left(\gamma - \overline{A}(X,X^*) \right). \tag{18}$$

For $M = B^d$, the second formula reduces to the equation for $\overline{U}(Z)$ given earlier.

Before we discuss (18) further, we mention a second approach to directional data. This is based on measure- or function-valued functionals which describe convex bodies K uniquely (up to translation). One such functional is the **area measure** $S_{d-1}(K,\cdot)$, a Borel measure on the unit sphere S^{d-1} which described the surface area of K in boundary points with prescribed outer normals (for a more detailed description of area measures, curvature measures and support measures, we refer to Section 2 of [34]). The reason that we consider only the $(d-1)$st area measure and not the other lower order ones lies in the fact that $S_{d-1}(K,\cdot)$ satisfies a (simple) translative formula (which is not the case for the other area measures),

$$\int_{\mathbb{R}^d} S_{d-1}(K \cap xM,\cdot)\,\lambda_d(dx) = V_d(M)S_{d-1}(K,\cdot) + V_d(K,\cdot)S_{d-1}(M,\cdot). \tag{19}$$

The iteration of (19) is obvious.

For each Borel set $B \subset S^{d-1}$, the real functional $K \mapsto S_{d-1}(K,B)$ satisfies the conditions of Theorem 2.3, it is additive, translation invariant, measurable and conditionally bounded (in general, for given B, it is not continuous on $\mathcal{K}$, but the measure-valued map $K \mapsto S_{d-1}(K,\cdot)$ is continuous in the weak topology). Thus, the densities $\overline{S}_{d-1}(Z,B)$ and $\overline{S}_{d-1}(X,B)$ exist and define finite Borel measures $\overline{S}_{d-1}(Z,\cdot)$ and $\overline{S}_{d-1}(X,\cdot)$ on S^{d-1}. For $\overline{S}_{d-1}(Z,\cdot)$ the nonnegativity may not be obvious, in fact one has to show first that the additive extension of the area measure to sets $K \in \mathcal{R}$ is nonnegative, since it is the image of the Hausdorff measure on ∂K under the spherical image map (for the latter, see [33, p. 78]). The following generalization of (12) is now easy to obtain,

$$\overline{S}_{d-1}(Z, \cdot) = e^{-\overline{V}_d(X)} \overline{S}_{d-1}(X, \cdot). \tag{20}$$

In fact, (20) is equivalent to the case $j = d - 1$ of Theorem 3.1, which reads

$$\overline{V}^{(0)}_{d-1,1}(Z, M) = e^{-\overline{V}_d(X)} \overline{V}^{(0)}_{d-1,1}(X, M),$$

for $M \in \mathcal{K}$. The connection follows from the equation

$$V^{(0)}_{d-1,1}(K, M) = \int_{S^{d-1}} h(M, -u) S_{d-1}(K, du),$$

which is a classical formula for mixed volumes of convex bodies K, M. Here, $h(M, \cdot)$ is the support function of M. Since area measures have centroid 0, we may replace $h(M, \cdot)$ by the **centred support function** $h^*(M, \cdot)$ (see [34], for details) and obtain

$$V^{(0)}_{d-1,1}(K, M) = \int_{S^{d-1}} h^*(M, -u) S_{d-1}(K, du). \tag{21}$$

A classical result from convex analysis tells us that the differences of centred support functions are dense in the Banach space of continuous functions f on S^{d-1}, which are centred in the sense that

$$\int_{S^{d-1}} u f(u) \, \sigma_{d-1}(du) = 0.$$

Therefore, the collection

$$\{V^{(0)}_{d-1,1}(K, M) : M \in \mathcal{K}\}$$

uniquely determines the measure $S_{d-1}(K, \cdot)$ and vice versa.

The (centred) support function $h^*(K, \cdot)$ is another directional functional of K, it determines K up to a translation and fulfills the assumptions of Theorem 2.3, namely it is translation invariant, additive, measurable and conditionally bounded (it is even continuous on $\mathcal{K}$). We therefore also have a density $\overline{h}(Z, \cdot)$ (we suppress the $*$ here). Since $h^*(K, \cdot)$ also satisfies an iterated translative formula (see Theorem 3.2 in [34]), we obtain a further formula for Z, either by copying the proof of the previous results (Theorem 2.9 or Theorem 3.1) or by using (21) in Theorem 3.1.

Theorem 3.2. *Let Z be a stationary Boolean model with convex grains. Then,*

$$\overline{h}(Z, \cdot) = e^{-\overline{V}_d(X)} \left(\overline{h}(X, \cdot) \right.$$

$$\left. - \sum_{s=2}^{d-1} \frac{(-1)^s}{s!} \sum_{\substack{m_1,\ldots,m_s=2 \\ m_1+\cdots+m_s=(s-1)d+1}}^{d-1} \overline{h}_{m_1,\ldots,m_s}(X, \ldots, X, \cdot) \right). \tag{22}$$

Here,

$$\overline{h}_{m_1,\dots,m_s}(X,\dots,X,\cdot)$$
$$:= \gamma^s \int_{\mathcal{K}_0} \cdots \int_{\mathcal{K}_0} h^*_{m_1,\dots,m_s}(K_1,\dots,K_s,\cdot)\mathbb{Q}(\mathrm{d}K_1)\cdots\mathbb{Q}(\mathrm{d}K_k), \quad (23)$$

where $h^*_{m_1,\dots,m_s}(K_1,\dots,K_s,\cdot)$ is the mixed (centred) support function occurring in [34, Theorem 3.2].

For $d = 2$, the formula reduces to

$$\overline{h}(Z,\cdot) = \mathrm{e}^{-\overline{A}(X)}\overline{h}(X,\cdot), \quad (24)$$

which does not give us further information since it is essentially equivalent to (20). In fact, both $S_{d-1}(K,\cdot)$ and $h^*(K;\cdot)$ determine K uniquely, up to translation, and therefore they also determine each other. In the plane, this connection can be made even more precise since

$$S_1(K,\cdot) = h^*(K,\cdot) + (h^*(K,\cdot))'', \quad (25)$$

in the sense of Schwartz distributions. If we look at (18) again, we can write it now as

$$\overline{A}(Z) = 1 - \mathrm{e}^{-\overline{A}(X)},$$
$$\overline{h}(Z,\cdot) = \mathrm{e}^{-\overline{A}(X)}\overline{h}(X,\cdot),$$
$$\overline{\chi}(Z) = \mathrm{e}^{-\overline{A}(X)}\left(\gamma - \overline{A}(X,X^*)\right).$$

These three formulas actually suffice to obtain an estimator for γ, if the left-hand densities are estimated. Namely, the first formula determines $\mathrm{e}^{-\overline{A}(X)}$, so the second gives us $\overline{h}(X,\cdot)$. From (25), we deduce

$$\overline{S}_1(X,\cdot) = \overline{h}(X,\cdot) + (\overline{h}(X,\cdot))''$$

and get $\overline{S}_1(X,\cdot)$. Equation (21) transfers to X as

$$\overline{V}^{(0)}_{d-1,1}(X,X) = \int_{S^{d-1}} \overline{h}(X,-u)\overline{S}_{d-1}(X,\mathrm{d}u),$$

hence

$$\overline{A}(X,X^*) = \frac{1}{2}\int_{S^1} \overline{h}(X,-u)\overline{S}_1(X,\mathrm{d}u)$$

is determined and so we get γ. Again, we emphasize that these results hold true for Boolean models Z with grains in $\mathcal{R}$ (under suitable additional integrability conditions).

Let us shortly discuss the case $d = 3$. From Theorem 3.1, we get four formulas for the mean values of mixed volumes. The first one is

$$\overline{V}(Z) = 1 - e^{-\overline{V}(X)},$$

which determines $e^{-\overline{V}(X)}$. The second one can be replaced by its measure version (20)

$$\overline{S}_2(Z, \cdot) = e^{-\overline{V}(X)} \overline{S}_2(X, \cdot),$$

and this gives us $\overline{S}_2(X, \cdot)$. By (21) and another denseness argument, the third equation for $\overline{V}_{1,2}^{(0)}(Z, M), M \in \mathcal{K}$, is equivalent to (and therefore can be replaced by)

$$\overline{h}(Z, \cdot) = e^{-\overline{V}(X)} \left(\overline{h}(X, \cdot) - \frac{1}{2} \overline{h}_{2,2}(X, X, \cdot) \right).$$

Using explicit formulas for $h_{2,2}(P, Q, \cdot)$, in the case of polytopes P, Q (see Theorem 3.2 in [34]), we obtain from (23) (and approximation of convex bodies by polytopes),

$$\overline{h}_{2,2}(X, X, \cdot) = \int_{S^2} \int_{S^2} f(u, v, \cdot) \overline{S}_2(X, du) \overline{S}_2(X, dv)$$

with a given geometric function f on $(S^2)^3$. Thus, $\overline{h}_{2,2}(X, X, \cdot)$ is determined by $\overline{S}_2(X, \cdot)$ and so the third equation gives us $\overline{h}(X, \cdot)$. The fourth equation reads

$$\overline{\chi}(Z) = e^{-\overline{V}_d(X)} \left(\gamma - \overline{V}_{1,2}^{(0)}(X, X) + \frac{1}{6} \overline{V}_{2,2,2}^{(0)}(X, X, X) \right).$$

We already know that

$$\overline{V}_{1,2}^{(0)}(X, X) = \int_{S^2} \overline{h}(X, -u) \overline{S}_2(X, du)$$

and similarly we get

$$\overline{V}_{2,2,2}^{(0)}(X, X, X) = \int_{S^2} \overline{h}_{2,2}(X, X, -u) \overline{S}_2(X, du).$$

Hence, both densities are determined by the quantities, which we already have. Therefore, we finally obtain γ.

For higher dimensions, the corresponding formulas become more and more complex and the question is open, whether the densities of mixed volumes of Z determine the intensity γ.

The above considerations show that, for $d = 2$ and $d = 3$, the densities for mixed volumes of Z, respectively their measure- and function-valued counterparts determine γ uniquely. Of course, the mean value formulas can be used to construct corresponding estimators for mean values of X and thus for γ. However, the question still remains whether these lead to applicable procedures in practice. For $d = 2$, this is the case as we shall show in a moment. First, we mention a third method to describe the directional behavior of random sets and particle processes, the method of associated bodies.

It has been a quite successful idea to associate convex bodies with the directional data of a random structure. In particular, this often allows to apply classical geometric inequalities (like the isoperimetric inequality) to these associated bodies and use this to formulate and solve extremal problems for the random structures. Such associated convex bodies even exist for random structures like processes of curves (fibre processes) or line processes. We only discuss here the case of Boolean models but hope that the general principle will be apparent.

The principle of associated bodies is based on the fact that certain mean measures of random structures are area measures of a convex body, uniquely determined up to translation, and similarly certain mean functions of random structures are (centred) support functions of a convex body, again determined up to translation. More precisely, any finite Borel measure ρ on S^{d-1} which has centroid 0 (and is not supported by a great sphere) is the area measure of a unique convex body K, $\rho = S_{d-1}(K, \cdot)$. This is Minkowski's existence theorem. Also, a continuous function h on S^{d-1} is the support function of a convex body K, $h = h(K, \cdot)$, if the positive homogeneous extension of h is convex on $\mathbb{R}^d$. The first principle can be used in nearly all situations in Stochastic Geometry, where measures on S^{d-1} occur. The second principle is only helpful in certain cases since the required convexity often fails. Using the first principle, we can define a convex body $B(Z)$ (the **Blaschke body** of Z) by

$$S_{d-1}(B(Z), \cdot) = \overline{S}_{d-1}(Z, \cdot)$$

and in the same way a Blaschke body $B(X)$ of X. Then, (20) becomes a formula between convex bodies,

$$B(Z) = e^{-\overline{V}_d(X)} B(X).$$

For a Boolean model with convex grains, the functions $\overline{h}_{m_1,\dots,m_s}(X, \dots, X, \cdot)$ in Theorem 3.2 are in fact support functions of a mixed convex body $M_{m_1,\dots,m_s}(X, \dots, X)$, but due to the alternating sign in (22) the function $\overline{h}(Z, \cdot)$ on the left-hand side is in general not a support function.

An exception is the case $d = 2$, where the convexity follows from (24). Here, $\overline{h}(Z, \cdot) = h(B(Z), \cdot)$ (and $\overline{h}(X, \cdot) = h(B(X), \cdot)$). The main statistical problem in the analysis of stationary, nonisotropic Boolean models Z in $\mathbb{R}^2$ thus consists in the estimation of the Blaschke body $B(Z)$. The latter and the volume fraction p immediately yield $B(X)$. Since $\overline{A}(X, X^*) = A(B(X), B(X)^*)$, simple formulas for the mixed area can then be used together with an empirical value for $\overline{\chi}(Z)$ to obtain γ. Several approaches for the estimation of $B(Z)$ (respectively $\overline{S}_1(Z, \cdot) = S_1(B(Z), \cdot)$) have been described in the literature. One uses the idea of convexification of a nonconvex set, another one is based on contact distributions (these will be discussed in Section 4).

Betke and Weil [3] proved the inequality

$$A(K, K^*) \leq \frac{\sqrt{3}}{18} U^2(K),$$

for planar convex bodies K, with equality if K is an equilateral triangle (the 'only if' part is open). Hence, for a planar Boolean model Z,

$$\begin{aligned}
\overline{\chi}(Z) &= e^{-\overline{A}(X)} \left(\gamma - \overline{A}(X, X^*) \right) \\
&= e^{-\overline{A}(X)} \left(\gamma - A(B(X), B(X)^*) \right) \\
&\geq e^{-\overline{A}(X)} \left(\gamma - \frac{\sqrt{3}}{18} U^2(B(X)) \right) \\
&= e^{-\overline{A}(X)} \left(\gamma - \frac{\sqrt{3}}{18} \overline{U}^2(X) \right),
\end{aligned}$$

with equality, if X consists of homothetic equilateral triangles. If the grain distribution is symmetric (that is, invariant under $K \mapsto K^*$), then $B(X) = B(X)^*$, hence $A(B(X), B(X)^*) = A(B(X)) \leq \frac{1}{4\pi} U^2(B(X))$, by the isoperimetric inequality (here, equality holds if and only if $B(X)$ is a ball). Thus, for a 'symmetric' Boolean model Z, we have the sharper inequality

$$\overline{\chi}(Z) \geq e^{-\overline{A}(X)} \left(\gamma - \frac{1}{4\pi} \overline{U}^2(X) \right),$$

with equality, if X is isotropic (but not only in this case).

In order to illustrate these results, we may start with a stationary Boolean model with homothetic equilateral triangles as grains. If we rotate the triangles randomly (and independently), this increases the specific Euler characteristic $\overline{\chi}(Z)$. If, instead of rotating, we reflect the particles randomly (and independently), we obtain an even larger $\overline{\chi}(Z)$.

Hints to the literature. Densities of mixed volumes were studied in [48]. Densities for area measures and support functions appear in [41], [45]. The intensity analysis for $d = 2$ was given in [42], the case $d = 3$ was discussed in [46]. In [48], a corresponding result was claimed also for $d = 4$. The given proof is however incomplete, since there is a term $\overline{V}_{2,2}^{(0)}(X, X)$ missing in the formula for $\overline{\chi}(Z)$ (there are also some constants missing in the formulas for $\overline{\chi}(Z)$, for $d = 2, 3, 4$). For the planar case, Rataj [31] and Kiderlen and Jensen [21] study methods to estimate the mean normal measure $\overline{S}_1(Z, \cdot)$. In higher dimensions, averages of planar sections can be used (see Kiderlen [20]).

Mean and Blaschke bodies for RACS and particle processes have been studied in [43], [44]. The use of associated convex bodies in stochastic geometry is demonstrated in Section 4.5 of [36].

It seems obvious that extremal properties of the specific Euler characteristic $\overline{\chi}(Z)$ of a planar stationary Boolean model Z should be related to the percolation threshold of Z. For a given grain distribution $\mathbb{Q}$, the latter is the critical value γ_0 of the intensity, such that Z a.s. contains an unbounded connected component, if $\gamma > \gamma_0$, and if $\gamma < \gamma_0$, then Z a.s. contains no unbounded connected component. For Boolean models with a deterministic (convex) grain K (i.e. $\mathbb{Q} = \delta_K$), Jonasson [18] has shown that γ_0 becomes minimal, if K is a triangle. In general, the exact value of γ_0 is not known and also relations between the percolation threshold and the specific Euler characteristic are not apparent. See [8, Section 12.10] and [38], for more information and a list of recent results on continuum percolation.

3.2 Local Densities

The mean values $\overline{\varphi}(Z)$ of additive, translation invariant functionals φ for a stationary RACS Z have been introduced as limits

$$\overline{\varphi}(Z) = \lim_{r \to \infty} \frac{\mathbb{E}\,\varphi(Z \cap rK_0)}{V_d(rK_0)}$$

over increasing sampling windows rK_0, $r \to \infty$. For a prospective extension to nonstationary RACS Z it would be helpful to have a local interpretation of these mean values. This is possible in the cases we treated so far, namely for the quermass densities $\overline{V}_j(Z)$ and the densities $\overline{S}_{d-1}(Z, \cdot)$ and $\overline{h}(Z, \cdot)$. The following considerations will also make clear why we speak of densities here. The basic idea is to replace the intrinsic volumes V_j by their local counterparts, the **curvature measures** $\Phi_j(K, \cdot), j = 0, ..., d-1$. These measures are defined for $K \in \mathcal{K}$, but have an additive extension to sets in the convex ring (which may be a finite signed measure). Since curvature measures are locally defined, they even extend to sets in $\mathcal{S}$ as signed Radon measures (set functions which are defined on bounded Borel sets). Hence, for a random $\mathcal{S}$-set Z, we are allowed to write $\Phi_j(Z, \cdot)$ and this is a random signed Radon measure (as always in these lectures, we skip some technical details like the measurability and integrability of $\Phi_j(Z, \cdot)$). If Z is stationary, $\Phi_j(Z, \cdot)$ is stationary, hence the (signed Radon) measure $\mathbb{E}\,\Phi_j(Z, \cdot)$ is translation invariant. Consequently, this measure is a multiple of λ_d, $\mathbb{E}\,\Phi_j(Z, \cdot) = d_j\,\lambda_d$. As it turns out, d_j equals the quermass density $\overline{V}_j(Z)$.

Theorem 3.3. *Let Z be a stationary random $\mathcal{S}$-set fulfilling* (14) *and $j \in \{0, ..., d\}$. Then,*

$$\mathbb{E}\,\Phi_j(Z, \cdot) = \overline{V}_j(Z)\,\lambda_d.$$

Proof. We copy the proof of Theorem 2.4 and modify it appropriately, but leave out some of the more technical details.

Let $A \subset \mathbb{R}^d$ be a bounded Borel set and K a convex body containing A in its interior. The stationarity of Z and the fact that the curvature measures are locally defined and translation covariant implies

$$\mathbb{E} \int_{\mathbb{R}^d} \Phi_j(Z \cap K \cap xrB^d, A) \, \lambda_d(dx) = \mathbb{E} \int_{\mathbb{R}^d} \Phi_j(Z \cap yK \cap rB^d, yA) \, \lambda_d(dy).$$

On the left-hand side, we can replace the integration over $\mathbb{R}^d$ by the invariant integral over G_d and apply the principal kinematic formula for curvature measures. We get

$$\sum_{k=j}^{d} c_{j,d}^{k,d+j-k} \mathbb{E} \, \Phi_k(Z, A) V_{d+j-k}(rB^d)$$

(here we have used that $\Phi_k(Z \cap K, A) = \Phi_k(Z, A)$). On the right-hand side, we apply the principal translative formula for curvature measures (see [34]) and use the fact that the mixed measures $\Phi_{k,d+j-k}^{(j)}(Z \cap rB^d, K; \mathbb{R}^d \times A)$ vanish, for $k > j$, since $A \subset \text{int } K$. The remaining summand, with $k = j$, is $\Phi_j(Z \cap rB^d, \mathbb{R}^d) \lambda_d(A)$. Hence we obtain

$$\sum_{k=j}^{d} c_{j,d}^{k,d+j-k} \mathbb{E} \, \Phi_k(Z, A) V_{d+j-k}(rB^d) = \mathbb{E} \, \Phi_j(Z \cap rB^d, \mathbb{R}^d) \lambda_d(A).$$

Dividing both sides by $V_d(rB^d)$ and letting $r \to \infty$, we obtain

$$\mathbb{E} \, \Phi_j(Z, A) = \overline{V}_j(Z) \lambda_d(A).$$

$\square$

For the directional densities $\overline{S}_{d-1}(Z, \cdot)$ and $\overline{h}(Z, \cdot)$, we obtain similar results. In the first case, we use the support measure $\Xi_{d-1}(K, \cdot)$, $K \in \mathcal{K}$, (see [34]) and show that

$$\mathbb{E} \, \Xi_{d-1}(Z, \cdot \times A) = \frac{1}{2} \overline{S}_{d-1}(Z, A) \lambda_d, \tag{26}$$

for each Borel set $A \subset S^{d-1}$ (the factor $1/2$ here comes from the different normalizations: $\Xi_{d-1}(K, \mathbb{R}^d \times S^{d-1}) = V_{d-1}(K)$, which is half the surface area $S_{d-1}(K, S^{d-1})$). (26) can be deduced as in the proof above and is based on the translative formula

$$\int_{\mathbb{R}^d} \Xi_{d-1}(K \cap xM, (B \cap xC) \times A) \, \lambda_d(dx)$$
$$= \Xi_{d-1}(K, B \times A) \lambda_d(M \cap C) + \Xi_{d-1}(M, C \times A) \lambda_d(K \cap B),$$

for $K, M \in \mathcal{K}$ and Borel sets $B, C \subset \mathbb{R}^d$, $A \subset S^{d-1}$, which follows from Theorem 1.2.7 in [35].

For $\overline{h}(Z, \cdot)$, we use the **support kernel** $\rho(K; u, \cdot)$, $K \in \mathcal{K}, u \in S^{d-1}$,

$$\rho(K; u, \cdot) := \Phi_{1,d-1}^{(0)}(K, u^+; \cdot \times \beta(u)),$$

here $u^+ := \{x \in \mathbb{R}^d : \langle x, u \rangle \geq 0\}$ and $\beta(u)$ is a ball in ∂u^+ of $(d-1)$-volume 1. The extension properties and the translative integral formula for mixed measures then show that

$$\mathbb{E}\,\rho(Z; u, \cdot) = \overline{h}(Z, u)\,\lambda_d,$$

again with a similar proof as above.

These results show also that the densities can be considered as Radon-Nikodym derivatives, in particular we have

$$\overline{V}_j(Z) = \lim_{r \to 0} \frac{\mathbb{E}\,\Phi_j(Z, rB^d)}{V_d(rB^d)}, \qquad j = 0, ..., d-1,$$

(note, however, that the relation $\overline{V}_j(Z) = \lim r \to 0\mathbb{E}\,V_j(Z \cap rB^d)/V_d(rB^d)$ is wrong).

For particle processes, we only formulate the corresponding results (and concentrate on convex particles). The proofs are similar but simpler because of Campbell's theorem.

Theorem 3.4. *Let X be a stationary process of convex particles. Then,*

$$\mathbb{E} \sum_{K \in X} \Phi_j(K, \cdot) = \overline{V}_j(X)\,\lambda_d, \qquad j = 0, ..., d,$$

$$\mathbb{E} \sum_{K \in X} \Xi_{d-1}(K, \cdot \times A) = \frac{1}{2}\overline{S}_{d-1}(X, A)\,\lambda_d, \qquad A \subset S^{d-1},$$

$$\mathbb{E} \sum_{K \in X} \rho(K; u, \cdot) = \overline{h}(X, u)\,\lambda_d, \qquad u \in S^{d-1}.$$

Hints to the literature. The interpretation of quermass densities as Radon-Nikodym derivatives goes back to Weil and Wieacker [49], Weil [39] and, for sets of positive reach, to Zähle [50]. The derivative interpretation of $\overline{S}_{d-1}(Z, \cdot)$ and $\overline{h}(Z, \cdot)$, and the corresponding results in Theorem 3.4 can be found at different places in the literature, in particular see [7].

3.3 Nonstationary Boolean Models

In this subsection, we discuss extensions of the previous results to nonstationary Boolean models Z. It is rather obvious that we have to require some regularity of Z, because with arbitrary intensity measures Θ of the underlying Poisson particle process X rather pathological union sets Z can arise. Starting with a Boolean model with compact grains, our basic assumption is that Θ allows a decomposition

$$\Theta(\mathcal{A}) = \int_{C_0} \int_{\mathbb{R}^d} 1_{\mathcal{A}}(x + K)f(K, x)\,\lambda_d(dx)\mathbb{Q}(dK), \qquad \mathcal{A} \in \mathcal{B}(\mathcal{C}'), \quad (27)$$

with a nonnegative and measurable function f on $\mathcal{C}_0 \times \mathbb{R}^d$ and a probability measure $\mathbb{Q}$ on $\mathcal{C}_0$. Theorem 1.3 shows that (27) means that Θ is absolutely continuous to a translation invariant measure Ω on $\mathcal{C}'$. Such a measure Ω is not uniquely determined and consequently the representation (27) for Θ is not unique, in general. There is however an important case, where uniqueness holds, and we will concentrate on this case, although most of the results in this subsection hold true in the more general situation (27).

Theorem 3.5. *Let Θ be a locally finite measure on $\mathcal{C}'$ and suppose that*

$$\Theta(\mathcal{A}) = \int_{\mathcal{C}_0} \int_{\mathbb{R}^d} \mathbf{1}_{\mathcal{A}}(x + K) f(x) \, \lambda_d(dx) \mathbb{Q}(dK), \qquad \mathcal{A} \in \mathcal{B}(\mathcal{C}'), \qquad (28)$$

with a nonnegative and measurable function f on $\mathbb{R}^d$ and a probability measure $\mathbb{Q}$ on $\mathcal{C}_0$. Then $\mathbb{Q}$ is uniquely determined and f is determined up to a set of λ_d-measure 0.

If we consider the image measure $\tilde{\Theta}$ of Θ under $K \mapsto (c(K), K - c(K))$, the theorem presents the obvious fact that a decomposition of the form $\tilde{\Theta} = (\int f \mathrm{d}\lambda_d) \otimes \mathbb{Q}$ is unique.

If Θ is now the intensity measure of a Poisson particle process X and Z the corresponding Boolean model, we call f the **intensity function** of Z and $\mathbb{Q}$ the **distribution of the typical grain**. We still have a simple interpretation of Z which can also be used for simulations. If we distribute points $x_1, x_2, \ldots$ according to a Poisson process with intensity function f and then add independent (convex) particles $Z_1, Z_2, \ldots$ with common distribution $\mathbb{Q}$, the resulting union set is equivalent (in distribution) to Z. The Boolean model Z is stationary (with intensity γ), if and only if $f \equiv \gamma$.

How strong are conditions like (27) or (28)? Since $\mathcal{C}$ and $\mathbb{R}^d$ are both Polish spaces and since Θ is assumed to be locally finite, a general decomposition principle in measure theory shows that a representation

$$\Theta(\mathcal{A}) = \int_{\mathcal{C}_0} \int_{\mathbb{R}^d} \mathbf{1}_{\mathcal{A}}(x + K) \rho(K, dx) \mathbb{Q}(dK), \qquad \mathcal{A} \in \mathcal{B}(\mathcal{C}'),$$

always exists, where ρ is a kernel (i.e, for each $K \in \mathcal{C}_0$, $\rho(K, \cdot)$ is a measure on $\mathbb{R}^d$ and, for each $B \in \mathcal{B}(\mathbb{R}^d)$, $\rho(\cdot, B)$ is a (nonnegative) measurable function on $\mathcal{C}_0$) and $\mathbb{Q}$ is a probability measure. Condition (27) thus requires, in addition, that the kernel ρ is absolutely continuous, that means, there are versions of ρ and $\mathbb{Q}$ (remember that ρ and $\mathbb{Q}$ are not uniquely determined), such that $\rho(K, \cdot)$ has a density $f(K, \cdot)$ with respect to λ_d, for each $K \in \mathcal{C}_0$. In contrast to this, (28) is a much stronger condition since it requires that $\rho(K, \cdot)$ is in addition independent of $K \in \mathcal{C}_0$.

Now we concentrate on Boolean models with convex grains again. Our next goal is to define densities of additive functionals like V_j, S_{d-1}, h^* and the mixed functionals $V_{m_1,\ldots,m_k}^{(j)}$ for the Poisson process X and the Boolean model Z.

Since X and Z are not assumed to be stationary anymore, we cannot expect a constant density but a quantity depending on the location z in space. The results of Section 3.2 motivate us to introduce densities as Radon-Nikodym derivatives of appropriate random measures with respect to λ_d. The main task is therefore to show that the random measures are absolutely continuous. Here, condition (28) is of basic importance. For the mixed functionals $V^{(j)}_{m_1,...,m_k}$, the corresponding random measures are measures on $(\mathbb{R}^d)^k$ and consequently the Radon-Nikodym derivatives will be taken with respect to λ_d^k and will be functions of k variables $z_1, ..., z_k$.

Theorem 3.6. *Let X be a Poisson process of convex particles with intensity measure satisfying (28). Then, for $j = 0, ..., d$, $k \in \mathbb{N}$, and $m_1, ..., m_k \in \{j, ..., d\}$ with $m_1 + \cdots + m_k = (k-1)d + j$, the signed measure*

$$\mathbb{E} \sum_{(K_1,...,K_k) \in X^k_{\neq}} \Phi^{(j)}_{m_1,...,m_k}(K_1, ..., K_k; \cdot)$$

is locally finite and absolutely continuous with respect to λ_d^k.

Its Radon-Nikodym derivative $\overline{V}^{(j)}_{m_1,...,m_k}(X, ..., X; \cdot)$ fulfills λ_d^k-almost everywhere

$$\overline{V}^{(j)}_{m_1,...,m_k}(X, ..., X; z_1, ..., z_k)$$
$$= \int_{\mathcal{K}_0} \cdots \int_{\mathcal{K}_0} \int_{(\mathbb{R}^d)^k} f(z_1 - x_1) \cdots f(z_k - x_k)$$
$$\times \Phi^{(j)}_{m_1,...,m_k}(K_1, ..., K_k; d(x_1, ..., x_k)) \mathbb{Q}(dK_1) \cdots \mathbb{Q}(dK_k)$$
$$= \lim_{r \to 0} \frac{1}{(V_d(rK))^k}$$
$$\times \mathbb{E} \sum_{(K_1,...,K_k) \in X^k_{\neq}} \Phi^{(j)}_{m_1,...,m_k}(K_1, ..., K_k; (z_1 + rK) \times \cdots \times (z_k + rK)),$$

for each $K \in \mathcal{K}$ with inner points.

For $k = 1$, we thus obtain the local quermass density $\overline{V}_j(X; \cdot)$ with

$$\overline{V}_j(X; z) = \int_{\mathcal{K}_0} \int_{\mathbb{R}^d} f(z - x) \Phi_j(M, dx) \mathbb{Q}(dM)$$
$$= \lim_{r \to 0} \frac{1}{(V_d(rK))^k} \mathbb{E} \sum_{M \in X} \Phi_j(M, (z + rK)), \qquad z \in \mathbb{R}^d.$$

For $j = d$, we get the volume density (or volume fraction)

$$V_d(X; z) = \int_{\mathcal{K}_0} \int_M f(z - x) \lambda_d(dx) \mathbb{Q}(dM).$$

This equation (with the proof given below) holds true for Poisson processes X on $\mathcal{C}'$ as well.

Apart from integrability considerations which are of a more technical nature, the basic tool in the proof of Theorem 3.6 is Campbell's theorem again which gives us

$$\mathbb{E} \sum_{(K_1,...,K_k) \in X_{\neq}^k} \Phi_{m_1,...,m_k}^{(j)}(K_1, ..., K_k; B)$$

$$= \int_{\mathcal{K}_0} \cdots \int_{\mathcal{K}_0} \int_{(\mathbb{R}^d)^k} \Phi_{m_1,...,m_k}^{(j)}(K_1 + x_1, ..., K_k + x_k; B) f(x_1) \cdots f(x_k)$$

$$\times \lambda_d^k(d(x_1, ..., x_k)) \mathbb{Q}(dK_1) \cdots \mathbb{Q}(dK_k),$$

since Θ^k is the intensity measure of $X_{\neq}^k$. The main step is then to show that

$$\int_{(\mathbb{R}^d)^k} \Phi_{m_1,...,m_k}^{(j)}(K_1 + x_1, ..., K_k + x_k; B) f(x_1) \cdots f(x_k) \lambda_d^k(d(x_1, ..., x_k))$$

$$= \int_B \int_{(\mathbb{R}^d)^k} f(z_1 - x_1) \cdots f(z_k - x_k) \Phi_{m_1,...,m_k}^{(j)}(K_1, ..., K_k; d(x_1, ..., x_k))$$

$$\times \lambda_d^k(d(z_1, ..., z_k)),$$

which follows from the translation covariance of the mixed measures and a simple change of variables.

Densities for other (mixed) functionals which occurred in previous sections, in the stationary case for X, can be treated in a similar way.

Now we consider the union set Z.

Theorem 3.7. *Let Z be a Boolean model with convex grains, where the intensity measure of the underlying Poisson process X satisfies (28), and let $j \in \{0, ..., d\}$. Then, the signed Radon measure*

$$\mathbb{E}\, \Phi_j(Z, \cdot)$$

is locally finite and absolutely continuous with respect to λ_d. Its Radon-Nikodym derivative $\overline{V}_j(Z; \cdot)$ fulfills

$$\overline{V}_j(Z; z) = \lim_{r \to 0} \frac{1}{V_d(rK)} \mathbb{E}\, \Phi_j(Z, z + rK)$$

for λ_d-almost all $z \in \mathbb{R}^d$ and each $K \in \mathcal{K}$ with inner points. Moreover, we have, for λ_d-almost all $z \in \mathbb{R}^d$,

$$\overline{V}_d(Z; z) = 1 - e^{-\overline{V}_d(X;z)}$$

and

$$\overline{V}_j(Z;z) = e^{-\overline{V}_d(X;z)} \left(\overline{V}_j(X;z) \right.$$

$$\left. + \sum_{k=2}^{d-j} \frac{(-1)^{k+1}}{k!} \sum_{\substack{m_1,\ldots,m_k=j \\ m_1+\ldots+m_k=(k-1)d+j}}^{d-1} \overline{V}_{m_1,\ldots,m_k}^{(j)}(X,\ldots,X;z,\ldots,z) \right),$$

for $j = 0, \ldots, d-1$.

Proof. The assertions on the volume density hold in greater generality. In fact, for a RACS Z, we have

$$\mathbb{E}\,\lambda_d(Z \cap A) = \int_A \mathbb{P}(z \in Z)\,\lambda_d(dz), \qquad A \in \mathcal{B}(\mathbb{R}^d).$$

Hence, the measure $\mathbb{E}\,\lambda_d(Z \cap \cdot)$ is absolutely continuous and has density $\overline{V}_d(Z;z) := \mathbb{P}(z \in Z)$. For a Boolean model Z with compact grains,

$$\mathbb{P}(z \in Z) = 1 - \mathbb{P}(\{z\} \cap Z = \emptyset) = 1 - e^{-\Theta(\mathcal{F}_{\{z\}})}$$

$$= 1 - \exp\left(-\int_{\mathcal{C}_0}\int_{\mathbb{R}^d} 1_{x+K}(z)f(x)\,\lambda_d(dx)\mathbb{Q}(dK) \right)$$

$$= 1 - \exp\left(-\int_{\mathcal{C}_0}\int_K f(z-y)\,\lambda_d(dy)\mathbb{Q}(dK) \right)$$

$$= 1 - e^{-\overline{V}_d(X;z)}. \tag{29}$$

Now we consider convex grains and $j \in \{0, \ldots, d-1\}$. Again, we leave out the technical parts concerning the local finiteness etc. We apply Theorem 2.1 with $\varphi(K) := \Phi_j(K,B)$, B a fixed bounded Borel set, and choose a convex body K_0 with B in its interior. Because of (28), we obtain

$$\mathbb{E}\,\Phi_j(Z,B) = \mathbb{E}\,\Phi_j(Z \cap K_0, B)$$

$$= \sum_{k=1}^{\infty} \frac{(-1)^{k+1}}{k!} \int_{\mathcal{K}_0}\cdots\int_{\mathcal{K}_0}\int_{\mathbb{R}^d}\cdots\int_{\mathbb{R}^d} \Phi_j(K_0 \cap x_1K_1 \cap \cdots \cap x_kK_k, B)$$

$$\times f(x_1)\cdots f(x_k)\,\lambda_d(dx_1)\cdots\lambda_d(dx_k)\mathbb{Q}(dK_1)\cdots\mathbb{Q}(dK_k).$$

The iterated translative formula for curvature measures, in its version (49) from [34], shows that

$$\int_{\mathbb{R}^d} \cdots \int_{\mathbb{R}^d} \Phi_j(K_0 \cap x_1 K_1 \cap \cdots \cap x_k K_k, B) f(x_1) \cdots f(x_k) \lambda_d(\mathrm{d}x_1) \cdots \lambda_d(\mathrm{d}x_k)$$

$$= \int_{\mathbb{R}^d} \cdots \int_{\mathbb{R}^d} \int_{\mathbb{R}^d} \mathbf{1}_B(z) f(x_1) \cdots f(x_k) \Phi_j(K_0 \cap x_1 K_1 \cap \cdots \cap x_k K_k, \mathrm{d}z)$$

$$\times \lambda_d(\mathrm{d}x_1) \cdots \lambda_d(\mathrm{d}x_k)$$

$$= \sum_{\substack{m_1,\ldots,m_k=j \\ m_1+\cdots+m_k=(k-1)d+j}}^{d} \int_B \int_{(\mathbb{R}^d)^k} f(z-x_1) \cdots f(z-x_k)$$

$$\times \Phi_{m_1,\ldots,m_k}^{(j)}(K_1,\ldots,K_k; \mathrm{d}(x_1,\ldots,x_k)) \, \lambda_d(\mathrm{d}z).$$

Therefore, we obtain from Theorem 3.6

$$\mathbb{E}\,\Phi_j(Z,B)$$

$$= \int_B \left(\sum_{k=1}^{\infty} \frac{(-1)^{k+1}}{k!} \sum_{\substack{m_1,\ldots,m_k=j \\ m_1+\cdots+m_k=(k-1)d+j}}^{d} \overline{V}_{m_1,\ldots,m_k}^{(j)}(X,\ldots,X;z,\ldots,z) \right) \lambda_d(\mathrm{d}z).$$

The remaining part of the proof follows now the standard procedure, since the decomposition properties for mixed measures carry over to the local densities. □

Analogous results hold for densities of mixed volumes.

There are also versions for local functionals, in particular for the surface area measure. The case $j = d-1$ of Theorem 3.7 yields

$$\overline{V}_{d-1}(Z;z) = e^{-\overline{V}_d(X;z)} \overline{V}_{d-1}(X;z), \tag{30}$$

with

$$\overline{V}_{d-1}(X;z) = \int_{\mathcal{K}_0} \int_{\mathbb{R}^d} f(z-x) \Phi_{d-1}(M,\mathrm{d}x) \mathbb{Q}(\mathrm{d}M).$$

The local densities of the surface area measure are defined as

$$\overline{S}_{d-1}(Z,A;z) := 2 \frac{\mathrm{d}\mathbb{E}\,\Xi_{d-1}(Z,\cdot \times A)}{\mathrm{d}\lambda_d}(z),$$

$$\overline{S}_{d-1}(X,A;z) := 2 \int_{\mathcal{K}_0} \int_{\mathbb{R}^d} f(z-x) \Xi_{d-1}(M,\mathrm{d}x \times A) \mathbb{Q}(\mathrm{d}M),$$

for Borel sets $A \subset S^{d-1}$. The local version of (30) then reads

$$\overline{S}_{d-1}(Z,\cdot;z) = e^{-\overline{V}_d(X;z)} \overline{S}_{d-1}(X,\cdot;z).$$

The main difference to the stationary situation is that $\overline{S}_{d-1}(X,\cdot;z)$ (and thus $\overline{S}_{d-1}(Z,\cdot;z)$) need not have center 0 anymore. Therefore, Minkowski's theorem is not applicable and there are no local Blaschke bodies $B(X;z)$ and $B(Z;z)$, in general.

However, for $d = 2$ and $d = 3$, an intensity analysis similar to the stationary situation can still be performed. Namely, for $d = 2$, we have

$$\overline{A}(Z; z) = 1 - e^{-\overline{A}(X; z)},$$

$$\overline{S}_1(Z, \cdot; z) = e^{-\overline{A}(X; z)} \overline{S}_1(X, \cdot; z),$$

$$\overline{\chi}(Z; z) = e^{-\overline{A}(X; z)} \left(\overline{\chi}(X; z) - \overline{A}(X, X^*; z, z) \right). \tag{31}$$

Formulas for mixed volumes (mixed functionals) in the plane can be used to show that

$$\overline{A}(X, X^*; z, z) = \frac{1}{2} \int_{\mathcal{K}_0} \int_{\mathcal{K}_0} \int_{\mathrm{Nor}\,M} \int_{\mathrm{Nor}\,K} f(z - x) f(z - y) \alpha(u, v) \sin \alpha(u, v)$$

$$\times \, \Xi_1(K, \mathrm{d}(x, u)) \Xi_{d-1}(M, \mathrm{d}(y, v)) \mathbb{Q}(\mathrm{d}K) \mathbb{Q}(\mathrm{d}M)$$

$$= \frac{1}{2} \int_{S^1} \int_{S^1} \alpha(u, v) \sin \alpha(u, v) \overline{S}_1(X, \mathrm{d}u; z) \overline{S}_1(X, \mathrm{d}v; z).$$

Here $\mathrm{Nor}\,K$ denotes the normal bundle of K and $\alpha(u, v)$ is the (smaller) angle between $u, v \in S^1$. We see that, as in the stationary case, the left-hand sides of (31) determine $\overline{\chi}(X; z)$ (which equals γ in the stationary case). This is a pointwise result, which holds for λ_2-almost all z (respectively for all z, if the intensity function f is continuous). But (for general d)

$$\overline{\chi}(X; z) = \int_{\mathcal{K}_0} \int_{\mathbb{R}^d} f(z - x) \Phi_0(M, \mathrm{d}x) \mathbb{Q}(\mathrm{d}M)$$

$$= (f * \tilde{\Phi}_0)(z),$$

where $*$ denotes convolution and $\tilde{\Phi}_0$ is the mean of Φ_0 w.r.t. $\mathbb{Q}$. In general, $f * \tilde{\Phi}_0$ does not determine f (not even in the planar case). Uniqueness holds, if f and $\tilde{\Phi}_0$ have compact support. Thus, if the intensity function f is continuous with compact support and if the grains of Z are uniformly bounded, the knowledge of the densities $\overline{A}(Z; z), \overline{S}_1(Z, \cdot; z)$ and $\overline{\chi}(Z; z)$, for all $z \in \mathbb{R}^2$, would determine the intensity function f uniquely.

The corresponding three-dimensional result is more involved and requires additional assumptions on Z.

Hints to the literature. Nonstationary Boolean models were first studied in detail by Fallert [5], [6], who also defined quermass densities and proved the basic Theorems 3.6 and 3.7. Corresponding results for mixed volumes are given in [48]. The intensity analysis for instationary Boolean models in dimensions $d = 2$ and $d = 3$ was performed in [47] and [7].

3.4 Sections of Boolean Models

If Z is a Boolean model in $\mathbb{R}^d$ and L any affine subspace (with dimension in $\{1, ..., d-1\}$), then the intersection $Z \cap L$ is a Boolean model. This follows from the fact that the intersections $K \cap L$, $K \in X$, build a Poisson process again, as can be easily checked. It is therefore natural to ask, how the characteristic quantities of Z and $Z \cap L$ are connected. Formulas of this kind are helpful since they can be used to estimate quantities of Z from lower dimensional sections. Such estimation problems are typical for the field of stereology.

It is, in general, quite difficult to get information about the grain distribution $\mathbb{Q}$ from sections. An exception is a Boolean model Z with balls as grains (the *Wicksell problem*). Here, an integral equation connecting the radii distributions of Z and $Z \cap L$ can be given and used for an inversion (which is however unstable and leads to an ill-posed problem). For the quermass densities, the Crofton formulas from integral geometry can be used. These lead to formulas which hold for RACS Z and particle processes X and therefore can be used for Boolean models as well. We only state some results here, the proofs are quite similar to the ones given previously for intersections of Z (or X) with sampling windows K_0.

Theorem 3.8. *Let Z be a stationary random $\mathcal{S}$-set fulfilling (14) and let X be a stationary process of particles in $\mathcal{K}'$. Let $L \in \mathcal{L}_k^d$ be a k-dimensional linear subspace, $k \in \{0, ..., d\}$, and let ϑ be a random rotation with distribution ν and independent of Z and X. Then,*

$$\mathbb{E}_\nu \overline{V}_j(Z \cap \vartheta L) = c_{j,d}^{k,d+j-k} \overline{V}_{d+j-k}(Z)$$

and

$$\mathbb{E}_\nu \overline{V}_j(X \cap \vartheta L) = c_{j,d}^{k,d+j-k} \overline{V}_{d+j-k}(X),$$

for $j = 0, ..., k$.

The cases $k = d$ and $k = 0$ are trivial and not of interest. If $j = k$ or if Z is isotropic, the expectation $\mathbb{E}_\nu$ can be omitted.

Hints to the literature. Crofton formulas are classical in stochastic geometry (see [37] or [36]). For the Wicksell problem, see also [37] and [36] and the literature cited there. Stereological applications of stochastic geometry are described in [37], see also the recent book by Beneš and Rataj [2].

4 Contact Distributions

In the previous sections, we mostly discussed properties of Boolean models Z which are based on intrinsic geometric quantities of the set Z. A corresponding statistical analysis of Z would require observations of Z from inside, for example by counting pixel points of Z in a bounded sampling window K_0, by considering curvatures in boundary points of Z, etc. In this last section, we shall consider quantities which are based on observations outside Z, for example by measuring the distance $d(x, Z)$ from a given point $x \notin Z$ to Z. The distribution of $d(x, Z)$ is a particular case of a contact distribution. Such contact distributions are often easier to estimate in practice and will also give us interesting information about Z.

4.1 Contact Distribution with Structuring Element

Contact distributions can be introduced for arbitrary RACS Z. In their general form, they are based on a given convex body $B \in \mathcal{K}'$ with $0 \in B$, the **structuring element**, and describe the distribution of the 'B-distance' from a point $x \notin Z$ to the set Z.

To give a more precise definition, we start with

$$d_B(x, A) := \inf\{r \geq 0 : (x + rB) \cap A \neq \emptyset\},$$

for $x \in \mathbb{R}^d$ and $A \in \mathcal{F}$. In this situation, we also call B a **gauge body**. The set on the right-hand side may be empty (for example, if $0 \in \partial B$ or if B is lower dimensional), then $d_B(x, Z) = \infty$. We also put $d_B(A, x) := d_{B^*}(x, A)$ and $d_B(x, y) := d_B(x, \{y\}), y \in \mathbb{R}^d$. If the gauge body has inner points and is symmetric with respect to 0, then $(x, y) \mapsto d_B(x, y)$ is a metric on $\mathbb{R}^d$ and hence $(\mathbb{R}^d, d_B)$ is a Minkowski space. Of course, for $B = B^d$ we obtain the Euclidean metric, we then write $d(x, A)$ instead of $d_{B^d}(x, A)$, etc.

Now we consider a RACS Z and $x \in \mathbb{R}^d$ and define the **contact distribution** H_B at x by

$$H_B(x, r) := \mathbb{P}((x + rB) \cap Z \neq \emptyset \,|\, x \notin Z)$$
$$= \mathbb{P}(d_B(x, Z) \leq r \,|\, x \notin Z), \qquad r \geq 0.$$

Thus, $H_B(x, \cdot)$ is the distribution function of the B-distance $d_B(x, Z)$ of x to Z, conditional to the event that $x \notin Z$ (it is of course also possible to consider the unconditional distribution function, but the above definition is the widely used one). Actually, $H_B(x, \cdot)$ is the contact distribution function and the contact distribution would be the corresponding probability measure on $[0, \infty]$, but it is now common use in the literature not to distinguish between these two notions. If $\mathbb{P}(x \notin Z) > 0$, then one can express $H_B(x, \cdot)$ in terms of the (local) volume density, namely

$$H_B(x, r) = \mathbb{P}((x + rB) \cap Z \neq \emptyset \,|\, x \notin Z)$$
$$= 1 - \frac{\mathbb{P}(x \notin Z + rB^*)}{\mathbb{P}(x \notin Z)}$$
$$= \frac{\mathbb{P}(x \in Z + rB^*) - \mathbb{P}(x \in Z)}{1 - \mathbb{P}(x \in Z)}$$
$$= \frac{\overline{V}_d(Z + rB^*, x) - \overline{V}_d(Z, x)}{1 - \overline{V}_d(Z, x)}.$$

The condition $\mathbb{P}(x \notin Z) > 0$ is always satisfied if Z is stationary and non-trivial, i.e. $Z \neq \mathbb{R}^d$ with positive probability.

Now we assume that Z is a Boolean model (with compact grains) and that the intensity measure Θ of the underlying Poisson particle process X fulfills (28). Then, (29) implies

$$H_B(x,r) = \frac{e^{-\overline{V}_d(X;x)} - e^{-\overline{V}_d(X+rB^*;x)}}{e^{-\overline{V}_d(X;x)}}$$

$$= 1 - \exp\left(-\int_{\mathcal{C}_0}\int_{(K+rB^*)\setminus K} f(x-y)\,\lambda_d(dy)\mathbb{Q}(dK)\right). \quad (32)$$

We first discuss the stationary situation and assume convex grains. Hence, $f \equiv \gamma$ and $H_B(x,r)$ is independent of x. We put $H_B(r) := H_B(0,r)$. Then,

$$H_B(r) = 1 - \exp\left(-\gamma\int_{\mathcal{K}_0} V_d((K+rB^*)\setminus K)\mathbb{Q}(dK)\right)$$

$$= 1 - \exp\left(-\gamma\sum_{i=0}^{d-1}\binom{d}{i}r^{d-i}\int_{\mathcal{K}_0} V(\underbrace{K,...,K}_{i},\underbrace{B^*,...,B^*}_{d-i})\mathbb{Q}(dK)\right)$$

$$= 1 - \exp\left(-\sum_{i=0}^{d-1}r^{d-i}\overline{V}^{(0)}_{i,d-i}(X,B)\right),$$

where we used the expansion of $V_d(K+rB^*)$ into mixed volumes (resp. mixed functionals). We thus get the following result.

Theorem 4.1. *Let Z be a stationary Boolean model with convex grains and B a gauge body. Then*

$$H_B(r) = 1 - \exp\left(-\sum_{i=0}^{d-1}r^{d-i}\overline{V}^{(0)}_{i,d-i}(X,B)\right), \qquad for\ r \geq 0.$$

Interesting special cases are $B = B^d$ (the **spherical contact distribution**) and $B = [0,u], u \in S^{d-1}$, (the **linear contact distribution**). For the spherical contact distribution, we get

$$H_{B^d}(r) = 1 - \exp\left(-\sum_{i=0}^{d-1}r^{d-i}\kappa_{d-i}\overline{V}_i(X)\right),$$

the result we already obtained at the end of Section 1 (for the capacity functional). For $H_{[0,u]}$, we observe that

$$V(\underbrace{K,...,K}_{i},\underbrace{[0,u],...,[0,u]}_{d-i}) = 0, \qquad for\ i = 0,...,d-2,$$

whereas $V(K,...,K,[0,u])$ is proportional to the $(d-1)$-volume of the orthogonal projection of K onto the hyperplane orthogonal to u,

$$V(K,...,K,[0,u]) = \frac{1}{d}V_{d-1}(K\,|\,u^\perp).$$

Therefore,

$$H_{[0,u]}(r) = 1 - \exp\left(-\gamma r \int_{\mathcal{K}_0} V_{d-1}(K \,|\, u^\perp)\mathbb{Q}(dK)\right).$$

From the formula

$$V_{d-1}(K \,|\, u^\perp) = \frac{1}{2}\int_{S^{d-1}} |\langle x, u\rangle| S_{d-1}(K, dx),$$

we obtain

$$H_{[0,u]}(r) = 1 - \exp\left(-\frac{r}{2}\int_{S^{d-1}} |\langle x, u\rangle| S_{d-1}(B(X), dx)\right)$$
$$= 1 - \exp\left(-rh(\Pi B(X), u)\right),$$

with the **projection body** $\Pi B(X)$ of the Blaschke body $B(X)$ of X. The linear contact distribution $H_{[0,u]}$ therefore determines the support value $h(\Pi B(X), u)$. If we know $H_{[0,u]}$ in all directions $u \in S^{d-1}$, the convex body $\Pi B(X)$ is determined. The corresponding integral transform

$$h(\Pi B(X), u) = \frac{1}{2}\int_{S^{d-1}} |\langle x, u\rangle| S_{d-1}(B(X), dx)$$

is called the **cosine transform**. It is injective on even measures, which means that $B(X)$ and therefore the mean area measure $\overline{S}_{d-1}(X, \cdot) = S_{d-1}(B(X), \cdot)$ are determined, provided they are symmetric. A sufficient condition for the latter is of course that all particles $K \in X$ are centrally symmetric. As we mentioned in Section 3.1, for the estimation of $\overline{S}_{d-1}(X, \cdot)$ in the nonsymmetric case, modifications of the linear contact distribution are necessary.

We take a small detour and discuss an application to the mean visible volume. For a stationary Boolean model Z with convex grains, let $S_0(Z)$ be the visible part outside Z, seen from 0, provided $0 \notin Z$. $S_0(Z)$ is the (open) set of all points $x \in \mathbb{R}^d$ such that $[0, x] \cap Z = \emptyset$. Since $S_0(Z)$ is star-shaped,

$$V_d(S_0(Z)) = \frac{1}{d}\int_{S^{d-1}} \rho(S_0(Z), u)^d \sigma_{d-1}(du),$$

where $\rho(M, \cdot)$ is the radial function of a star-shaped set M. We are interested in the mean visible volume outside Z, that is, in the conditional expectation

$$\mathbb{E}(V_d(S_0(Z)) \,|\, 0 \notin Z).$$

Fubini's theorem gives

$$\mathbb{E}(V_d(S_0(Z)) \,|\, 0 \notin Z) = \frac{1}{d}\int_{S^{d-1}} \mathbb{E}(\rho(S_0(Z), u)^d \,|\, 0 \notin Z)\sigma_{d-1}(du).$$

We have $\rho(S_0(Z), u) \le r$, if and only if $d_{[0,u]}(Z) \le r$. Hence,

$$\mathbb{P}(\rho(S_0(Z), u) \le r \,|\, 0 \notin Z) = H_{[0,u]}(r) = 1 - \exp\left(-rh(\Pi B(X), u)\right),$$

an exponential distribution. This implies

$$\mathbb{E}(\rho(S_0(Z), u)^d \mid 0 \notin Z) = d! h(\Pi B(X), u)^{-d} = d! \rho(\Pi^\circ B(X), u)^d,$$

where $\Pi^\circ B(X)$ is the polar of $\Pi B(X)$ (the **polar projection body** of the Blaschke body). We arrive at

$$\mathbb{E}(V_d(S_0(Z)) \mid 0 \notin Z) = (d-1)! \int_{S^{d-1}} \rho(\Pi^\circ B(X), u)^d \sigma_{d-1}(\mathrm{d}u)$$
$$= d! V_d(\Pi^\circ B(X)),$$

which gives us the following result.

Theorem 4.2. *For a stationary Boolean model Z with convex grains (and underlying particle process X), the mean volume outside Z visible from a point $x \notin Z$ equals*

$$d! V_d(\Pi^\circ B(X)).$$

Since

$$V_1(\Pi B(X)) = 2\overline{V}_{d-1}(X),$$

the inequality

$$V_d(M^\circ) \geq c_d V_1(M)^{-d}$$

(with a given constant c_d and with equality, if and only if M is a ball) yields the following result.

Corollary 4.1. *Given the surface area density $2\overline{V}_{d-1}(X)$ of a stationary Poisson process on $\mathcal{K}'$, the mean visible volume outside the union set Z is minimal, iff $\Pi B(X)$ is a ball; for example, if X is isotropic.*

We come back to the discussion of contact distributions and consider nonstationary Boolean models (but still with convex grains). Then, (32) shows that we need a general Steiner-type formula for

$$\int_{(K+rB^*)\setminus K} f(x-y) \, \lambda_d(\mathrm{d}y).$$

This can be given using **relative support measures** $\Xi_{j;d-j}(K; B^*; \cdot)$ and results in

$$\int_{(K+rB^*)\setminus K} f(x-y) \, \lambda_d(\mathrm{d}y)$$
$$= \sum_{j=0}^{d-1} (d-j)k_{d-j} \int_0^r \int_{\mathbb{R}^d \times \mathbb{R}^d} t^{d-1-j} f(x-tu-y) \Xi_{j;d-j}(K; B^*; \mathrm{d}(y,u)) \mathrm{d}t.$$

The definition of relative support measures requires that K and B^* are in general relative position, which means that for all support sets $K(u), B^*(u)$ and $(K + B^*)(u)$, $u \in S^{d-1}$, of K, B^* and $K + B^*$ we have

$$\dim(K + B^*)(u) = \dim K(u) + \dim B^*(u).$$

The condition of general relative position is satisfied, for example, if one of the bodies K, B^* is strictly convex.

Theorem 4.3. *Let Z be a Boolean model with convex grains satisfying (28) and let B be a gauge body. Assume that K and B^* are in general relative position, for $\mathbb{Q}$-almost all K. Then*

$$H_B(x,r) = 1 - \exp\left(- \int_0^r \lambda_B(x,t)dt \right), \qquad \text{for } r \geq 0,$$

with

$$\lambda_B(x,t) := \sum_{j=0}^{d-1} (d-j)k_{d-j}t^{d-1-j} \int_{\mathcal{K}_0} \int_{\mathbb{R}^d \times \mathbb{R}^d} f(x - tu - y)$$

$$\times \, \Xi_{j;d-j}(K; B^*; d(y,u))\mathbb{Q}(dK).$$

If $B = B^d$, the measure $\Xi_{j;d-j}(K; B^*; \cdot)$ is the (ordinary) support measure $\Xi_j(K, \cdot)$ of K.

Theorem 4.3 shows that $H_B(x, \cdot)$ is differentiable. In particular, we get

$$\frac{\partial}{\partial r} H_{B^d}(x,r)|_{r=0} = \lambda_{B^d}(x,0)$$

$$= 2 \int_{\mathcal{K}_0} \int_{\mathbb{R}^d} f(x-y)\Phi_{d-1}(K, dy)\mathbb{Q}(dK)$$

$$= 2\overline{V}_{d-1}(X; x),$$

and thus

$$2\overline{V}_{d-1}(Z; x) = (1 - \overline{V}_d(Z; x))\frac{\partial}{\partial r} H_{B^d}(x,r)|_{r=0}.$$

Hints to the literature. For stationary (or stationary and isotropic) Boolean models, the results of this section are classical and can be found in [37] and [36]. For the nonstationary case, [12] and [14] are the main references. A survey on contact distributions with further references is given in [16]. Relative support measures are introduced and studied in [14] and [22]. Note that in [12], [16] and [22] different normalizations of the (relative) support measures are used.

There are further, interesting inequalities for the mean visible volume, mainly in connection with intersection densities of the boundaries of the particles. Such results go back to Wieacker and are described in [36, Section 4.5].

4.2 Generalized Contact Distributions

The results of the last section can be generalized in various directions.

First we can replace the point x by a convex body L and thus measure the B-distance from L to Z, provided that $L \cap Z = \emptyset$. Hence, we define

$$d_B(L, A) := \inf\{r \geq 0 : (L + rB) \cap A \neq \emptyset\},$$

for $A \in \mathcal{F}$, and

$$H_B(L, r) := \mathbb{P}(L + rB \cap Z \neq \emptyset \mid L \cap Z = \emptyset)$$
$$= \mathbb{P}(d_B(L, Z) \leq r \mid L \cap Z = \emptyset).$$

Then we have to work with a corresponding more general notion of general relative position for three convex bodies L, K, B, use **mixed relative support measures** $\Xi_{i,j;d-i-j}(L, K^*; B; \cdot)$, and get the following generalization of Theorem 4.3.

Theorem 4.4. *Let Z be a Boolean model with convex grains satisfying (28) and let B be a gauge body. Let $L \in \mathcal{K}'$ be such that L, K and B^* are in general relative position, for $\mathbb{Q}$-almost all K. Then*

$$H_B(L, r) = 1 - \exp\left(-\int_0^r \lambda_B(L, t)\mathrm{d}t\right), \qquad \text{for } r \geq 0,$$

with

$$\lambda_B(L, t) := \sum_{i,j,k=0}^{d-1} c_{dijk} t^k \int_{\mathcal{K}_0} \int_{\mathbb{R}^d \times \mathbb{R}^d \times \mathbb{R}^d} f(x + tu + y)$$
$$\times \Xi_{i,j;k+1}(L, K^*; B; \mathrm{d}(x, y, u))\mathbb{Q}(\mathrm{d}K).$$

Here the coefficient c_{dijk} can be given explicitly, it is non-zero only for $i + j + k = d - 1$.

If Z is stationary, we can use

$$\Xi_{i,j;k+1}(L, K^*; B; \mathbb{R}^d \times \mathbb{R}^d \times \mathbb{R}^d) = a_{dijk}V(L[i], K^*[j], B[k + 1])$$

and get

$$H_B(L, r) = 1 - \exp\left(-\sum_{i,j,k=0}^{d-1} b_{dijk} r^{k+1}\overline{V}(L[i], X^*[j], B[k+1])\right),$$

with given constants a_{dijk} and b_{dijk}.

As a further generalization, we may include directions. As one can show, if $L \cap Z = \emptyset$ (and under the condition of general relative position), with probability one, the pair of points $(p_B(L, Z), q_B(L, Z)) \in \partial L \times \partial Z$ which realizes the B-distance,

$$d_B(L, Z) = d_B(p_B(L, Z), q_B(L, Z)),$$

is uniquely determined. Let $u_B(L, Z) \in \partial B$ be the 'direction' from $p_B(L, Z)$ to $q_B(L, Z)$, thus

$$q_B(L, Z) = p_B(L, Z) + d_B(L, Z)u_B(L, Z).$$

We now may even add some geometric information $l_B(L, Z) = \rho_Z(q_B(L, Z))$ in the boundary point $q_B(L, Z)$, which is translation covariant and 'local' in the sense that it depends only on Z in an arbitrarily small neighborhood of $q_B(L, Z)$. For example, in the plane and for a Boolean model with smooth convex grains, $\rho_Z(y)$ could be the curvature of Z in $y \in \partial Z$. A quite general version of Theorem 4.4 then yields the formula

$$\mathbb{E}\left(\mathbf{1}\{d_B(L, Z) < \infty\}g(d_B(L, Z), u_B(L, Z), p_B(L, Z), l_B(L, Z)) \mid Z \cap L = \emptyset\right)$$

$$= \sum_{i,j,k=0}^{d-1} c_{dijk} \int_0^\infty t^k(1 - H_B(L, t)) \int_{\mathcal{K}_0} \int_{\mathbb{R}^d \times \mathbb{R}^d \times \mathbb{R}^d} g(t, u, x, \rho_{y+K^*}(0))$$

$$\times f(x + tu + y)\Xi_{i,j;k+1}(L, K^*; B; \mathrm{d}(x, y, u))\mathbb{Q}(\mathrm{d}K)\mathrm{d}t,$$

for any measurable nonnegative function g.

We sketch the proof of this general result (assuming that the sets under consideration are in general position). We consider a measurable enumeration $X = \{K_1, K_2, ...\}$ of X. Then, the distance $d_B(L, Z)$ is a.s. realized in exactly one grain K_i,

$$d_B(L, Z) = d_B(L, K_i).$$

Let

$$\tilde{g}(Z) := g(d_B(L, Z), u_B(L, Z), p_B(L, Z), l_B(L, Z)).$$

Then,

$$\mathbb{E}\left(\mathbf{1}\{0 < d_B(L, Z) < \infty\}\tilde{g}(Z)\right)$$

$$= \mathbb{E}\left(\sum_{i=1}^\infty \mathbf{1}\{0 < d_B(L, K_i) < \infty\}\mathbf{1}\{d_B(L, Z \setminus K_i) > d_B(L, K_i)\}\tilde{g}(K_i)\right)$$

$$= \mathbb{E}\int \tilde{g}(K)\mathbf{1}\{0 < d_B(L, K) < \infty\}\mathbf{1}\{d_B(L, Z \setminus K) > d_B(L, K)\}X(\mathrm{d}K)$$

$$= \int_{\mathcal{K}'} \int \tilde{g}(K)\mathbf{1}\{0 < d_B(L, K) < \infty\}\mathbf{1}\{d_B(L, Z_N) > d_B(L, K)\}$$

$$\times \mathbb{P}_X^K(\mathrm{d}N)\Theta(\mathrm{d}K),$$

by the refined Campbell theorem. Here $\mathbb{P}_X^K$ is the Palm distribution of X at $K \in X$ and Z_N is the union of the particles in N. For a Poisson process X,

$$\mathbb{P}_X^K = \mathbb{P}_X * \Delta_K$$

(see the contribution of A. Baddeley [1], for notation and result), hence

$$\mathbb{E}\left(\mathbf{1}\{0 < d_B(L, Z) < \infty\}\tilde{g}(Z)\right)$$

$$= \int_{\mathcal{K}'} \tilde{g}(K)\mathbf{1}\{0 < d_B(L, K) < \infty\}\mathbb{P}(d_B(L, Z) > d_B(L, K))\Theta(dK)$$

$$= \mathbb{P}(L \cap Z = \emptyset) \int_{\mathcal{K}_0} \int_{\mathbb{R}^d} \tilde{g}(z + K)\mathbf{1}\{0 < d_B(L, z + K) < \infty\}$$

$$\times (1 - H_B(L, d_B(L, z + K)))\, f(z)\lambda_d(dz)\mathbb{Q}(dK).$$

For the inner integral, the generalized Steiner formula can be used, which finally yields the result.

The above arguments show that a similar formula holds for general germ-grain models (with an additional assumption on the second moment measure of the underlying particle process), if the Palm distributions are used in the statement.

We illustrate the last result with an example. Let $Z \subset \mathbb{R}^2$ be a stationary Boolean model of circles with intensity γ and radii distribution $\mathbb{S}$. For each $x \notin Z$, we measure the (Euclidean) distance $d(x, Z)$ and the radius $r(x, Z)$ in the point $q(x, Z) = x + d(x, Z)u(x, Z)$. Then,

$$\mathbb{P}(d(x, Z) \in A, r(x, Z) \le s \mid x \notin Z)$$

$$= \gamma \int_0^s \int_A (c_0 r + c_1 t)(1 - H(t))dt\, \mathbb{S}(dr),$$

with constants c_0, c_1, which implies in particular that the radii distribution $\mathbb{S}$ can be determined.

This is only a special case of more general results which show that the generalized contact distributions determine, in certain cases, the grain distribution $\mathbb{Q}$.

As a further, far-reaching generalization, we consider Boolean models with arbitrary compact grains and use a general Steiner formula (see Theorem 2.6 in [34]). We only formulate a corresponding result in the stationary case.

Theorem 4.5. *Let Z be a stationary Boolean model with compact grains. Then*

$$H(r) = 1 - \exp\left(-\int_0^r \lambda(t)dt\right), \qquad \text{for } r \ge 0,$$

with

$$\lambda(t)$$

$$:= \sum_{j=0}^{d-1} d_{dj} t^{d-1-j} \gamma \int_{\mathcal{C}_0} \int_{\mathbb{R}^d \times S^{d-1}} \mathbf{1}\{t < \delta(C, x, u)\}\Xi_j(C, d(x, u))\mathbb{Q}(dC).$$

Here, the $\Xi_j(C, \cdot)$ are the (signed) **support measures** (defined on the **generalized normal bundle** of C) and $\delta(C, \cdot)$ is the **reach function**.

The general Steiner formula also yields results for (generalized) contact distributions of arbitrary stationary RACS Z. Namely, let

$$H(t, A) := \mathbb{P}(d(0, Z) \leq t, u(0, Z) \in A \,|\, 0 \notin Z), \quad t \geq 0, A \in \mathcal{B}(S^{d-1}).$$

Then,

$$(1 - p)H(t, A) = \sum_{i=0}^{d-1} c_{di} \int_0^t s^{d-1-i} \Gamma_i(A \times (s, \infty]) \mathrm{d}s$$

with constants c_{di} and

$$\Gamma_i(\cdot) := \mathbb{E}\left(\int_{C^d \times S^{d-1}} \mathbf{1}\{(u, \delta(Z, x, u)) \in \cdot\} \Xi_i(Z, \mathrm{d}(x, u)) \right).$$

Thus, $H(\cdot, A)$ is absolutely continuous and we have an explicit formula for the density.

Moreover, $H(\cdot, A)$ is differentiable with exception of at most countably many points, but it need not be differentiable at 0. If

$$\mathbb{E}|\Xi_i|(Z, B \times S^{d-1}) < \infty, \quad \text{for some } B \in \mathcal{B}(\mathbb{R}^d), \lambda_d(B) > 0,$$

(which excludes fractal behavior, for example), then

$$\lim_{t \to 0+} t^{-1}(1 - p)H(t, A) = \overline{S}_{d-1}(Z, A),$$

for $A \in \mathcal{B}(S^{d-1})$, where

$$\overline{S}_{d-1}(Z, A) := 2 \, \mathbb{E} \, \Xi_{d-1}(Z; C^d \times A) < \infty.$$

In particular, we have, for such RACS,

$$(1 - p)H'(0) = 2\overline{V}_{d-1}(Z) := \overline{S}_{d-1}(Z, S^{d-1}).$$

Hence, for a stationary RACS Z fulfilling the expectation condition above, the surface area density is defined and, even more, a mean surface area measure $\overline{S}_{d-1}(Z, \cdot)$ exists. The normalized measure

$$R(Z, \cdot) = \frac{\overline{S}_{d-1}(Z, \cdot)}{\overline{S}_{d-1}(Z, S^{d-1})}$$

is called the **rose of directions** of Z. It is the distribution of the (outer) normal in a typical point of ∂Z.

Hints to the literature. The results of this section were taken from [15] (see also [16]).

4.3 Characterization of Convex Grains

For a stationary Boolean model Z with convex grains and a gauge body B, Theorem 4.1 shows that

$$-\ln(1 - H_B(r))$$

is a polynomial in $r \geq 0$ (of degree d). This is an important property for statistical purposes, in particular, for suitable gauge bodies B, it yields simple estimators for the intensity γ, the quermass densities $\overline{V}_j(X)$ and other mean functionals like $\overline{S}_{d-1}(X, \cdot)$. For nonconvex grains, the occurrence of the reach function in Theorem 4.5 shows that we cannot expect a polynomial behavior of contact distributions anymore, in general. This can be made more precise and yields a possibility to check the convexity of the grains.

Starting with a stationary Boolean model with compact grains, we assume that the grains of Z are $\mathbb{Q}$-a.s. **regular** (i.e. the closure of their interior) and fulfill a more general integrability assumption, namely

$$\int_{\mathcal{C}_0} V_d(\text{conv } K + B^d)\mathbb{Q}(\mathrm{d}K) < \infty. \tag{33}$$

We define the **ALLC-function** (average logarithmic linear contact distribution) L of Z by

$$L(r) := -\int_{S^{d-1}} \ln(1 - H_{[0,u]}(r))\sigma_{d-1}(\mathrm{d}u), \quad r \geq 0.$$

Theorem 4.6. *Let Z be a stationary Boolean model with regular compact grains satisfying (33). Then, the ALLC-function L of Z is linear, if and only if the grains are a.s. convex.*

We sketch the *proof* of the non-obvious direction. Thus, we assume that L is linear. Then

$$f := \overline{V}_d(X) + L$$

is a polynomial and

$$f(r) = \gamma \int_{S^{d-1}} \int_{\mathcal{C}_0} \lambda_d(K + r[0, u])\mathbb{Q}(\mathrm{d}K)\sigma_{d-1}(\mathrm{d}u).$$

We have

$$f(r) \leq \tilde{f}(r) := \gamma \int_{S^{d-1}} \int_{\mathcal{C}_0} \lambda_d(K_u + r[0, u])\mathbb{Q}(\mathrm{d}K)\sigma_{d-1}(\mathrm{d}u)$$

$$= \gamma \int_{S^{d-1}} \int_{\mathcal{C}_0} \left(\lambda_d(K_u) + \lambda_{d-1}(K \mid u^{\perp})r\right)\mathbb{Q}(\mathrm{d}K)\sigma_{d-1}(\mathrm{d}u)$$

$$= \tilde{a} + \tilde{b}r,$$

where K_u is the convexification of K in direction u (for each line l in direction u, we replace $K \cap l$ by its convex hull). Hence,

$$f(r) = a + br, \qquad \text{with} \qquad a \leq \tilde{a}, b \leq \tilde{b}.$$

For large enough r, we have $K + r[0, u] = K_u + r[0, u]$, uniformly in u. This implies $f = \tilde{f}$, for large r and therefore for all $r \geq 0$. But then $K = K_u$, for all u, which implies convexity. (Notice that in some of these arguments the regularity of the grains is used.) $\qquad \square$

There is a corresponding result for the two-dimensional unit disc B which concerns the **ALDC-function** (average logarithmic disc contact distribution) D of Z,

$$D(r) := -\int_{SO_d} \ln(1 - H_{\vartheta B}(r))\nu(d\vartheta), \quad r \geq 0.$$

Instead of (33), we need the stronger assumption of uniformly bounded grains.

Theorem 4.7. *Let Z be a stationary Boolean model with regular, uniformly bounded compact grains. Then, the ALDC-function D of Z is a polynomial, if and only if the grains are a.s. convex.*

The *proof* is more complicated and is based on the following steps.

First, by Fubini's theorem, it is sufficient to show the result for $d = 2$ and a fixed (regular) grain $K \in \mathcal{C}'$. More precisely, it is sufficient to show that $K \subset \mathbb{R}^2$ is convex, if $\lambda_2(K + rB^2)$ is a polynomial in $r \geq 0$.

As in the linear case, we compare $\lambda_2(K(r)) = \sum_{i=0}^m a_i r^i, K(r) := K + rB^2$, with the volume of the convex hull $\bar{K}(r) := \bar{K} + rB^2, \bar{K} := \text{conv } K$,

$$\lambda_2(\bar{K}(r)) = A(\bar{K}) + 2rA(\bar{K}, B^2) + r^2 A(B^2).$$

Since $\lambda_2(rB^2) \leq \lambda_2(K(r)) \leq \lambda_2(\bar{K}(r))$, we obtain

$$\lambda_2(K(r)) = a_0 + a_1 r + a_2 r^2, \qquad r \geq 0,$$

with $a_2 = A(B^2)$ and $a_0 + a_1 r \leq A(\bar{K}) + 2rA(\bar{K}, B^2)$.

For λ_1-almost all $r \geq 0$, we have

$$\frac{d}{dr} A(K(r)) = \int_{\partial K(r)} h(B^2, u_{K(r)}(x))\mathcal{H}^1(dx),$$

where $u_{K(r)}$ is the ($\mathcal{H}^1$-a.e. existing) outer normal in x and $h(M, \cdot)$ is the support function of a convex body M. Moreover, $K(r)$ is star-shaped, for r large enough. Hence,

$$\int_{\partial K(r)} h(B^2, u_{K(r)}(x))\mathcal{H}^1(dx) \leq \int_{\partial \bar{K}(r)} h(B^2, u_{\bar{K}(r)}(x))\mathcal{H}^1(dx),$$

which, under the spherical image map, transforms to

$$\int_{S^1} h(B^2, u)S_1(K(r), du) \leq \int_{S^1} h(B^2, u)S_1(\bar{K}(r), du). \qquad (34)$$

Here, the image measure $S_1(\bar{K}(r), \cdot)$ is the surface area measure of $\bar{K}(r)$ and $S_1(K(r), \cdot)$ is, by Minkowski's theorem, also the surface area measure of some convex body $\tilde{K}(r)$, the **convexification** of $K(r)$.

As one can show, $K(r) \subset \tilde{K}(r)$, after a suitable translation, and therefore $\bar{K}(r) \subset \tilde{K}(r)$. This implies $h(\bar{K}(r), \cdot) \leq h(\tilde{K}(r), \cdot)$. On the other hand, (34) and the symmetry of mixed areas yield

$$\int_{S^1} h(\bar{K}(r), u)\sigma_1(\mathrm{d}u) = \int_{S^1} h(B^2, u)S_1(\bar{K}(r), \mathrm{d}u)$$

$$\geq \int_{S^1} h(B^2, u)S_1(K(r), \mathrm{d}u)$$

$$= \int_{S^1} h(B^2, u)S_1(\tilde{K}(r), \mathrm{d}u)$$

$$= \int_{S^1} h(\tilde{K}(r), u)\sigma_1(\mathrm{d}u).$$

Therefore $h(\bar{K}(r), \cdot) = h(\tilde{K}(r), \cdot)$, hence $\bar{K}(r) = \tilde{K}(r)$, which implies that $K(r)$ and the convex hull $\bar{K}(r)$ have the same boundary length, $\mathcal{H}^1(\partial K(r)) = \mathcal{H}^1(\partial \bar{K}(r))$. For a planar star-shaped set this implies $K(r) = \bar{K}(r)$. Consequently, $K = \bar{K} = \mathrm{conv}\, K$. $\qquad \square$

Theorem 4.7 holds in a more general version, with the unit disc B replaced by a smooth planar body, and the proof is essentially the same.

However, a corresponding result for three-dimensional gauge bodies B is wrong. In particular, for $d = 3$ and $B = B^3$, $\ln(1 - H)$ can be a polynomial, without that Z has convex grains! An example is given by a Boolean model Z, the primary grain of which is the union of two touching balls (of equal radius).

Hints to the literature. The results in this section can be found in [11] and [17].

References

1. Baddeley, A.: Spatial point processes and their applications. In this volume
2. Beneš, V., Rataj, J.: Stochastic Geometry: Selected Topics. Kluwer Academic Publ., Boston Dortrecht New York London (2004)
3. Betke, U., Weil, W.: Isoperimetric inequalities for the mixed area of plane convex sets. Arch. Math., **57**, 501–507 (1991)
4. Daley, D., Vere-Jones, D.: An Introduction to the Theory of Point Processes. Springer, New York Berlin Heidelberg (1988)
5. Fallert, H.: Intensitätsmaße und Quermaßdichten für (nichtstationäre) zufällige Mengen und geometrische Punktprozesse. Doctoral Thesis, Universität Karlsruhe (1992)
6. Fallert, H.: Quermaßdichten für Punktprozesse konvexer Körper und Boolesche Modelle. Math. Nachr., **181**, 165–184 (1996)

7. Goodey, P., Weil, W.: Representations of mixed measures with applications to Boolean models. Rend. Circ. Mat. Palermo (2) Suppl., **70**, 325–346 (2002)
8. Grimmett, G.: Percolation. Second edition, Springer, Berlin Heidelberg New York (1999)
9. Hall, P.: Introduction to the Theory of Coverage Processes. Wiley, New York (1988)
10. Heinrich, L.: On existence and mixing properties of germ-grain models. Statistics, **23**, 271–286 (1992)
11. Heveling, M., Hug, D., Last, G.: Does parallel volume imply convexity? Math. Ann., **328**, 469–479 (2004)
12. Hug, D.: Contact distributions of Boolean models. Rend. Circ. Mat. Palermo (2) Suppl, **65**, 137–181 (2000)
13. Hug, D.: Random mosaics. In this volume
14. Hug, D., Last, G.: On support measures in Minkowski spaces and contact distributions in stochastic geometry. Ann. Probab., **28**, 796–850 (2000)
15. Hug, D., Last, G., Weil, W.: Generalized contact distributions of inhomogeneous Boolean models. Adv. in Appl. Probab., **34**, 21–47 (2002)
16. Hug, D., Last, G., Weil, W.: A survey on contact distributions. In: Mecke, K., Stoyan, D. (eds) Morphology of Condensed Matter, Physics and Geometry of Spatially Complex Systems. Lect. Notes in Physics, **600**, Springer, Berlin Heidelberg New York (2002)
17. Hug, D., Last, G., Weil, W.: Polynomial parallel volume, convexity and contact distributions of random sets. Probab. Theory Related Fields, (in print)
18. Jonasson, J.: Optimization of shape in continuum percolation. Ann. Probab., **29**, 624–635 (2001)
19. Kendall, D.G.: Foundations of a theory of random sets. In: Harding, E.F., Kendall, D.G. (eds) Stochastic Geometry. A Tribute to the Memory of Rollo Davidson. Wiley, London (1974)
20. Kiderlen, M.: Determination of the mean normal measure from isotropic means of flat sections. Adv. in Appl. Probab., **34**, 505–519 (2002)
21. Kiderlen, M., Jensen, E.B. Vedel: Estimation of the directional measure of planar random sets by digitization. Adv. in Appl. Probab., **35**, 583–602 (2003)
22. Kiderlen, M., Weil, W.: Measure-valued valuations and mixed curvature measures of convex bodies. Geom. Dedicata, **76**, 291–329 (1999)
23. König, D., Schmidt, V.: Zufällige Punktprozesse. Teubner, Stuttgart (1992)
24. Matheron, G.: Ensembles fermeés aléatoires, ensembles semi-markoviens et polyèdres poissoniens. Adv. in Appl. Probab., **4**, 508–541 (1972)
25. Matheron, G.: Random Sets and Integral Geometry. Wiley, New York (1975)
26. Mecke, K.: Additivity, convexity and beyond: Applications of Minkowski functionals in statistical physics. In: Mecke, K., Stoyan, D. (eds) Statistical Physics and Spatial Statistics. The Art of Analyzing and Modeling Spatial Structures and Pattern Formation. Lect. Notes in Physics, **554**, Springer, Berlin Heidelberg New York (2000)
27. Molchanov, I.S.: Statistics of the Boolean Model for Practitioners and Mathematicians. Wiley, New York (1997)
28. Molchanov, I.: Theory of Random Sets. Springer, London (2005)
29. Molchanov, I., Stoyan, D.: Asymptotic properties of estimators for parameters of the Boolean model. Adv. in Appl. Probab., **26**, 301–323 (1994)
30. Nguyen, X.X., Zessin, H.: Ergodic theorems for spatial processes. Z. Wahrsch. Verw. Gebiete, **48**, 133–158 (1979)

31. Rataj, J.: Estimation of oriented direction distribution of a planar body. Adv. in Appl. Probab., **28**, 349–404 (1996)

32. Schmidt, V., Spodarev, E.: Joint estimators for the specific intrinsic volumes of stationary random sets. Stochastic Process. Appl., **115**, 959–981 (2005)

33. Schneider, R.: Convex Bodies: the Brunn–Minkowski Theory. Encyclopedia of Mathematics and Its Applications, **44**, Cambridge University Press, Cambridge (1993)

34. Schneider, R.: Integral geometric tools for stochastic geometry. In this volume

35. Schneider, R., Weil, W.: Integralgeometrie. Teubner, Stuttgart (1992)

36. Schneider, R., Weil, W.: Stochastische Geometrie. Teubner, Stuttgart (2000)

37. Stoyan, D., Kendall, W.S., Mecke, J.: Stochastic Geometry and Its Applications. Second edition, Wiley, New York (1995)

38. Torquato, S.: Random Heterogeneous Materials. Microstructure and Macroscopic Properties. Interdisciplinary Applied Mathematics, **16**, Springer, New York Berlin Heidelberg (2002)

39. Weil, W.: Densities of quermassintegrals for stationary random sets. In: Ambartzumian, R., Weil, W. (eds) Stochastic Geometry, Geometric Statistics, Stereology. Proc. Conf. Oberwolfach, 1983. Teubner, Leipzig (1984)

40. Weil, W.: Iterations of translative integral formulae and non-isotropic Poisson processes of particles. Math. Z., **205**, 531–549 (1990)

41. Weil, W.: Support functions on the convex ring in the plane and support densities for random sets and point processes. Rend. Circ. Mat. Palermo (2) Suppl., **35**, 323–344 (1994)

42. Weil, W.: The estimation of mean shape and mean particle number in overlapping particle systems in the plane. Adv. in Appl. Probab., **27**, 102–119 (1995)

43. Weil, W.: On the mean shape of particle processes. Adv. in Appl. Probab., **29**, 890–908 (1997)

44. Weil, W.: Mean bodies associated with random closed sets. Rend. Circ. Mat. Palermo (2) Suppl., **50**, 387–412 (1997)

45. Weil, W.: The mean normal distribution of stationary random sets and particle processes. In: Jeulin, D. (ed) Advances in Theory and Applications of Random Sets. Proc. International Symposium, Oct. 9-11, 1996, Fontainebleau, France. World Scientific, Singapore (1997)

46. Weil, W.: Intensity analysis of Boolean models. Pattern Recognition, **32**, 1675–1684 (1999)

47. Weil, W.: A uniqueness problem for non-stationary Boolean models. Rend. Circ. Mat. Palermo (2) Suppl., **65**, 329–344 (2000)

48. Weil, W.: Densities of mixed volumes for Boolean models. Adv. in Appl. Probab., **33**, 39–60 (2001)

49. Weil, W., Wieacker, J.A.: Densities for stationary random sets and point processes. Adv. in Appl. Probab., **16**, 324–346 (1984)

50. Zähle, M.: Curvature measures and random sets, II. Probab. Theory Related Fields, **71**, 37–58 (1986)

Random Mosaics

Daniel Hug

Mathematisches Institut, Albert-Ludwigs-Universität
Eckerstr. 1, D-79104 Freiburg, Germany
e-mail: daniel.hug@math.uni-freiburg.de

Mosaics arise frequently as models in materials science, biology, geology, telecommunications, and in data and point pattern analysis. Very often the data from which a mosaic is constructed is random, hence we arrive at a random mosaic. Such a random mosaic can be considered as a special particle process or as a random closed set having a special structure. Random mosaics provide a convenient and important model for illustrating some of the general concepts discussed in previous parts of these lecture notes. In particular, the investigation of random mosaics naturally requires a combination of probabilistic, geometric and combinatorial ideas.

In this brief survey of selected results and methods related to random mosaics, we will mainly concentrate on classical topics which are contained in the monographs [11], [14] and [16]. A survey with a different emphasis is provided in [12]. Some of the more recent developments related to modelling of communication networks, iterated constructions of random mosaics and related central limit theorems cannot be discussed here. But there is also interesting recent research which is concerned with obtaining quantitative information about various shape characteristics of random mosaics. In the final part of this appendix, we will report on work carried out in this direction.

1 General Results

In this section, we collect general information which is available for arbitrary mosaics. Later we will concentrate on results for special types of mosaics such as hyperplane, Voronoi and Delaunay mosaics.

1.1 Basic Notions

In a purely deterministic setting, a mosaic can be defined in a rather general way. For instance, we may request a mosaic (tessellation) in $\mathbb{R}^d$ to be a system of closed sets which cover $\mathbb{R}^d$ and have no common interior points. It turns out, however, that for many applications this definition is unnecessarily general. Therefore we use the following more restricted notion of mosaic which is sufficient for our present introductory purposes. Thus we define a **mosaic** in $\mathbb{R}^d$ as a locally finite system of compact convex sets (cells) with nonempty interiors which cover $\mathbb{R}^d$ and have mutually no common interior points. Let $\mathcal{M}$ denote the set of mosaics.

Lemma 1.1. *The cells of a mosaic* m *are convex polytopes.*

Proof. This follows from basic separation properties of convex sets. In fact, if $K \in$ m and $K_1, \ldots, K_k \in$ m $\setminus \{K\}$ satisfy $K \cap K_i \neq \emptyset$ and $\partial K = \bigcup_{i=1}^{k}(K \cap K_i)$, then $K = \bigcap_{i=1}^{k} H_i^+$. Here, for $i = 1, \ldots, k$, H_i^+ is a halfspace which contains K and is bounded by a hyperplane which separates K and K_i.   □

As a consequence of the lemma, it is natural to consider the facial structure of the cells. A k-face of m $\in \mathcal{M}$ is a k-face of a cell of m. These faces contain additional explicit information about the metric and combinatorial structure of the mosaic. For the mathematical analysis of mosaics and their faces, it is often convenient to consider more restricted classes of tessellations which enjoy additional properties. For instance, we will exclude the possibility that the intersection of two faces of different cells is not a face of both cells. Thus we use the following terminology. For $k = 0, \ldots, d$ and a polytope $P \subset \mathbb{R}^d$, let $\mathcal{S}_k(P)$ denote the set of k-faces of P, and let $\mathcal{S}(P)$ be the set of faces of P (of any dimension). Then a mosaic m $\in \mathcal{M}$ is **face-to-face** if $P \cap P' \in (\mathcal{S}(P) \cap \mathcal{S}(P')) \cup \{\emptyset\}$, for all $P, P' \in$ m. We write $\mathcal{S}_k($m$) := \bigcup_{P \in m} \mathcal{S}_k(P)$ for the set of all k-faces of m. Let $\mathcal{M}_s$ denote the set of face-to-face mosaics. A mosaic m $\in \mathcal{M}_s$ is called **normal** if each k-face lies in exactly $d - k + 1$ cells.

All mosaics considered here will be face-to-face. A line mosaic is an example of a mosaic which is not normal. It easily follows from the definition that if m is a normal mosaic, then each j-face of m lies in exactly $\binom{d-j+1}{d-k+1} = \binom{d-j+1}{k-j}$ different k-faces of m.

1.2 Random Mosaics

In the following, let $(\Omega, \mathbf{A}, \mathbb{P})$ denote an underlying probability space. Further, let $\mathsf{N}(\mathcal{F}')$ denote the space of locally finite subsets of $\mathcal{F}'$, where $\mathcal{F}'$ is the set of all nonempty closed subsets of $\mathbb{R}^d$. Then a **random mosaic** X in $\mathbb{R}^d$ is a particle process (i.e. a measurable map) $X : (\Omega, \mathbf{A}, \mathbb{P}) \to \mathsf{N}(\mathcal{F}')$ satisfying $X \in \mathcal{M}_s$ with probability one. Put

$$X^{(k)} := \mathcal{S}_k(X), \qquad \Theta^{(k)} := \mathbb{E}X^{(k)}, \qquad k = 0, \ldots, d.$$

The **intensity measure** $\Theta^{(k)}$ of the **process of k-faces** $X^{(k)}$ is assumed to be locally finite. If X is stationary, then $X^{(k)}$ is stationary, and hence we obtain the decomposition

$$\Theta^{(k)} = \gamma^{(k)} \int_{\mathcal{K}_0} \int_{\mathbb{R}^d} \mathbf{1}\{x + K \in \cdot\} \lambda_d(\mathrm{d}x) \mathbb{P}_0^{(k)}(\mathrm{d}K),$$

where $\mathcal{K}_0 := \{K \in \mathcal{K} : c(K) = 0\}$ with a centre function c, $\mathbb{P}_0^{(k)}$ is a probability measure, and the factor $\gamma^{(k)}$ is the **intensity** of $X^{(k)}$. Any random polytope Z_k with distribution $\mathbb{P}_0^{(k)}$ is called **typical k-face** of X. The **quermass densities** (the densities of the intrinsic volumes) of $X^{(k)}$ are

$$d_j^{(k)} := \overline{V}_j(X^{(k)}) := \gamma^{(k)} \int_{\mathcal{K}_0} V_j(K) \mathbb{P}_0^{(k)}(\mathrm{d}K).$$

For these densities alternative representations and various relationships are known. For instance, Theorem 14 in Wolfgang Weil's contribution in this volume implies

$$\overline{V}_j(X) = \lim_{r \to \infty} \frac{1}{V_d([-r,r]^d)} \mathbb{E} \sum_{K \in X} V_j(K \cap [-r,r]^d),$$

and hence $d_d^{(d)} = \overline{V}_d(X) = 1$, which shows that $\mathbb{E}V_d(Z) = 1/\gamma^{(d)}$. Moreover, the intensities of the processes of i-faces satisfy the Euler relation

$$\sum_{i=0}^{d} (-1)^i \gamma^{(i)} = 0,$$

which is the special case $j = 0$ of the following theorem.

Theorem 1.1. *Let X be a stationary random mosaic in $\mathbb{R}^d$. Then, for $j \in \{0, \ldots, d-1\}$,*

$$\sum_{i=j}^{d} (-1)^i d_j^{(i)} = 0.$$

Proof. For $\mathrm{m} \in \mathcal{M}$, let $S_1, \ldots, S_p$ be the cells of m hitting the unit ball B^d with centre at the origin. By additivity,

$$V_j(B^d) = V_j \left(B^d \cap \bigcup_{l=1}^{p} S_l \right) = \sum_{r=1}^{p} (-1)^{r-1} \sum_{|I|=r} V_j \left(\bigcap_{l \in I} S_l \cap B^d \right)$$

$$= \sum_{i=j}^{d} \sum_{F \in \mathcal{S}_i(\mathrm{m})} V_j(F \cap B^d) \sum_{r=1}^{p} (-1)^{r-1} \nu(F, r), \tag{1}$$

where

$$\nu(F, r) = \mathrm{card}\left(\{I \subset \{1, \ldots, p\} : |I| = r, \bigcap_{l \in I} S_l = F\}\right).$$

Below we will show that

$$\sum_{r=1}^{p}(-1)^{r-1}\nu(F,r) = (-1)^{d-\dim(F)}. \tag{2}$$

From (1) and (2) with ρB^d instead of B^d, we get

$$V_j(\rho B^d)/V_d(\rho B^d) = \sum_{i=j}^{d}(-1)^{d-i} \sum_{F\in \mathcal{S}_i(X)} V_j(F\cap \rho B^d)/V_d(\rho B^d). \tag{3}$$

By Theorem 14 in Wolfgang Weil's contribution and since $j < d$,

$$0 = \lim_{\rho\to\infty} \frac{V_j(\rho B^d)}{V_d(\rho B^d)} = \lim_{\rho\to\infty} \frac{1}{V_d(\rho B^d)}\mathbb{E}\sum_{i=j}^{d}(-1)^{d-i} \sum_{F\in X^{(i)}} V_j(F\cap \rho B^d)$$

$$= \sum_{i=j}^{d}(-1)^{d-i}\overline{V}_j(X^{(i)}).$$

We verify (2) in the special case $F = \{x\}$, the general case can be deduced from the special one. Let $T_1,\dots,T_q$ be the cells containing x, choose a polytope $P \subset \mathbb{R}^d$ with $x \in \mathrm{int}(P)$ and $P \subset \mathrm{int}(T_1\cup\dots\cup T_q)$. Then, for $|I| = r$,

$$\chi\left(\bigcap_{l\in I}T_l\cap \partial P\right) = \begin{cases} 0, & \bigcap_{l\in I}T_l = \{x\}, \\ 1, & \text{otherwise,}\end{cases}$$

and hence

$$\sum_{r=1}^{p}(-1)^{r-1}\nu(\{x\},r) = \sum_{r=1}^{q}(-1)^{r-1}\sum_{|I|=r}\left(1-\chi\left(\bigcap_{l\in I}T_l\cap \partial P\right)\right)$$

$$= 1 - \chi(\partial P)$$

$$= (-1)^d,$$

which completes the proof. $\square$

1.3 Face-Stars

In order to admit a finer analysis of the combinatorial relations (incidences) between faces of different dimensions, we use the notion of a **face-star**. For a j-face $T \in \mathcal{S}_j(\mathsf{m})$, let

$$\mathcal{S}_k(T,\mathsf{m}) := \begin{cases} \mathcal{S}_k(T - c(T)), & k \le j, \\ \{S - c(T) : S \in \mathcal{S}_k(\mathsf{m}), T \subset S\}, & k > j.\end{cases}$$

Then we call $(T,\mathcal{S}_k(T,\mathsf{m}))$ a (j,k)-**face-star**. The special centering used here turns out to be appropriate in view of the symmetric form of the basic relation provided by Theorem 1.2. Let X be a stationary random mosaic X. From X we derive a stationary marked particle process by considering the (j, k)-face-stars of the j-faces of X. For its intensity measure we obtain the decomposition

$$\mathbb{E} \sum_{T \in \mathcal{S}_j(X)} h(T, \mathcal{S}_k(T, X))$$

$$= \gamma^{(j,k)} \int_{\mathbb{R}^d} \int_{\mathcal{K}_0 \times \mathcal{F}(\mathcal{K}')} h(x + T, \mathcal{S}) \mathbb{P}_0^{(j,k)}(\mathrm{d}(T, \mathcal{S})) \lambda_d(\mathrm{d}x),$$

where h is a nonnegative measurable function. The probability measure $\mathbb{P}_0^{(j,k)}$ is called the shape-mark distribution of X. A random (j, k)-face-star with this distribution is said to be a typical (j, k)-face star. Specialization leads to

$$\gamma^{(j,k)} = \gamma^{(j)} \qquad \text{and} \qquad \mathbb{P}_0^{(j,k)}(\cdot \times \mathcal{F}(\mathcal{K}')) = \mathbb{P}_0^{(j)}.$$

Moreover, the characteristics

$$v_i^{(j,k)} := \int_{\mathcal{K}_0 \times \mathcal{F}(\mathcal{K}')} \sum_{S \in \mathcal{S}} V_i(S) \mathbb{P}_0^{(j,k)}(\mathrm{d}(T, \mathcal{S})), \qquad d_i^{(j,k)} := \gamma^{(j)} v_i^{(j,k)}$$

are useful. For instance, the mean number of k-faces of a typical (j, k)-face-star is given by

$$n_{jk} := v_0^{(j,k)} = \int_{\mathcal{K}_0 \times \mathcal{F}(\mathcal{K}')} \mathrm{card}(\mathcal{S}) \, \mathbb{P}_0^{(j,k)}(\mathrm{d}(T, \mathcal{S})).$$

If $k \leq j$, then

$$n_{jk} = \int_{\mathcal{K}_0} \mathrm{card}(\mathcal{S}_k(T)) \, \mathbb{P}_0^{(j)}(\mathrm{d}T),$$

and hence

$$\sum_{k=0}^{j} (-1)^k n_{jk} = 1, \qquad j = 0, \ldots, d, \tag{4}$$

which is an Euler relation for random mosaics. A general symmetry result is provided by the following theorem.

Theorem 1.2. *Let X be a stationary random mosaic in $\mathbb{R}^d$, let $f : \mathcal{K} \times \mathcal{K} \to \mathbb{R}$ be a nonnegative, measurable, translation invariant function. Then, for $j, k \in \{0, \ldots, d\}$,*

$$\gamma^{(j)} \int_{\mathcal{K}_0 \times \mathcal{F}(\mathcal{K}')} \sum_{S \in \mathcal{S}} f(S, T) \mathbb{P}_0^{(j,k)}(\mathrm{d}(T, \mathcal{S}))$$

$$= \gamma^{(k)} \int_{\mathcal{K}_0 \times \mathcal{F}(\mathcal{K}')} \sum_{T \in \mathcal{T}} f(S, T) \mathbb{P}_0^{(k,j)}(\mathrm{d}(S, \mathcal{T})).$$

For the proof one uses an extension of Theorem 14 in Wolfgang Weil's contribution in this volume for stationary marked particle processes and the symmetry relation

$$\{(S + c(T), T) : T \in \mathcal{S}_j(\mathsf{m}), S \in \mathcal{S}_k(T, \mathsf{m})\}$$
$$= \{(S, T + c(S)) : S \in \mathcal{S}_k(\mathsf{m}), T \in \mathcal{S}_j(S, \mathsf{m})\}.$$

The usefulness of Theorem 1.2 can be seen from the consequence

$$\gamma^{(j)} n_{jk} = \gamma^{(k)} n_{kj}, \tag{5}$$

which will be used later. Further, if X is normal and $j \le k$, then we have $n_{jk} = \binom{d+1-j}{k-j}$, and hence

$$\sum_{j=0}^{k} (-1)^j \binom{d+1-j}{k-j} \gamma^{(j)} = \sum_{j=0}^{k} (-1)^j \gamma^{(j)} n_{jk} = \gamma^{(k)} \sum_{j=0}^{k} (-1)^j n_{kj} = \gamma^{(k)},$$

where (4) was used for the last equality. Thus we arrive at

$$\left(1 - (-1)^k\right) \gamma^{(k)} = \sum_{j=0}^{k-1} (-1)^j \binom{d+1-j}{k-j} \gamma^{(j)},$$

which is another Euler-type relation for the intensities of the processes of faces of the given random mosaic.

1.4 Typical Cell and Zero Cell

The **typical cell** Z of a stationary random mosaic X can be considered as a spatial average, since its distribution can be written in the form

$$\mathbb{P}_0^{(d)} = \lim_{r \to \infty} \frac{\mathbb{E} \sum_{K \in X} \mathbf{1}\{K \subset rB^d\} \mathbf{1}\{K - c(K) \in \cdot\}}{\mathbb{E} \sum_{K \in X} \mathbf{1}\{K \subset rB^d\}}.$$

Instead of such an average shape we may also consider a specific cell such as the **zero cell** Z_0, i.e.

$$Z_0 = \bigcup_{K \in X} \mathbf{1}\{0 \in \text{int}(K)\} K.$$

Is there any relation between the sizes of Z and Z_0? For a first answer, observe that by **Campbell's theorem**

$$\mathbb{E} f(Z_0) = \gamma^{(d)} \mathbb{E} \left[f(Z) V_d(Z) \right], \quad \gamma^{(d)} = (\mathbb{E} V_d(Z))^{-1}$$

for any nonnegative, measurable, translation invariant function $f : \mathcal{K} \to \mathbb{R}$. The Cauchy-Schwarz inequality then implies

$$\mathbb{E} V_d(Z_0) = \mathbb{E} V_d(Z)^2 / \mathbb{E} V_d(Z) \ge \mathbb{E} V_d(Z).$$

A stronger result is given below.

Theorem 1.3. *Let X be a stationary random mosaic in $\mathbb{R}^d$. Let F_0, F denote the distribution functions of $V_d(Z_0)$, $V_d(Z)$. Then $F_0(x) \leq F(x)$ for $x \geq 0$, and thus $\mathbb{E}V_d(Z_0)^k \geq \mathbb{E}V_d(Z)^k$, $k \in \mathbb{N}$.*

Proof. The first assertion follows by verifying

$$(F(x) - F_0(x)) \, \mathbb{E}V_d(Z)$$
$$= F(x) \int_x^\infty (1 - F(t))\mathrm{d}t + (1 - F(x)) \int_0^x F(t)\mathrm{d}t \geq 0,$$

the second then is clear from

$$\mathbb{E}V_d(Z)^k = k \int_0^\infty x^{k-1}(1 - F(x))\mathrm{d}x,$$

which completes the proof.   □

2 Voronoi and Delaunay Mosaics

The two most important mosaics are the Voronoi and the Delaunay mosaics. First, we introduce them separately, but then we also point out their combinatorial equivalence.

2.1 Voronoi Mosaics

Let $A \subset \mathbb{R}^d$ be locally finite. We will write $\|\cdot\|$ for the Euclidean norm in $\mathbb{R}^d$. Then, for $x \in A$, the **Voronoi cell** of A with centre x is defined by

$$C(x, A) := \{z \in \mathbb{R}^d : \|z - x\| = d(z, A)\},$$

where $d(z, A) := \min\{\|z - a\| : a \in A\}$. It is clear from the definition that the Voronoi cells are always polyhedral sets.

Lemma 2.1. *If* $\mathrm{conv}(A) = \mathbb{R}^d$, *then* $\mathcal{V}(A) := \{C(x, A) : x \in A\}$ *is a mosaic which is face-to-face.*

If $\tilde{X}$ is a stationary point process in $\mathbb{R}^d$, then $X := \mathcal{V}(\tilde{X})$ is a stationary random mosaic, the **Voronoi mosaic** of $\tilde{X}$. If $\tilde{X}$ is a Poisson process, we call X the **Poisson-Voronoi mosaic** induced by $\tilde{X}$.

It will be useful to have a characterization of the flats spanned by the faces of a Voronoi mosaic. To describe this, let $A \subset \mathbb{R}^d$ be a locally finite set in general position. The latter means that any $k + 1$ points of A do not lie in a $(k - 1)$-flat $(k = 1, \ldots, d + 1)$. Then, for $m \in \{1, \ldots, d\}$ and $x_0, \ldots, x_m \in A$, let $B^m(x_0, \ldots, x_m)$ denote the m-ball having $x_0, \ldots, x_m$ in its boundary. Let $z(x_0, \ldots, x_m)$ be the centre of this ball, $F(x_0, \ldots, x_m)$ the orthogonal flat through it, and write $S(x_0, \ldots, x_m; A)$ for the set of all $y \in F(x_0, \ldots, x_m)$ satisfying $\|y - x_0\| = d(y, A)$. Then

$$F(x_0, \ldots, x_m) = \text{aff}(S) \quad \text{for some } S \in \mathcal{S}_{d-m}(\mathcal{V}(A))$$

if and only if

$$S(x_0, \ldots, x_m; A) \neq \emptyset.$$

If either condition is satisfied, then $S = S(x_0, \ldots, x_m; A)$. Moreover, any face $S \in \mathcal{S}_{d-m}(\mathcal{V}(A))$ can be obtained in this way.

Theorem 2.1. *Let X be the Poisson-Voronoi mosaic induced by $\tilde{X}$ with intensity $\tilde{\gamma}$. Then X is a normal mosaic. For $0 \le k \le j \le d$,*

$$d_k^{(j,k)} = \binom{d - k + 1}{j - k} d_k^{(k)} \quad \text{and} \quad d_k^{(k)} = c(d, k) \, \tilde{\gamma}^{\frac{d-k}{d}},$$

where $c(d, k)$ is an explicitly known constant.

Proof. By Theorem 1.2, we obtain

$$d_k^{(j,k)} = \gamma^{(k)} \int\limits_{\mathcal{K}_0 \times \mathcal{F}(\mathcal{K}')} V_k(S) N_j(S, T) \, \mathbb{P}_0^{(k,j)}(\mathrm{d}(S, T))$$

$$= \binom{d - k + 1}{j - k} d_k^{(k)},$$

where $N_j(S, T)$ is the number of j-faces in the (k, j)-face-star (S, T). Thus it remains to calculate $d_k^{(k)}$. For this task, we will use the (extended) **Slivnyak-Mecke formula**

$$\mathbb{E} \sum_{(x_1, \ldots, x_m) \in \tilde{X}_{\neq}^m} f(\tilde{X}; x_1, \ldots, x_m)$$

$$= \tilde{\gamma}^m \int_{\mathbb{R}^d} \cdots \int_{\mathbb{R}^d} \mathbb{E} f(\tilde{X} \cup \{x_1, \ldots, x_m\}; x_1, \ldots, x_m) \lambda_d(\mathrm{d}x_1) \ldots \lambda_d(\mathrm{d}x_m),$$

which holds for any nonnegative, measurable function f (cf. Theorems 3.1 and 3.2 in Adrian Baddeley's contribution in this volume, [11, Proposition 4.1], [13, Theorem 3.3]). Hence we get

$$d_k^{(k)} = \overline{V}_k(X^{(k)}) = \frac{1}{\kappa_d} \mathbb{E} \sum_{S \in X^{(k)}} V_k(S \cap B^d)$$

$$= \frac{1}{\kappa_d (d - k + 1)!} \mathbb{E} \sum_{(x_0, \ldots, x_{d-k}) \in \tilde{X}_{\neq}^{d-k+1}} V_k(S(x_0, \ldots, x_{d-k}; \tilde{X}) \cap B^d)$$

$$= \frac{\tilde{\gamma}^{d-k+1}}{\kappa_d (d - k + 1)!} \int \mathbb{E} V_k(S(x_0, \ldots, x_{d-k}; \tilde{X} \cup \{x_0, \ldots, x_{d-k}\}) \cap B^d)$$

$$\times \lambda_d^{\otimes(d-k+1)}(\mathrm{d}(x_0, \ldots, x_{d-k})).$$

For $x_0, \ldots, x_{d-k} \in \mathbb{R}^d$ in general position, we obtain

$$\mathbb{E} V_k(S(x_0, \ldots, x_{d-k}; \tilde{X} \cup \{x_0, \ldots, x_{d-k}\}) \cap B^d)$$

$$= \iint \mathbf{1}\{y \in F(x_0, \ldots, x_{d-k}) \cap B^d\}\mathbf{1}\{\|y - x_0\| = d(y, \tilde{X})\}\lambda_k(\mathrm{d}y)\mathrm{d}\mathbb{P}$$

$$= \int_{F(x_0, \ldots, x_{d-k}) \cap B^d} \mathbb{P}(\tilde{X} \cap \mathrm{int}(B^d(y, \|y - x_0\|)) = \emptyset)\lambda_k(\mathrm{d}y)$$

$$= \int_{F(x_0, \ldots, x_{d-k}) \cap B^d} e^{-\tilde{\gamma}\kappa_d\|y - x_0\|^d}\lambda_k(\mathrm{d}y).$$

Thus we have expressed $d_k^{(k)}$ in purely geometric terms. Now we can use integral geometric transformation formulae due to Blaschke, Petkantschin and Miles (see [15]) to complete the proof. □

2.2 Delaunay Mosaics

Let $A \subset \mathbb{R}^d$ be locally finite and assume that $\mathrm{conv}(A) = \mathbb{R}^d$. We put $\mathsf{m} := \mathcal{V}(A)$. For any vertex $e \in \mathcal{S}_0(\mathsf{m})$, the **Delaunay cell** of e is defined by

$$D(e, A) := \mathrm{conv}\{x \in A : e \in \mathcal{S}_0(C(x, A))\}.$$

In the following, we assume that A is in general position and that any $d + 2$ points of A do not lie on a sphere (in this case, A is said to be in general quadratic position). Then the Delaunay cells are simplices which can be defined without reference to the Voronoi mosaic. The convex hull of $d + 1$ points of A is a Delaunay simplex if and only if its circumscribed sphere does not contain any other point of A in its interior.

Lemma 2.2. *If $A \subset \mathbb{R}^d$ is locally finite and $\mathrm{conv}(A) = \mathbb{R}^d$, then*

$$\mathsf{d} := \mathcal{D}(A) := \{D(e, A) : e \in \mathcal{S}_0(\mathsf{m})\} \in \mathcal{M}_s.$$

If A is in general quadratic position, then the Delaunay cells are simplices and

$$\Sigma : \mathcal{S}(\mathsf{m}) \to \mathcal{S}(\mathsf{d}), \quad S(x_0, \ldots, x_{d-m}; A) \mapsto \mathrm{conv}\{x_0, \ldots, x_{d-m}\},$$

is a combinatorial anti-isomorphism.

If $\tilde{X}$ is a stationary point process in $\mathbb{R}^d$, then $Y := \mathcal{D}(\tilde{X})$ is a stationary random mosaic, the **Delaunay mosaic** of $\tilde{X}$. In case $\tilde{X}$ is a Poisson process, we call Y the induced **Poisson-Delaunay mosaic**. Moreover, we write $Y^{(j)}$ and $\beta^{(j)}$ for the process of j-faces and its intensity, respectively. The corresponding notation for $X := \mathcal{V}(\tilde{X})$ is $X^{(j)}$ and $\gamma^{(j)}$. If $\tilde{X}$ is a Poisson process, then $\tilde{X}$ is in general quadratic position almost surely. This yields the first assertion of the next theorem.

Theorem 2.2. *A Poisson-Delaunay mosaic Y is simplicial and $\beta^{(j)} = \gamma^{(d-j)}$.*

Proof. The following brief argument is based on [16, Theorem 4.3.1]. There one can also find useful information about the intensities of stationary particle processes and associated marked point processes with respect to general centre functions. Consider the stationary marked point process

$$\tilde{X}^{(d-j)} := \{(z(x_0, \ldots, x_j), S(x_0, \ldots, x_j; \tilde{X}) - z(x_0, \ldots, x_j)) :$$
$$(x_0, \ldots, x_j) \in \tilde{X}^{j+1}, S(x_0, \ldots, x_j; \tilde{X}) \neq \emptyset\}.$$

The associated particle process is $X^{(d-j)}$ with intensity $\gamma^{(d-j)}$. Hence,

$$\tilde{Y}^{(j)} := \{z(x_0, \ldots, x_j) : (x_0, \ldots, x_j) \in \tilde{X}^{j+1}, S(x_0, \ldots, x_j; \tilde{X}) \neq \emptyset\}$$

has the same intensity. But $\tilde{Y}^{(j)}$ consists of the circumcentres of the polytopes in $Y^{(j)}$.   □

Let $\triangle^{(d)}$ be the set of d-simplices in $\mathbb{R}^d$. Let $z(S)$ denote the centre of the circumscribed sphere of $S \in \triangle^{(d)}$. The distribution of the **typical cell** of a Poisson-Delaunay mosaic Y is the shape distribution $\mathbb{Q}_0$ of Y on the space $\triangle_0^{(d)} := \{S \in \triangle^{(d)} : z(S) = 0\}$ with respect to the centre function z. Hence, for a measurable set $\mathcal{A} \subset \triangle_0^{(d)}$, we have

$$\mathbb{Q}_0(\mathcal{A}) = \mathbb{E} \sum_{S \in Y} \mathbf{1}\{z(S) \in [0,1]^d\}\mathbf{1}\{S - z(S) \in \mathcal{A}\}$$
$$= \lim_{r \to \infty} \frac{\operatorname{card}\{S \in Y : z(S) \in [0,r]^d, S - z(S) \in \mathcal{A}\}}{\operatorname{card}\{S \in Y : z(S) \in [0,r]^d\}}.$$

The first equation is true by definition, the second holds almost surely and follows from the ergodic properties of Poisson-Delaunay mosaics (see [16, Section 6.4]).

Theorem 2.3. *Let Y be a Poisson-Delaunay mosaic associated with a stationary Poisson point process of intensity $\tilde{\gamma}$. Then, for a measurable set $\mathcal{A} \subset \triangle_0^{(d)}$,*

$$\mathbb{Q}_0(\mathcal{A}) = c(d)\tilde{\gamma}^d \int_0^\infty \int_{S^{d-1}} \cdots \int_{S^{d-1}} \mathbf{1}\{r\operatorname{conv}\{u_0, \ldots, u_d\} \in \mathcal{A}\}$$
$$\times \exp(-\tilde{\gamma}\kappa_d r^d)r^{d^2-1}\Delta_d(u_0, \ldots, u_d)\sigma(du_0) \ldots \sigma(du_d)dr,$$

where $\Delta_d(u_0, \ldots, u_d) := \lambda_d(\operatorname{conv}\{u_0, \ldots, u_d\})$ and $c(d)$ is an explicitly known constant.

Proof. Since $\beta^{(d)} = \gamma^{(0)} = c(d)^{-1}\tilde{\gamma}$ by Theorems 2.2 and 2.1, we can again use the Slivnyak-Mecke formula to get

$$Q_0(\mathcal{A}) = \frac{c(d)\tilde{\gamma}^{-1}}{(d+1)!} \mathbb{E} \sum_{(x_0,\ldots,x_d)\in\tilde{X}_{\neq}^{d+1}} \mathbf{1}\{\mathrm{conv}\{x_0,\ldots,x_d\} - z(x_0,\ldots,x_d) \in \mathcal{A}\}$$

$$\times \mathbf{1}\{z(x_0,\ldots,x_d) \in [0,1]^d\}\mathbf{1}\{\tilde{X} \cap \mathrm{int}(B^d(x_0,\ldots,x_d)) = \emptyset\}$$

$$= \frac{c(d)\tilde{\gamma}^d}{(d+1)!} \int_{\mathbb{R}^d}\cdots\int_{\mathbb{R}^d} \mathbf{1}\{\mathrm{conv}\{x_0,\ldots,x_d\} - z(x_0,\ldots,x_d) \in \mathcal{A}\}$$

$$\times \mathbf{1}\{z(x_0,\ldots,x_d) \in [0,1]^d\}\,\mathbb{P}(\tilde{X} \cap \mathrm{int}(B^d(x_0,\ldots,x_d)) = \emptyset)$$

$$\times \lambda_d(dx_0)\ldots\lambda_d(dx_d).$$

Finally, an application of an integral geometric transformation formula due to Miles (see [16, Satz 7.2.2]) completes the proof. $\qquad\square$

3 Hyperplane Mosaics

Let $\hat{X}$ be a stationary **hyperplane process** in $\mathbb{R}^d$. Each realization of $\hat{X}$ decomposes $\mathbb{R}^d$ into polyhedral cells. If all cells are bounded, we obtain a **stationary hyperplane mosaic** X. We always assume that X (i.e. $\hat{X}$) is in **general position**. The j-th **intersection process** of $\hat{X}$, which is a process of $(d-j)$-flats, is denoted by $\hat{X}_j$. We collect some information about the quermass densities of the processes of faces of the given hyperplane mosaic.

Theorem 3.1. *Let X be a stationary hyperplane mosaic in $\mathbb{R}^d$. Assume that the intersection processes of $\hat{X}$ have finite intensity. Then, for $0 \le j \le k \le d$,*

$$d_j^{(k)} = \binom{d-j}{d-k}d_j^{(j)}, \qquad \gamma^{(k)} = \binom{d}{k}\gamma^{(0)}, \qquad n_{kj} = 2^{k-j}\binom{k}{j}.$$

Proof. Define $\nu_k := \mathrm{card}\,(\{F \in \mathcal{S}_k(X) : F \cap rB^d \neq \emptyset\})$, $\alpha_0 := 0$ and, for $1 \le j \le d$,

$$\alpha_j := \frac{1}{j!}\mathrm{card}\,(\{(H_1,\ldots,H_j) \in \hat{X}_{\neq}^j : H_1 \cap \ldots \cap H_j \cap rB^d \neq \emptyset\}).$$

Miles has shown that

$$\nu_k = \sum_{j=d-k}^{d} \binom{j}{d-k}\alpha_j.$$

Hence $\mathbb{E}\nu_k < \infty$ follows from the assumption of finite intersection densities. Let $0 \le j < k \le d-1$ and $r > 0$. For $E \in \hat{X}_{d-k}$, a stationary random mosaic is induced in E by $\hat{X} \cap E$ for which, by applying (3) in E, we get

$$\sum_{E\in\hat{X}_{d-k}} V_j(E \cap rB^d) = \sum_{i=j}^{k}(-1)^{k-i} \sum_{E\in\hat{X}_{d-k}} \sum_{F\in X^{(i)},F\subset E} V_j(F \cap rB^d)$$

$$= \sum_{i=j}^{k}(-1)^{k-i}\binom{d-i}{d-k} \sum_{F\in X^{(i)}} V_j(F \cap rB^d).$$

We estimate

$$\frac{1}{V_d(rB^d)}\mathbb{E}\sum_{E\in\hat{X}_{d-k}}V_j(E\cap rB^d) \le \frac{1}{V_d(rB^d)}\mathbb{E}\sum_{E\in\hat{X}_{d-k}}r^jV_j(B^k)\chi(E\cap rB^d)$$

$$\le \frac{r^jV_j(B^k)}{V_d(rB^d)}\kappa_{d-k}r^{d-k}\hat{\gamma}_{d-k}\to 0$$

as $r\to\infty$. Thus, for $0\le j<k\le d$, we arrive at

$$\sum_{i=j}^{k}(-1)^{k-i}\binom{d-i}{d-k}d_j^{(i)}=0.$$

It remains to solve this triangular system of linear equations.

As to the third assertion, as a consequence of Theorem 1.2 we obtained relation (5), i.e. $\gamma^{(k)}n_{kj}=\gamma^{(j)}n_{jk}$. Here we have $n_{jk}=2^{k-j}\binom{d-j}{d-k}$. Combining this with the second assertion, the required result follows. $\quad\square$

Let $\hat{X}$ be a stationary hyperplane process in $\mathbb{R}^d$. Let $\langle\cdot,\cdot\rangle$ denote the scalar product in $\mathbb{R}^d$ and put $H(u,t):=\{x\in\mathbb{R}^d:\langle x,u\rangle=t\}$ whenever $u\in S^{d-1}$ and $t\in\mathbb{R}$. Then the translation invariant intensity measure Θ of $\hat{X}$ can be written in the form

$$\Theta:=\mathbb{E}\hat{X}=\hat{\gamma}\int_{S^{d-1}}\int_{\mathbb{R}}\mathbf{1}\{H(u,t)\in\cdot\}\mathrm{d}t\,\tilde{\mathbb{P}}(\mathrm{d}u),$$

where $\hat{\gamma}$ is called the **intensity** and $\tilde{\mathbb{P}}$ the **direction distribution** (an even probability measure on S^{d-1}) of $\hat{X}$. We call $\hat{X}$ **non-degenerate** if $\tilde{\mathbb{P}}$ is not concentrated on a great subsphere. For $K\in\mathcal{C}^d$, let $\mathcal{H}_K:=\{E\in\mathcal{E}_{d-1}^d:K\cap E\ne\emptyset\}$.

A **Poisson hyperplane process** $\hat{X}$ satisfies, for $K\in\mathcal{C}^d$ and $k\in\mathbb{N}_0$,

$$\mathbb{P}(\hat{X}(\mathcal{H}_K)=k)=\frac{\Theta(\mathcal{H}_K)^k}{k!}\mathrm{e}^{-\Theta(\mathcal{H}_K)}.$$

It is in general position and induces a **Poisson hyperplane mosaic** X.

The probabilistic information contained in the direction distribution of a stationary hyperplane process $\hat{X}$ can be expressed in geometric terms by associating with $\hat{X}$ a suitable convex body. More precisely, the **associated zonoid** $\Pi_{\hat{X}}$ of $\hat{X}$ is defined by its support function

$$h(\Pi_{\hat{X}},\cdot)=\frac{\hat{\gamma}}{2}\int_{S^{d-1}}|\langle\cdot,v\rangle|\,\tilde{\mathbb{P}}(\mathrm{d}v).$$

Introducing geometric tools and methods for describing hyperplane processes turns out to be a fruitful strategy. We mention two of the simpler applications which are suggested by this method. A different geometric idea used for other purposes will be described in the next section.

Let X be a stationary Poisson hyperplane mosaic in $\mathbb{R}^d$, derived from a stationary Poisson hyperplane process $\hat{X}$ with intensity $\hat{\gamma}$.

- In the special case of a Poisson hyperplane mosaic, the formulas of Theorem 3.1 take a more specific form. In particular, they can be interpreted by means of geometric functionals. Thus known inequalities from geometry become available for the study of extremal problems. For $0 \leq j \leq k \leq d$, we have

$$d_j^{(k)} = \binom{d-j}{d-k} V_{d-j}(\Pi_{\hat{X}}), \quad \gamma^{(k)} = \binom{d}{k} V_d(\Pi_{\hat{X}}).$$

If X is isotropic, then even

$$d_j^{(k)} = \binom{d-j}{d-k}\binom{d}{j} \frac{\kappa_{d-1}^{d-j}}{d^{d-j}\kappa_d^{d-1-j}\kappa_j} \hat{\gamma}^{d-j}, \quad \gamma^{(k)} = \binom{d}{k} \frac{\kappa_{d-1}^d}{d^d\kappa_d^{d-1}} \hat{\gamma}^d.$$

- Let Z_0 be the zero cell of X. Then

$$\mathbb{E}V_d(Z_0) = 2^{-d}d!V_d((\Pi_{\hat{X}})^\circ) \geq d!\kappa_d \left(\frac{2\kappa_{d-1}}{d\kappa_d}\hat{\gamma}\right)^{-d}$$

with equality if and only if $\hat{X}$ is isotropic. Here $(\Pi_{\hat{X}})^\circ$ denotes the polar body of $\Pi_{\hat{X}}$. For a direct approach to this inequality one can proceed as follows. The **radial function** $\rho(Z_0, v)$ of Z_0 at $v \in S^{d-1}$, which is the distance from the origin to the boundary point of Z_0 in direction v, is exponentially distributed, that is

$$\mathbb{P}(\rho(Z_0, v) \leq r) = \mathbb{P}(\hat{X}(\mathcal{F}_{[0,rv]}) > 0) = 1 - \exp\left(-2rh(\Pi_{\hat{X}}, v)\right),$$

where we use that

$$\mathbb{E}\hat{X}(\mathcal{F}_{[0,rv]}) = 2\hat{\gamma} \int_{S^{d-1}} \int_0^\infty 1\{H(u,t) \cap [0,rv] \neq \emptyset\}dt\,\tilde{\mathbb{P}}(du)$$

$$= \hat{\gamma}r \int_{S^{d-1}} |\langle u, v\rangle|\,\tilde{\mathbb{P}}(du) = 2rh(\Pi_{\hat{X}}, v).$$

Hence, by Jensen's inequality

$$\mathbb{E}V_d(Z_0) = \frac{1}{d} \int_{S^{d-1}} \mathbb{E}\rho(Z_0, v)^d \sigma(dv) = \frac{1}{d} \int_{S^{d-1}} \frac{d!}{2^d} h(\Pi_{\hat{X}}, v)^{-d} \sigma(dv),$$

$$\geq \frac{d!\kappa_d(d\kappa_d)^d}{2^d} \left(\int_{S^{d-1}} h(\Pi_{\hat{X}}, u)\sigma(du)\right)^{-d}$$

$$= \frac{d!\kappa_d(d\kappa_d)^d}{2^d} \left(\frac{\hat{\gamma}}{2} \int_{S^{d-1}} \int_{S^{d-1}} |\langle u, v\rangle|\sigma(du)\tilde{\mathbb{P}}(dv)\right)^{-d},$$

which yields the required estimate.

4 Kendall's Conjecture

In this section, we will consider the shape of large cells in Poisson hyperplane, in Poisson-Voronoi and in Poisson-Delaunay mosaics. We start with the description of a classical problem which was first formulated in the 1940s in a more restricted framework. For the purpose of introduction, let $\hat{X}$ be a stationary isotropic Poisson line process in $\mathbb{R}^2$. Further, let Z_0 denote the zero cell (Crofton cell) of the associated Poisson-line mosaic. Then D. Kendall's classical conjecture (Kendall's problem) can be stated as follows.

The conditional law for the shape of Z_0, given a lower bound for $A(Z_0)$, converges weakly, as that lower bound tends to infinity, to the law concentrated at the circular shape.

Recently, various contributions to this subject have been made by R. Miles (heuristically) [10], I. Kovalenko [8], [9], A. Goldman [2], P. Calka [1], and in the papers [3], [5], [4]. For a more detailed description of the relevant literature we refer to [3], [4].

In order to motivate a crucial ingredient for the solution of Kendall's problem, we start with a related, but much simpler extremal problem. Let $\hat{X}$ be a stationary Poisson hyperplane process with intensity measure

$$\Theta = 2\hat{\gamma} \int_{S^{d-1}} \int_0^\infty \mathbf{1}\{H(u,t) \in \cdot\} \mathrm{d}t\, \tilde{\mathbb{P}}(\mathrm{d}u).$$

Now we wish to know which convex bodies $K \subset \mathbb{R}^d$ with $0 \in K$ and fixed $V_d(K) > 0$ maximize the inclusion probability $\mathbb{P}(K \subset Z_0)$.

By **Minkowski's existence theorem** there is a unique centred convex body $B \subset \mathbb{R}^d$ such that $\tilde{\mathbb{P}} = S_{d-1}(B, \cdot)$, where $S_{d-1}(B, \cdot)$ is the surface area measure of order $d-1$ of B. The convex body B, the **direction body** associated with the given hyperplane process, thus is another example of a convex auxiliary body which can be used to translate the given probabilistic information into geometric terms. Thus tools and results from convex geometry become available in the present setting. Hence, for any $K \in \mathcal{K}$, we obtain

$$\Theta(\mathcal{H}_K) = 2\hat{\gamma} \int_{S^{d-1}} h(K, u) S_{d-1}(B, \mathrm{d}u) = 2d\hat{\gamma} V_1(B, K).$$

At this point an application of **Minkowski's inequality** yields

$$\mathbb{P}(K \subset Z_0) = \exp\left(-2d\hat{\gamma} V_1(B, K)\right) \le \exp\left(-2d\hat{\gamma} V_d(K)^{1/d} V_d(B)^{(d-1)/d}\right).$$

This shows that the inclusion probability is maximized precisely if K is homothetic to the direction body B and has the prescribed volume.

4.1 Large Cells in Poisson Hyperplane Mosaics

We turn to the solution of Kendall's problem in a more general framework. Let $\hat{X}$ be a stationary hyperplane process with intensity $\hat{\gamma}$, direction distribution $\tilde{\mathbb{P}}$ and direction body B. Now we ask for the **limit shape** of the zero cell Z_0, given the volume of the zero cell is big. In view of the previous extremal problem, the shape of the direction body B is a natural candidate for the limit shape of Z_0, given "Z_0 is big".

An important initial step in the solution of the problem is to find a precise formulation of the type of result one is aiming at. For instance, we have to clarify what we mean by the shape of a convex body, and we have to specify the type of convergence which is used. In the following, we identify homothetic convex bodies and determine distances up to homotheties. In other words, we work in the shape space

$$\mathcal{S}^* := \{K \in \mathcal{K} : 0 \in K, R(K) = 1\},$$

where $R(K)$ is the radius of the circumscribed ball of K, and measure distances between (the shape of) a convex body K and the (shape of the) direction body B by means of the deviation measure

$$r_B(K) := \min\left\{\frac{r_2}{r_1} - 1 : r_1 B + z \subset K \subset r_2 B + z, 0 < r_1 \leq r_2, z \in \mathbb{R}^d\right\}.$$

Then we say that B^* is the limit shape of Z_0 with respect to the size functional V_d, which is used to measure the size of the zero cell, if

$$\mu_a := \mathbb{P}\left(Z_0^* \in \cdot \mid V_d(Z_0) \geq a\right) \to \delta_{B^*} \tag{6}$$

weakly on $\mathcal{S}^*$, as $a \to \infty$. By a compactness argument, (6) is equivalent to

$$\mathbb{P}(r_B(Z_0) \geq \epsilon \mid V_d(Z_0) \geq a) \to 0,$$

as $a \to \infty$, for all $\epsilon > 0$. The following theorem contains a much more explicit and quantitative result. As a very special case it yields a solution of Kendall's problem in generalized form, in arbitrary dimensions and, more importantly, without the assumption of isotropy.

Theorem 4.1. *Let $\hat{X}$ be a Poisson hyperplane process for the data $\hat{\gamma}$ and B. There is a constant $c_0 = c_0(B)$ such that the following is true. If $\epsilon \in (0,1)$ and $I = [a,b]$ with $a^{1/d}\hat{\gamma} \geq \sigma_0 > 0$, then*

$$\mathbb{P}(r_B(Z_0) \geq \epsilon \mid V_d(Z_0) \in I) \leq c \exp\left(-c_0 \epsilon^{d+1} a^{1/d} \hat{\gamma}\right),$$

where $c = c(B, \epsilon, \sigma_0)$.

It turns out that Theorem 4.1 remains true if the zero cell is replaced by the typical cell.

Theorem 4.2. *The assertion of Theorem 4.1 remains true if the zero cell Z_0 is replaced by the typical cell Z.*

The proof of these two results is quite involved and technical. Instead of trying to give a sketch of proof, we have to content ourselves with providing some further motivation for the strategy of proof. The basic task consists in estimating the conditional probability

$$\mathbb{P}(r_B(Z_0) \geq \epsilon \mid V_d(Z_0) \geq a) = \frac{\mathbb{P}(r_B(Z_0) \geq \epsilon, V_d(Z_0) \geq a)}{\mathbb{P}(V_d(Z_0) \geq a)}.$$

We estimate numerator and denominator separately.

- To obtain a lower bound for the denominator, we define the convex body $K := (a/V_d(B))^{1/d}B$. Then

$$\mathbb{P}(V_d(Z_0) \geq a) \geq \mathbb{P}(\hat{X}(\mathcal{H}_K) = 0) = \exp\left(-2dV_1(B, K)\hat{\gamma}\right)$$

$$= \exp\left(-2dV_d(B)^{(d-1)/d}a^{1/d}\hat{\gamma}\right).$$

This simple argument has to be refined considerably to yield a *local* improvement.

- Next we consider an upper bound for the numerator. **Minkowski's inequality** states that

$$V_1(B, K) \geq V_d(B)^{(d-1)/d}V_d(K)^{1/d}$$

with equality if and only if K and B are homothetic. If we know that $r_B(K) \geq \epsilon$, then in fact a stronger stability result

$$V_1(B, K) \geq (1 + f(\epsilon))V_d(B)^{(d-1)/d}V_d(K)^{1/d}$$

is true involving a nonnegative and explicitly known stability function f. Hence, if the deterministic convex body K satisfies

$$r_B(K) \geq \epsilon \qquad \text{and} \qquad V_d(K) \geq a,$$

then

$$\mathbb{P}(\hat{X}(\mathcal{H}_K) = 0) = \exp\left(-2dV_1(B, K)\hat{\gamma}\right)$$

$$\leq \exp\left(-2d(1 + f(\epsilon))V_d(B)^{(d-1)/d}a^{1/d}\hat{\gamma}\right).$$

- In a bold analogy, we might hope that a similar result holds for the random cell Z_0 instead of a deterministic convex body K. In other words, we would like to have an estimate of the form

$$\mathbb{P}(r_B(Z_0) \geq \epsilon, V_d(Z_0) \geq a) \leq c\exp\left(-2d(1 + g(\epsilon))V_d(B)^{(d-1)/d}a^{1/d}\hat{\gamma}\right)$$

with some explicitly given, nonnegative function g. Such an estimate and the lower bound from the first step would essentially imply the theorem in the special case considered.

- However, things are not so easy. The previous analogy cannot be justified in a straightforward way. We first consider $V_d(Z_0) \in a[1, 1+h]$ for small $h > 0$ and, in addition, we classify the random convex body Z_0 according to its degree of elongation.

The method of proof can be modified to yield results about asymptotic distributions.

Theorem 4.3. *For $\hat{X}$ and Z_0 as in Theorem 4.1,*

$$\lim_{r \to \infty} a^{-1/d} \ln \mathbb{P}(V_d(Z_0) \geq a) = -2dV_d(B)^{(d-1)/d}\hat{\gamma}.$$

The special case $d = 2$ and $B = B^2$ (isotropic case) had previously been established by Goldman [2] using different techniques.

4.2 Large Cells in Poisson-Voronoi Mosaics

We now describe an analogue of Kendall's problem for Poisson-Voronoi mosaics. Let $\tilde{X}$ be a stationary Poisson process with intensity $\tilde{\gamma}$. We write $\mathcal{K}_*$ for the space of convex bodies containing 0, and $X := \mathcal{V}(\tilde{X})$ for the Poisson-Voronoi mosaic induced by $\tilde{X}$. The **typical cell** of X is a random convex body Z having distribution $\mathbb{Q}$, where

$$\mathbb{Q}(\mathcal{A}) := \frac{1}{\tilde{\gamma}} \mathbb{E} \sum_{x \in \tilde{X}} 1\{x \in B\} 1\{C(x, \tilde{X}) - x \in \mathcal{A}\},$$

for measurable sets $B \subset \mathbb{R}^d$ with $\lambda_d(B) = 1$ and $\mathcal{A} \subset \mathcal{K}_*$. This mean value can be rewritten by means of the **Slivnyak-Mecke formula**. Thus we obtain

$$\mathbb{Q}(\mathcal{A}) = \int_{\mathbb{R}^d} \mathbb{E} 1\{x \in B\} 1\{C(x, \tilde{X} \cup \{x\}) - x \in \mathcal{A}\}\lambda_d(dx)$$

$$= \int_{\mathbb{R}^d} \mathbb{E} 1\{x \in B\} 1\{C(0, \tilde{X} \cup \{0\}) \in \mathcal{A}\}\lambda_d(dx)$$

$$= \mathbb{P}(C(0, \tilde{X} \cup \{0\}) \in \mathcal{A}).$$

Hence we can specify

$$Z = C(0, \tilde{X} \cup \{0\}) = \bigcap_{x \in \tilde{X}} H^-(x),$$

i.e. $Z = Z_0(X^*)$ with $X^* := \{H(x) : x \in \tilde{X}\}$. Here X^* is a nonstationary (but isotropic) Poisson hyperplane process with intensity measure

$$\mathbb{E}X^* = 2^d \tilde{\gamma} \int_{S^{d-1}} \int_0^\infty 1\{H(u, t) \in \cdot\} t^{d-1} dt\, \sigma(du).$$

Hence, for $K \in \mathcal{K}_*$, we get $\mathbb{E}X^*(\mathcal{H}_K) = 2^d \tilde{\gamma} U(K)$ with

$$U(K) = \lambda_d(\Phi(K)) = \frac{1}{d}\int_{S^{d-1}} h(K,u)^d \sigma(\mathrm{d}u),$$

where

$$\Phi(K) = \{y \in \mathbb{R}^d : H(2y) \cap K \neq \emptyset\} = \bigcup\{B^d(x, \|x\|) : 2x \in K\}.$$

The strategy of proof for Theorem 4.1 can be adapted and extended to yield a corresponding result for large typical cells in Poisson-Voronoi mosaics. On the geometric side, however, we need a new stability result for the geometric functional U, which states that

$$U(K) \geq \left(1 + \gamma(d)\vartheta(K)^{(d+3)/2}\right) \kappa_d^{1-d/k} v_k(K)^{d/k}.$$

The much weaker estimate $U(K) \geq \kappa_d^{1-d/k} v_k(K)^{d/k}$ is easy to derive from results available in the literature. But again the main point is to obtain a quantitative improvement involving the nonnegative deviation measure ϑ which measures the deviation of the shape of K from the shape of a Euclidean ball. For details we refer to [4].

Theorem 4.4. *Let $\tilde{X}$ be a Poisson process with intensity $\tilde{\gamma}$. Let Z denote the typical cell of $X = \mathcal{V}(\tilde{X})$ and $k \in \{1,\dots,d\}$. There is some $c_0 = c_0(d)$ such that the following is true. If $\epsilon \in (0,1)$ and $I = [a,b]$ with $a^{d/k}\tilde{\gamma} \geq \sigma_0 > 0$, then*

$$\mathbb{P}(\vartheta(Z) \geq \epsilon \mid v_k(Z) \in I) \leq c\exp\left(-c_0\epsilon^{(d+3)/2}a^{d/k}\tilde{\gamma}\right),$$

where $c = c(d, \epsilon, \sigma_0)$.

The following result provides an analogue of Theorem 4.3.

Theorem 4.5. *For $\tilde{X}$, X and Z as in Theorem 4.4,*

$$\lim_{a\to\infty} a^{-d/k}\ln\mathbb{P}(v_k(Z) \geq a) = -2^d \kappa_d^{1-d/k}\tilde{\gamma}.$$

Similar results can be established for other size functionals as well, e.g., for the inradius of the typical cell.

4.3 Large Cells in Poisson-Delaunay Mosaics

To complete the picture we finally address a version of Kendall's problem for Poisson-Delaunay mosaics. Let $\tilde{X}$ be a stationary Poisson process with intensity $\tilde{\gamma}$. Let Z denote the typical cell of $Y = \mathcal{D}(\tilde{X})$ with distribution $\mathbb{Q}_0$. For d-simplices $S_1, S_2 \subset \mathbb{R}^d$, we write $\eta(S_1, S_2)$ for the smallest number η such that for any vertex p of one of the simplices, $B^d(p,\eta)$ contains a vertex q of the other simplex. Let $\mathcal{T}^d$ be the set of regular simplices inscribed to S^{d-1}. For a d-simplex inscribed to a ball with centre z and radius r, we put

$$\rho(S) := \min\left\{\eta(r^{-1}(S-z), T) : T \in \mathcal{T}^d\right\}.$$

Then again we can state a large deviation result for the shape of the typical Poisson-Delaunay cell given it has large volume.

Theorem 4.6. *Let $\tilde{X}$, $\tilde{\gamma}$, Y and Z be as described above. There is some constant $c_0 = c_0(d)$ such that the following is true. If $\epsilon \in (0,1)$ and $I = [a,b]$ with $a\tilde{\gamma} \geq \sigma_0 > 0$, then*

$$\mathbb{P}(\rho(Z) \geq \epsilon \mid V_d(Z) \in I) \leq c\exp\left(-c_0\epsilon^2 a\tilde{\gamma}\right),$$

where $c = c(d, \epsilon, \sigma_0)$.

Similar results hold for the zero cell Z_0 of Y and if $V_d(Z)$ is replaced by the inradius of Z. These results for Poisson-Delaunay mosaics can be obtained by more direct estimates than in the previous cases. This is mainly due to the explicit formula for the distribution of the typical cell which is available in the present case (cf. Theorem 2.3). In addition, one needs sharp estimates of isoperimetric type. A simple and well known geometric extremal result states that a simplex S inscribed to S^{d-1} has maximal volume among all simplices inscribed to S^{d-1} if and only if S is regular. Let T^d be such a regular simplex with volume τ_d. In the proof of Theorem 4.6 we need an improved version of such a uniqueness result. It can be summarized in the estimate

$$V_d(S) \leq (1 - c\,\rho(S))V_d(T^d).$$

It should be emphasized that corresponding uniqueness results are completely unknown in dimension $d \geq 3$ for such basic functionals as the mean width or the surface area.

There is also a result on the asymptotic distribution of the volume of the typical Poisson-Delaunay mosaic.

Theorem 4.7. *For $\tilde{X}$, $\tilde{\gamma}$, Y and Z as in Theorem 4.6,*

$$\lim_{a\to\infty} a^{-1}\ln\mathbb{P}(V_d(Z) \geq a) = -\frac{\kappa_d}{\tau_d}\tilde{\gamma}.$$

More recently, asymptotic shapes and limit shapes of random polytopes (related to random mosaics) have been investigated in [6] and [7]. There are fundamental connections between inequalities of isoperimetric type and the shape of large cells in random mosaics are discovered. In particular, these papers contribute to Kendall's problem in a very general framework.

References

1. Calka, P.: The distributions of the smallest disks containing the Poisson-Voronoi typical cell and the Crofton cell in the plane. Adv. in Appl. Probab., **34**, 702–717 (2002)
2. Goldman, A.: Sur une conjecture de D.G. Kendall concernant la cellule de Crofton du plan et sur sa contrepartie brownienne. Ann. Probab., **26**, 1727–1750 (1998)

3. Hug, D., Reitzner, M., Schneider, R.: The limit shape of the zero cell in a stationary Poisson hyperplane tessellation. Ann. Probab., **32**, 1140–1167 (2004)

4. Hug, D., Reitzner, M., Schneider, R.: Large Poisson-Voronoi cells and Crofton cells. Adv. in Appl. Probab., **36**, 667–690 (2004)

5. Hug, D., Schneider, R.: Large cells in Poisson-Delaunay tessellations. Discrete Comput. Geom., **31**, 503–514 (2004)

6. Hug, D., Schneider, R.: Large typical cells in Poisson-Delaunay mosaics. Rev. Roumaine Math. Pures Appl., (to appear)

7. Hug, D., Schneider, R.: Asymptotic shapes of large cells in random tessellations. Preprint

8. Kovalenko, I.: A simplified proof of a conjecture of D.G. Kendall concerning shapes of random polygons. J. Appl. Math. Stochastic Anal., **12**, 301–310 (1999)

9. Kovalenko, I.: An extension of a conjecture of D.G. Kendall concerning shapes of random polygons to Poisson Voronoï cells. In: Engel, P. et al. (eds) Voronoï's Impact on Modern Science. Book I. Transl. from the Ukrainian. Kyiv: Institute of Mathematics. Proc. Inst. Math. Natl. Acad. Sci. Ukr., Math. Appl., **212**, 266–274 (1998)

10. Miles, R.: A heuristic proof of a long-standing conjecture of D.G. Kendall concerning the shapes of certain large random polygons. Adv. in Appl. Probab., **27**, 397–417 (1995)

11. Møller, J.: Lectures on Random Voronoi Tessellations. Lect. Notes Statist., **87**, Springer, New York (1994)

12. Møller, J.: Topics in Voronoi and Johnson-Mehl tessellations. In: Barndorff-Nielsen, O.E., Kendall, W.S., van Lieshout, M.N.M. (eds) Stochastic Geometry, Likelihood and Computation. Monographs on Statistics and Applied Probability, **80**, Chapman & Hall/CRC Press, Boca Raton, FL (1999)

13. Møller, J., Waagepetersen, R.P.: Statistical Inference and Simulation for Spatial Point Processes. Monographs on Statistics and Applied Probability, **100**, Chapman & Hall/CRC Press, Boca Raton, FL (2004)

14. Okabe, A., Boots, B., Sugihara, K., Chiu, S.N.: Spatial Tessellations. Wiley, Chichester (2002)

15. Schneider, R., Weil, W.: Integralgeometrie. Teubner, Stuttgart (1992)

16. Schneider, R., Weil, W.: Stochastische Geometrie. Teubner, Stuttgart (2000)

On the Evolution Equations of Mean Geometric Densities for a Class of Space and Time Inhomogeneous Stochastic Birth-and-growth Processes

Vincenzo Capasso[1] and Elena Villa[2]

[1] Department of Mathematics, University of Milan
via Saldini 50, 20133 Milano, Italy
e-mail: vincenzo.capasso@mat.unimi.it
[2] Department of Mathematics, University of Milan
via Saldini 50, 20133 Milano, Italy
e-mail: villa@mat.unimi.it

1 Introduction

Many real phenomena may be modelled as birth-and-growth processes. Our aim is to provide evolution equations for relevant quantities describing the geometric process $\{\Theta^t, t \in \mathbb{R}_+\}$ associated with the birth-and-growth process.

A birth-and-growth process in d-dimensional space $\mathbb{R}^d$ is driven by a spatially marked point process $N = \{(T_i, X_i)\}_{i \in \mathbb{N}}$, where $T_i \in \mathbb{R}_+$ represents the random time of birth of the i-th germ, and $X_i \in \mathbb{R}^d$ its random spatial location [17, 26]. Once born, each germ generates a grain which grows at the surface (growth front), with a speed $G(t, x) > 0$ which should, in general, be assumed space-time dependent.

Application areas include crystallization processes (see [10], and references therein; see also [27] for the crystallization processes on sea shells); tumor growth [23] and angiogenesis [24]; spread of fires in the woods, spread of a pollutant in an environment; etc. Because of the coupling with an underlying field such as temperature, nutrients, etc., all this kind of phenomena may include a space-time structured random process of birth (nucleation), and a growth process that, as a first approximation, we consider deterministic.

In Section 2 we introduce basic ingredients of a birth-and-growth process, and a *normal growth* model for the growth of grains.

A quantitative description of the evolution of the resulting random set Θ^t can be obtained in terms of local mean densities of the set itself, and of its free surface $\partial \Theta^t$, with respect to the standard Lebesgue measure on $\mathbb{R}^d$. Given a sufficiently regular random set $\Theta_n \subset \mathbb{R}^d$, with integer Hausdorff dimension $n \leq d$, in Section 3 we introduce a concept of *geometric density* of the set, with respect to the standard Lebesgue measure λ_d on $\mathbb{R}^d$, in terms of a suitable class of linear functionals. Correspondingly, in the stochastic case, we consider *mean geometric densities*.

In Section 4 we derive the announced evolution equation. We show how the evolution equations for mean densities can be given in terms of the hazard function of the birth-and-growth process, thus extending, to the non homogeneous case, known results from the literature.

For a specific case, i.e. for a Poisson birth process, we provide explicit expressions for the hazard rate in terms of the birth and growth rate fields. As shown in [22] and [12], once we know the hazard function of the geometric process, it is also possible to derive evolution equations for mean densities of n-facets associated with the morphology of stochastic tessellations produced by birth-and-growth processes.

2 Birth-and-growth Processes

2.1 The Nucleation Process

For the model presented here we refer to [10, 11]. Consider the measurable space $(\mathbb{R}^d, \mathcal{B}(\mathbb{R}^d))$ endowed with the usual Lebesgue measure λ_d. Consider further an underlying probability space $(\Omega, \mathcal{A}, \mathbb{P})$.

A birth-and-growth process is a dynamic germ-grain model whose birth process is modelled as a marked point process (MPP). Consider a Borel set $E \subset \mathbb{R}^d$, $d \geq 2$, endowed with its Borel σ-algebra $\mathcal{E}$. A marked point process (MPP) N on $\mathbb{R}_+$, with marks in E, is a point process on $\mathbb{R}_+ \times E$ with the property that the marginal process $\{N(B \times E) : B \in \mathcal{B}(\mathbb{R}_+)\}$ is itself a point process. So, it is defined as a random measure given by

$$N = \sum_{n=1}^{\infty} \varepsilon_{T_n, X_n},$$

where

- T_n is an $\mathbb{R}_+$-valued random variable representing the time of birth of the n-th nucleus,
- X_n is an E-valued random variable representing the spatial location of the nucleus born at time T_n,
- $\varepsilon_{t,x}$ is the Dirac measure on $\mathcal{B}(\mathbb{R}_+) \times \mathcal{E}$ such that for any $t_1 < t_2$ and $A \in \mathcal{E}$,

$$\varepsilon_{t,x}([t_1, t_2] \times A) = \begin{cases} 1 & \text{if } t \in [t_1, t_2], \ x \in A, \\ 0 & \text{otherwise.} \end{cases}$$

Hence, in particular, for any $B \in \mathcal{B}(\mathbb{R}_+)$ and $A \in \mathcal{E}$ bounded, we have

$$N(B \times A) = \text{card} \{T_n \in B, X_n \in A\} < \infty, \tag{1}$$

i.e. it is the (random) number of germs born in the region A, during time B.

We assume that the nucleation process N is such that the marginal process is **simple** (i.e. $N(dt \times E) \leq 1$ for every infinitesimal time interval dt), and, for any fixed t, its mark distribution is absolutely continuous with respect to the d-dimensional Lebesgue measure λ_d on E.

It is well known [6, 20] that, under general conditions, a marked point process (MPP) is characterized by its compensator (or stochastic intensity), say $\nu(dt \times dx)$, with respect to the internal history of the process. Besides, if $\tilde{\nu}(dt)$ is the compensator of the marginal process, there exists a stochastic kernel k from $\Omega \times \mathbb{R}_+$ to E such that

$$\nu(dt \times dx) = k(t, dx)\tilde{\nu}(dt).$$

In many applications it is supposed that further nuclei cannot be born in an already crystallized zone. When we want to emphasize this, denoting by Θ^t the crystallized region at time t, we shall write

$$\nu(dt \times dx) = k(t, dx)\tilde{\nu}(dt) = k_0(t, dx)\tilde{\nu}(dt)(1 - \mathbf{1}_{\Theta^{t-}}(x)),$$

where $\nu_0(dt \times dx) = k_0(t, dx)\tilde{\nu}(dt)$ is the compensator of the process N_0, called the **free process**, in which nuclei can be born anywhere. (See also [13]).

The Poisson Case

In a great number of applications, it is supposed that N_0 is a marked Poisson process. In this case it is well known that its compensator is deterministic.

In particular, it is assumed that the MPP N_0 is a space-time inhomogeneous Poisson process with a given **deterministic** intensity $\alpha(t, x), t \geq 0, x \in E$, where α is a real valued measurable function on $\mathbb{R}_+ \times E$ such that $\alpha(t, \cdot) \in \mathcal{L}^1(E)$, for all $t > 0$ and such that

$$0 < \int_0^T \int_E \alpha(t, x) dx \, dt < \infty$$

for any $0 < T < \infty$.

If we want to exclude germs which are born within the region already occupied by Θ^t, we shall consider the thinned **stochastic** intensity

$$\nu(dt \times dx) = \alpha(t, x)(1 - \mathbf{1}_{\Theta^{t-}}(x))dtdx.$$

In this respect we shall call

$$\nu_0(dt \times dx) = \alpha(t, x)dtdx$$

the **free space intensity**.

2.2 The Growth Process

Let $\Theta_{T_n}^t(X_n)$ be the random closed set (RACS) obtained as the evolution up to time $t \geq T_n$ of the germ born at time T_n in X_n, according to some growth model; this will be the **grain** associated with the **germ** (T_n, X_n).

We call **birth-and-growth process** the family of RACS given by

$$\Theta^t = \bigcup_{T_n \leq t} \Theta_{T_n}^t(X_n), \qquad t \in \mathbb{R}_+.$$

In order to complete the definition of the birth-and-growth process we need to propose a growth model for each individual grain $\Theta_{t_0}^t(x_0)$ born at some time t_0, at some location x_0.

We denote by $\dim_{\mathcal{H}}$ and $\mathcal{H}^n$ the Hausdorff dimension, and the n-dimensional Hausdorff measure, respectively.

We assume here the **normal growth model** (see, e.g., [7, 9, 5]), according to which, at almost every time t, at every point of the actual grain surface, i.e. at every $x \in \partial \Theta_{t_0}^t(x_0)$, growth occurs with a given strictly positive normal velocity

$$v(t, x) = G(t, x)n(t, x),$$

where $G(t, x)$ is a given sufficiently regular deterministic **strictly positive growth field**, and $n(t, x)$ is the unit outer normal at point $x \in \partial \Theta_{t_0}^t(t_0)$. In particular, from now on we will assume that G is bounded and continuous on $\mathbb{R}_+ \times \mathbb{R}^d$ with $G_0 \leq G(t, x) < \infty$, for some constant $G_0 > 0$.

The well-posedness of the **evolution problem** for the growth front $\partial \Theta_{t_0}^t(x_0)$ requires further regularity. We refer to [7] (see also [8]) to state that, subject to the initial condition that each initial germ is a spherical ball of infinitesimal radius, under suitable regularity on the growth field $G(t, x)$, each grain $\Theta_{t_0}^t(x_0)$ is such that the following inclusion holds

$$\Theta_{t_0}^s(x_0) \subset \Theta_{t_0}^t(x_0), \text{ for } s < t, \tag{2}$$

and

$$\partial \Theta_{t_0}^s(x_0) \cap \partial \Theta_{t_0}^t(x_0) = \emptyset, \text{ for } s < t.$$

Moreover, for almost every $t \in \mathbb{R}_+$, we have $\dim_{\mathcal{H}} \partial \Theta_{t_0}^t(x_0) = d - 1$, and $\mathcal{H}^{d-1}(\partial \Theta_{t_0}^t(x_0)) < \infty$.

Under the regularity assumptions taken above, and since the whole germ-grain process Θ^t is a finite union of individual grains by (1), we may assume that for almost every $t \in \mathbb{R}_+$, there exists the tangent hyperplane to Θ^t at $\mathcal{H}^{d-1}$- a.e. $x \in \partial \Theta^t$. As a consequence Θ^t and $\partial \Theta^t$ are finite unions of rectifiable sets [15] and satisfy

$$\lim_{r \to 0} \frac{\mathcal{H}^d(\Theta^t \cap B_r(x))}{b_d r^d} = 1 \quad \text{for } \mathcal{H}^d\text{-a.e. } x \in \Theta^t,$$

$$\lim_{r \to 0} \frac{\mathcal{H}^{d-1}(\partial \Theta^t \cap B_r(x))}{b_{d-1} r^{d-1}} = 1 \quad \text{for } \mathcal{H}^{d-1}\text{-a.e. } x \in \partial \Theta^t,$$

where $B_r(x)$ is the d-dimensional open ball centered in x with radius r, and b_n denotes the volume of the unit ball in $\mathbb{R}^n$.

Further we assume that $G(t,x)$ is sufficiently regular so that, at almost any time $t > 0$, the following holds (see also [21, 25])

$$\lim_{r \to 0} \frac{\mathcal{H}^d(\Theta^t_{\oplus r}) - \mathcal{H}^d(\Theta^t)}{r} = \mathcal{H}^{d-1}(\partial\Theta^t), \tag{3}$$

where we have denoted by $\Theta^t_{\oplus r}$ the **parallel set** of Θ^t at distance $r \geq 0$ (i.e. the set of all points $x \in \mathbb{R}^d$ with distance from Θ^t at most r).

Note that the limit above implies that $\partial\Theta^t$ admits a finite $(d-1)$-dimensional Minkowski content (see [16]).

Definition 2.1. *Let Θ_n be a subset of $\mathbb{R}^d$ with Hausdorff dimension n. If Θ is $(\mathcal{H}^n, n)$-rectifiable and $\mathcal{H}^n$-measurable (see [15]), we say that Θ_n is n-**regular**.*

In particular we have that $\mathcal{H}^n(\Theta_n) < \infty$.

Note that, according to the above definition, Θ^t and $\partial\Theta^t$ are d-regular and $(d-1)$-regular closed sets, respectively.

3 Closed Sets as Distributions

3.1 The Deterministic Case

In order to pursue our analysis of the germ-grain model associated with a birth-and-growth process, we find it convenient to represent n-regular closed sets in $\mathbb{R}^d$ as distributions, in terms of their *geometric densities*.

Consider an n-regular closed set Θ_n in $\mathbb{R}^d$. Then we have

$$\lim_{r \to 0} \frac{\mathcal{H}^n(\Theta_n \cap B_r(x))}{b_d r^d} = \lim_{r \to 0} \frac{\mathcal{H}^n(\Theta_n \cap B_r(x))}{b_n r^n} \frac{b_n r^n}{b_d r^d} = \begin{cases} \infty & \mathcal{H}^n\text{-a.e. } x \in \Theta_n, \\ 0 & \forall x \notin \Theta_n. \end{cases}$$

By analogy with the delta function $\delta_{x_0}(x)$ associated with a point x_0, for $x \in \mathbb{R}^d$ we define the generalized density of Θ_n as

$$\delta_{\Theta_n}(x) := \lim_{r \to 0} \frac{\mathcal{H}^n(\Theta_n \cap B_r(x))}{b_d r^d}.$$

The density $\delta_{\Theta_n}(x)$ (delta function of the set Θ_n) can be seen as a linear functional defined by a measure in a similar way as the classical delta function $\delta_{x_0}(x)$ of a point x_0. We proceed as follows.

Let $\mu_{\Theta_n}^{(r)}$ be the measure on $\mathbb{R}^d$ with density function

$$\delta_{\Theta_n}^{(r)}(x) := \frac{\mathcal{H}^n(\Theta_n \cap B_r(x))}{b_d r^d},$$

and let μ_{Θ_n} be the measure defined by

$$\mu_{\Theta_n}(A) := \mathcal{H}^n(\Theta_n \cap A), \qquad A \in \mathcal{B}(\mathbb{R}^d). \tag{4}$$

With an abuse of notations, we may introduce the linear functionals $\delta_{\Theta_n}^{(r)}$ and δ_{Θ_n} associated with the measure $\mu_{\Theta_n}^{(r)}$ and μ_{Θ_n}, respectively, as follows:

$$(\delta_{\Theta_n}^{(r)}, f) := \int_{\mathbb{R}^d} f(x)\mu_{\Theta_n}^{(r)}(dx),$$

$$(\delta_{\Theta_n}, f) := \int_{\mathbb{R}^d} f(x)\mu_{\Theta_n}(dx),$$

for any bounded continuous $f : \mathbb{R}^d \to \mathbb{R}$. (We denote by $C_b(\mathbb{R}^d, \mathbb{R})$ the set of all such functions).

We may prove the following result (see [14]).

Proposition 3.1. *For any measurable subset A of $\mathbb{R}^d$ such that*

$$\mathcal{H}^n(\Theta_n \cap \partial A) = 0,$$

it holds

$$(\delta_{\Theta_n}, \mathbf{1}_A) := \mathcal{H}^n(\Theta_n \cap A) = \lim_{r \to 0} \int_A \frac{\mathcal{H}^n(\Theta_n \cap B_r(x))}{b_d r^d} dx =: \lim_{r \to 0} (\delta_{\Theta_n}^{(r)}, \mathbf{1}_A).$$

As a consequence it follows that (see [2]):

Corollary 3.1. *For all $f \in C_b(\mathbb{R}^d, \mathbb{R})$*

$$\lim_{r \to 0} \int_{\mathbb{R}^d} f(x)\mu_{\Theta_n}^{(r)}(dx) = \int_{\mathbb{R}^d} f(x)\mu_{\Theta_n}(dx).$$

Equivalently, we may claim that the sequence of measure $\mu_{\Theta_n}^{(r)}$ converges weakly to the measure μ_{Θ_n}, or, in other words, the sequence of linear functionals $\delta_{\Theta_n}^{(r)}$ converges weakly* to the linear functional δ_{Θ_n}, i.e.

$$(\delta_\Theta, f) = \lim_{r \to 0} (\delta_{\Theta_n}^{(r)}, f) \qquad \forall f \in C_b(\mathbb{R}^d, \mathbb{R}).$$

Remark 3.1. Note that, by definition, $\delta_{\Theta_n}(x) = \lim_{r \to 0} \delta_{\Theta_n}^{(r)}(x)$, and, if we denote by $C_0(\mathbb{R}^d, \mathbb{R})$ the set of all continuous functions $f : \mathbb{R}^d \to \mathbb{R}$ with compact support, we have $C_0(\mathbb{R}^d, \mathbb{R}) \subset C_b(\mathbb{R}^d, \mathbb{R})$. Thus, in analogy with the classical Dirac delta, if we restrict our attention to $C_0(\mathbb{R}^d, \mathbb{R})$, we may regard the continuous linear functional δ_{Θ_n} as a generalized function on the usual test space $C_0(\mathbb{R}^d, \mathbb{R})$, and, in accordance with the usual representations of distributions in the theory of generalized functions, we formally write

$$\int_{\mathbb{R}^d} f(x)\delta_{\Theta_n}(x)dx := (\delta_{\Theta_n}, f).$$

Further, we may like to notice that the classical Dirac delta $\delta_{x_0}(x)$ associated to a point x_0 now follows as a particular case.

In terms of the above arguments, we may state that $\delta_{\Theta_n}(x)$ is the (generalized) density of the measure μ_{Θ_n}, defined above, with respect to the standard Lebesgue measure λ_d on $\mathbb{R}^d$ and, formally, we may define

$$\frac{\mathrm{d}\mu_{\Theta_n}}{\mathrm{d}\lambda_d} := \delta_{\Theta_n}.$$

3.2 The Stochastic Case

Suppose now that Θ_n is an n-regular random closed set in $\mathbb{R}^d$ on a suitable probability space $(\Omega, \mathcal{A}, \mathbb{P})$, with $\mathbb{E}[\mathcal{H}^n(\Theta_n)] < \infty$.

Now μ_{Θ_n} defined as above by (4) is a random measure, and correspondingly δ_{Θ_n} is a random linear functional.

Consider now the linear functional $\mathbb{E}[\delta_{\Theta_n}]$ defined on $C_b(\mathbb{R}^d, \mathbb{R})$ by the measure $\mathbb{E}[\mu_{\Theta_n}](A) := \mathbb{E}[\mathcal{H}^n(\Theta_n \cap A)]$, i.e. by

$$(\mathbb{E}[\delta_{\Theta_n}], f) := \int_{\mathbb{R}^d} f(x) \mathbb{E}[\mu_{\Theta_n}](\mathrm{d}x), \qquad f \in C_b(\mathbb{R}^d, \mathbb{R}).$$

It can be shown (see [14]) that the expected linear functional $\mathbb{E}[\delta_{\Theta_n}]$ so defined is such that, for any $f \in C_b(\mathbb{R}^d, \mathbb{R})$,

$$(\mathbb{E}[\delta_{\Theta_n}], f) = \mathbb{E}[(\delta_{\Theta_n}, f)],$$

which corresponds to the expected linear functional in the sense of Gelfand-Pettis [1]. For a discussion about measurability of (δ_{Θ_n}, f) we refer to [4, 28].

As for the deterministic case, we may formally define the mean density

$$\frac{\mathrm{d}\mathbb{E}[\mu_{\Theta_n}]}{\mathrm{d}\lambda_d} := \mathbb{E}[\delta_{\Theta_n}].$$

Note that, even though for any realization θ_n of Θ_n, the measure μ_{θ_n} may be singular, the expected measure $\mathbb{E}[\mu_{\Theta_n}]$ may be absolutely continuous with respect to λ_d.

For $n = d$, it is easily seen that $\delta_{\Theta_d}(x) = \mathbf{1}_{\Theta_d}(x)$, λ_d-a.s.; which directly implies

$$\mathbb{E}[\delta_{\Theta_d}(x)] = \mathbb{P}(x \in \Theta_d), \quad \lambda_d\text{-a.s.},$$

and we know that $\mathbb{P}(x \in \Theta_d)$ is the Radon-Nikodym density of the measure $\mathbb{E}[\lambda_d(\Theta_d \cap \cdot)]$, since (see also [19])

$$\mathbb{E}[\lambda_d(\Theta_d \cap A)] = \mathbb{E}\left(\int_{\mathbb{R}^d} \mathbf{1}_{\Theta_d \cap A}(x)\mathrm{d}x\right)$$

$$= \mathbb{E}\left(\int_A \mathbf{1}_{\Theta_d}(x)\mathrm{d}x\right)$$

$$= \int_A \mathbb{E}(\mathbf{1}_{\Theta_d}(x))\mathrm{d}x$$

$$= \int_A \mathbb{P}(x \in \Theta_d)\mathrm{d}x,$$

so that in this case we may really exchange

$$\int_A \mathbb{E}[\delta_{\Theta_d}(x)]\mathrm{d}x = \mathbb{E}\left[\int_A \delta_{\Theta_d}(x)\mathrm{d}x\right].$$

The density

$$V_V(x) := \mathbb{E}[\delta_{\Theta_d}(x)] = \mathbb{P}(x \in \Theta_d)$$

in material science is known as the **degree of crystallinity**. The complement to 1 of the crystallinity, is known as **porosity**

$$p_x = 1 - V_V(x) = \mathbb{P}(x \notin \Theta_d).$$

4 The Evolution Equation of Mean Densities for the Stochastic Birth-and-growth Process

In a birth-and-growth process the RACS Θ^t evolves in time, so that the question arises about *WHEN* a point $x \in E$ is reached (**captured**) by this growing RACS; or vice versa up to when a point $x \in E$ survives capture?

In this respect the degree of crystallinity (now also depending on time) $V_V(t, x) = \mathbb{P}(x \in \Theta^t)$ may be seen as a probability of capture of the point x, by time t. In this sense the porosity

$$p_x(t) = 1 - V_V(t, x) = \mathbb{P}(x \notin \Theta^t)$$

is the **survival function** of the point x at time t, i.e. the probability that the point x is not yet covered by the random set Θ^t.

With reference to our birth-and-growth process $\{\Theta^t\}$, we may introduce the random variable $T(x)$, representing the **time of capture** of a given point $x \in E$, i.e.

$$\{x \in \Theta^t\} = \{T(x) \le t\}.$$

Denoting by $\mathrm{int}\Theta^t$ the interior of the set Θ^t, we may observe that

$$\delta_{\Theta^t}(x) = \begin{cases} 1 \ \forall x \in \mathrm{int}\Theta^t \\ 0 \ \forall x \notin \Theta^t, \end{cases}$$

and that $x \in \partial\Theta^{T(x)}$. Besides, by (2), $\Theta^t \subset \Theta^{t+s}$ for any $s > 0$.

So it follows that for any $t > T(x)$, $x \in \mathrm{int}\Theta^t$; hence

$$\delta_{\Theta^t}(x) = \begin{cases} 1 \text{ if } t > T(x) \\ 0 \text{ if } t < T(x), \end{cases}$$

and its time derivative coincides with the derivative of the Heaviside function $H_{T(x)}(t) = 1_{[T(x),+\infty]}$. From the theory of distributions we know that

$$\frac{\partial}{\partial t}\delta_{\Theta^t}(x) = \delta_{T(x)}(t). \tag{5}$$

Under our modelling assumptions, it follows that $T(x)$ is a continuous random variable [13]. Denoting by $f_x(t)$ its probability density function, as a consequence of (5), we have [14]

$$\mathbb{E}\left[\frac{\partial}{\partial t}\delta_{\Theta^t}(x)\right] = f_x(t).$$

So we may directly state that

$$\mathbb{E}\left[\frac{\partial}{\partial t}\delta_{\Theta^t}(x)\right] = f_x(t) = \frac{\partial}{\partial t}\mathbb{P}(x \in \Theta^t) = \frac{\partial}{\partial t}\mathbb{E}[\delta_{\Theta^t}(x)].$$

Hence, $\mathbb{E}\left[\frac{\partial}{\partial t}\delta_{\Theta^t}(x)\right]$ and $\frac{\partial}{\partial t}\mathbb{E}[\delta_{\Theta^t}(x)]$ coincide as functions, and by the equation above, we have the exchange between derivative and expectation.

Remark 4.1. Even though for any realization θ^t of Θ^t, $\frac{\partial}{\partial t}\delta_{\theta^t}(x)$ is a singular generalized function, when we consider its expectation we obtain a regular generalized function, i.e. a real integrable function. In particular the derivative is the usual derivative of functions.

Under our modelling assumptions, and in particular under the assumption (3), we may state the following [14]

Theorem 4.1. *Under the assumptions above, for any d-regular set A such that*

$$\mathbb{P}(\mathcal{H}^{d-1}(\partial\Theta^t \cap \partial A) > 0) = 0,$$

the following holds

$$\lim_{\Delta t \to 0} \int_A \frac{\mathbb{E}[\delta_{\Theta^{t+\Delta t}}(x)] - \mathbb{E}[\delta_{\Theta^t}(x)]}{\Delta t}\, dx = \int_A G(t,x)\mathbb{E}[\delta_{\partial\Theta^t}(x)]dx.$$

Moreover, regarding $\mathbb{E}[\delta_{\Theta^t}(x)]$ and $\mathbb{E}[\delta_{\partial\Theta^t}(x)]$ as generalized functions on $C_0^1(\mathbb{R} \times \mathbb{R}^d, \mathbb{R})$, i.e. the space of all functions $f : \mathbb{R} \times \mathbb{R}^d \to \mathbb{R}$ of class C^1 with compact support, we have

$$\left(\frac{\partial}{\partial t}\mathbb{E}[\delta_{\Theta^t}(x)] - G(t,x)\mathbb{E}[\delta_{\partial\Theta^t}(x)], f(t,x)\right) = 0 \qquad \forall f \in C_0^1(\mathbb{R} \times \mathbb{R}^d);$$

which is the weak form of the following evolution equation for the mean densities

$$\frac{\partial}{\partial t}\mathbb{E}[\delta_{\Theta^t}(x)] = G(t,x)\mathbb{E}[\delta_{\partial\Theta^t}(x)].$$

The expected value of $\delta_{\partial\Theta^t}(x)$ is what is usually called **mean free surface density**,

$$S_V(t,x) := \mathbb{E}[\delta_{\partial\Theta^t}(x)].$$

As a consequence of the previous theorem we can state that

$$\frac{\partial}{\partial t}V_V(t,x) = G(t,x)S_V(t,x), \tag{6}$$

to be taken, as usual, in a weak form.

In some cases (such as in the case of Poisson birth processes, as we shall later discuss as an example), it can be of interest to consider the following extended mean densities.

Definition 4.1. *We call* **mean extended volume density** *at point x and time t the quantity $V_{ex}(t,x)$ such that, for any $B \in \mathcal{B}(\mathbb{R}^d)$,*

$$\mathbb{E}[\sum_{T_j \le t} \nu^d(\Theta_{T_j}^t(X_j) \cap B)] = \int_B V_{ex}(t,x)\mathrm{d}x.$$

It represents the mean of the sum of the volume densities, at time t, of the grains supposed free to be born and grow [11].

Correspondingly we give the following definition.

Definition 4.2. *We call* **mean extended surface density** *at point x and time t the quantity $S_{ex}(t,x)$ such that, for any $B \in \mathcal{B}(\mathbb{R}^d)$,*

$$\mathbb{E}[\sum_{T_j \le t} \nu^{d-1}(\partial\Theta_{T_j}^t(X_j) \cap B)] = \int_B S_{ex}(t,x)\mathrm{d}x.$$

It represents the mean of the sum of the surface densities, at time t, of the grains supposed free to be born and grow.

Under our assumptions for the growth model, we can also claim, by linearity arguments, that

$$\frac{\partial}{\partial t}V_{ex}(t,x) = G(t,x)S_{ex}(t,x), \tag{7}$$

to be taken, as usual, in weak form.

4.1 Hazard Function and Causal Cone

Given a point $x \in E$, it is convenient to introduce a **hazard function** $h(t,x)$, defined as the rate of capture by the process Θ^t, i.e.

$$h(t,x) = \lim_{\Delta t \to 0} \frac{\mathbb{P}(x \in \Theta^{t+\Delta t} | x \notin \Theta^t)}{\Delta t}.$$

Under our modelling assumptions, the time of capture $T(x)$ admits a probability density function $f_x(t)$ such that

$$f_x(t) = \frac{d}{dt}(1 - p_x(t)) = \frac{\partial V_V(t,x)}{\partial t}$$

and

$$f_x(t) = p_x(t)h(t,x),$$

from which we immediately obtain

$$\frac{\partial V_V(t,x)}{\partial t} = (1 - V_V(t,x))h(t,x). \tag{8}$$

This is an extension of the well known Avrami-Kolmogorov formula [18, 3], proven for space homogeneous birth and growth rates; our expression holds whenever a mean volume density and an hazard function are well defined.

In order to obtain explicit evolution equations for the relevant densities of the birth-and-growth process, we need to make explicit their dependence upon the kinetic parameters of the processes of birth (such as the intensity of the corresponding marked point process) and growth (the growth rate). This can be done by means of the hazard function $h(t,x)$.

In general cases (e.g. not Poissonian cases) the expressions for the survival and the hazard functions are quite complicated, because they must take into account all the previous history of the process (cf. (9)). So, it can be useful to estimate the hazard function, directly. For example, under our assumptions, the following holds (see [13])

$$h(t,x) = G(t,x)\frac{\partial}{\partial r}H_{S,\Theta^t}(r,x)|_{r=0},$$

where H_{S,Θ^t} is the local spherical contact distribution function of Θ^t.

For the birth-and-growth model defined above, we may introduce a **causal cone** associated with a point $x \in E$ and a time $t > 0$, as defined by Kolmogorov [18],

$$A(t,x) := \{(y,s) \in E \times [0,t] \mid x \in \Theta_s^t(y)\}.$$

We denote by $C_s(t,x)$ the section of the causal cone at time $s < t$,

$$C_s(t,x) := \{y \in E \mid (y,s) \in A(t,x)\} = \{y \in E \mid x \in \Theta_s^t(y)\}.$$

It is possible to obtain (see [13]) expressions for the survival and the hazard functions in terms of the birth process N and of the causal cone.

For example, under our modelling assumptions, if N has continuous intensity measure, then for any fixed t

$$p_x(t) = \exp\left\{-\int_0^t \mathbb{E}(\tilde{\nu}(\mathrm{d}s)k(s, C_s(t,x)) \mid N[A(s, C_s(t,x))] = 0)\right\}. \tag{9}$$

In the Poisson case, the independence property of increments makes this expression simpler.

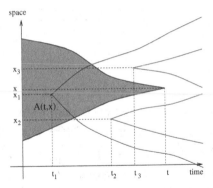

Fig. 1. The causal cone $A(t, x)$ of point x at time t: it is the space-time region where a nucleation has to take place so that point x is reached by a growing grain by time t.

The Poisson Case

If N is a marked Poisson process, it is easily seen that

$$p_x(t) = \mathbb{P}(x \notin \Theta^t) = \mathbb{P}(N(A(t, x)) = 0) = e^{-\nu_0(A(t, x))},$$

where $\nu_0(A(t, x))$ is the volume of the causal cone with respect to the free space intensity of the Poisson process:

$$\nu_0(A(t, x)) = \int_{A(t, x)} \alpha(s, y) \mathrm{d}s \mathrm{d}y.$$

So, we have

$$h(t, x) = -\frac{\partial}{\partial t} \ln p_x(t) = \frac{\partial}{\partial t} \nu_0(A(t, x)).$$

The following theorem holds [8] whose proof requires sophisticate techniques of PDE's in a geometric measure theoretic setting.

Theorem 4.2. *Under our modelling assumptions on birth and on growth, which make the evolution problem well posed, the following equality holds*

$$\frac{\partial}{\partial t} \nu_0(A(t, x)) = G(t, x) \int_0^t \int_{\mathbb{R}^d} K(t_0, x_0; t, x) \alpha(t_0, x_0) \mathrm{d}x_0 \, \mathrm{d}t_0. \tag{10}$$

with

$$K(t_0, x_0; t, x) := \int_{\{z \in \mathbb{R}^d | \tau(t_0, x_0; z) = t\}} \delta(z - x) \mathrm{d}a(z).$$

Here δ is the Dirac function, $\mathrm{d}a(z)$ is a $(d-1)$-dimensional surface element, and $\tau(t_0, x_0; z)$ is the solution of the eikonal problem

$$\left| \frac{\partial \tau}{\partial x_0}(t_0, x_0, x) \right| = \frac{1}{G(t_0, x_0)} \frac{\partial \tau}{\partial t_0}(t_0, x_0, x),$$

$$\left| \frac{\partial \tau}{\partial x}(t_0, x_0, x) \right| = \frac{1}{G(\tau(t_0, x_0, x), x)}.$$

In our case it has been shown [12] that the volume of the causal cone can be expressed in terms of the extended volume density.

Theorem 4.3. *Under the previous modelling assumptions on birth and on growth, the following equality holds*

$$\nu_0(A(t, x)) = V_{\text{ex}}(t, x). \tag{11}$$

As a consequence

$$h(t, x) = \frac{\partial}{\partial t} \nu_0(A(t, x)) = \frac{\partial}{\partial t} V_{\text{ex}}(t, x)$$

so that

$$\frac{\partial}{\partial t} V_V(t, x) = (1 - V_V(t, x)) \frac{\partial}{\partial t} V_{\text{ex}}(t, x).$$

This equation is exactly the Kolmogorov-Avrami formula extended to a birth-and-growth process with space-time inhomogeneous parameters of birth and of growth [18, 3].

Finally, by direct comparison between (11), (10) and (7), we may claim that

$$S_{\text{ex}}(t, x) = \int_0^t \int_{\mathbb{R}^d} K(t_0, x_0; t, x) \alpha(t_0, x_0) \mathrm{d}x_0 \, \mathrm{d}t_0.$$

and

$$h(t, x) = G(t, x) S_{\text{ex}}(t, x).$$

Consequently, by remembering (6) and (8), we obtain

$$S_V(t, x) = (1 - V_V(t, x)) S_{\text{ex}}(t, x).$$

The availability of an explicit expression for the hazard function, in terms of the relevant kinetic parameters of the process, makes it possible to obtain evolution equations for all facet densities of the Johnson-Mehl tessellation associated with our birth-and-growth process (see [17],[22],[11]).

A major problem arises when the birth-and-growth process is strongly coupled with an underlying field (such as temperature in a crystallization process); in such a case the kinetic parameters of the stochastic birth-and-growth process are themselves stochastic, depending on the history of the crystallization process. As a consequence the previous theory for obtaining evolution equations for densities cannot be applied. The interested reader may refer to [10] and [8], for an approximate solution of the problem by means of hybrid models that take into account the possible multiple scale structure of the problem.

Acknowledgements. It is a pleasure to acknowledge the contribution of M. Burger in Linz and A. Micheletti in the development of joint research projects relevant for this presentation.

References

1. Araujo, A., Giné, E.: The Central Limit Theorem for Real and Banach Valued Random Variables. John Wiley & Sons, New York (1980)
2. Ash, R.B.: Real Analysis and Probability. Academic Press, New York (1972)
3. Avrami A.: Kinetic of phase change. Part I. J. Chem. Phys., **7**, 1103–112 (1939)
4. Baddeley, A.J., Molchanov, I.S.: On the expected measure of a random set. In: Jeulin, D. (ed) Advances in Theory and Applications of Random Sets, Proc. International Symposium, Oct. 9-11, 1996, Fontainebleau, France. World Sci. Publ., Singapore (1997)
5. Barles, G., Soner, H.M., Souganidis, P.E.: Front propagation and phase-field theory. SIAM J. Control Optim., **31**, 439–469 (1993)
6. Brémaud, P.: Point Processes and Queues, Martingale Dynamics. Springer, New York (1981)
7. Burger, M.: Growth fronts of first-order Hamilton-Jacobi equations. SFB Report 02-8, J. Kepler University, Linz (2002)
8. Burger, M., Capasso, V., Pizzocchero, L.: Mesoscale averaging of nucleation and growth models. Submitted
9. Burger, M., Capasso, V., Salani, C.: Modelling multi-dimensional crystallization of polymers in interaction with heat transfer. Nonlinear Anal. Real World Appl., **3**, 139–160 (2002)
10. Capasso, V. (ed): Mathematical Modelling for Polymer Processing. Polymerization, Crystallization, Manufacturing. Mathematics in Industry Series, **2**, Springer, Heidelberg (2003)
11. Capasso, V., Micheletti, A.: Stochastic geometry of spatially structured birth-and-growth processes. Application to crystallization processes. In: Merzbach, E. (ed) Topics in Spatial Stochastic Processes. Lecture Notes in Mathematics (CIME Subseries), **1802**, Springer, Heidelberg (2002)
12. Capasso, V., Micheletti, A.: On the hazard function for inhomogeneous birth-and-growth processes. Submitted
13. Capasso, V., Villa, E.: Survival functions and contact distribution functions for inhomogeneous, stochastic geometric marked point processes. Stoch. Anal. Appl., **23**, 79–96 (2005).
14. Capasso, V., Villa, E.: On the stochastic geometric densities of time dependent random closed sets. In preparation
15. Federer, H.: Geometric Measure Theory. Springer, Berlin (1996)
16. Hug, D., Last, G., Weil, W.: A local Steiner-type formula for general closed sets and applications. Math. Z., **246**, 237–272 (2004)
17. Johnson, W.A., Mehl, R.F.: Reaction Kinetics in processes of nucleation and growth. Trans. A.I.M.M.E., **135**, 416–458 (1939)
18. Kolmogorov, A.N.: On the statistical theory of the crystallization of metals. Bull. Acad. Sci. USSR, Math. Ser., **1**, 355–359 (1937)
19. Kolmogorov, A.N.: Foundations of the Theory of Probability. Chelsea Pub. Co., New York (1956)
20. Last, G., Brandt, A.: Marked Point Processes on the Real Line. The Dynamic Approach. Springer, New York (1995)
21. Lorenz, T.: Set valued maps for image segmentation. Comput. Visual. Sci., **4**, 41–57 (2001)
22. Møller, J.: Random Johnson-Mehl tessellations. Adv. in Appl. Probab., **24**, 814–844 (1992)

23. Morale, D.: A stochastic particle model for vasculogenesis: a multiple scale approach. In: Capasso, V. (ed.) Mathematical Modelling and Computing in Biology and Medicine. The MIRIAM Project Series, ESCULAPIO Pub. Co., Bologna (2003)
24. Serini, G., et al.: Modeling the early stages of vascular network assembly. EMBO J., **22**, 1771–1779 (2003)
25. Sokolowski, J., Zolesio, J.-P.: Introduction to Shape Optimization. Shape Sensitivity Analysis. Springer, Berlin (1992)
26. Stoyan, D., Kendall, W.S., Mecke, J.: Stochastic Geometry and its Application. John Wiley & Sons, New York (1995)
27. Ubukata, T.: Computer modelling of microscopic features of molluscan shells. In: Sekimura, T., et al. (eds) Morphogenesis and Pattern Formation in Biological Systems. Springer, Tokyo (2003)
28. Zähle, M.: Random processes of Hausdorff rectifiable closed sets. Math. Nachr., **108**, 49–72 (1982)

List of Participants

1. Aletti Giacomo
   University of Milano, Italy
   giacomo.aletti@unimi.it
2. Azarina Svetlana
   Voronezh State University, Russia
   azarinas@mail.ru
3. Baddeley Adrian
   University of Western Australia,
   Australia
   adrian@maths.uwa.edu.au (**lecturer**)
4. Bárány Imre
   Rényi Institute of Mathematics
   Budapest, Hungary
   barany@renyi.hu (**lecturer**)
5. Berchtold Maik
   Swiss Federal Institute of Techn.,
   Switzerland
   berchtld@stat.math.ethz.ch
6. Bianchi Annamaria
   University of Milano, Italy
   abianchi@mat.unimi.it
7. Bianchi Gabriele
   University of Florence, Italy
   gabriele.bianchi@unifi.it
8. Bobyleva Olga
   Moscow State University, Russia
   o_bobyleva@mail.ru
9. Bongiorno Enea
   University of Milano, Italy
   bongiorno.enea@tiscali.it
10. Capasso Vincenzo
    University of Milano, Italy
    Vincenzo.Capasso@unimi.it
11. Cerdàn Ana
    Universidad de Alicante, Spain
    aacs@alu.ua.es

12. Cerny Rostislav
    Charles University Prague, Czech
    Republic
    rostislav.cerny@karlin.mff.cuni.cz
13. Connor Stephen
    University of Warwick, UK
    s.b.connor@warwick.ac.uk
14. Fleischer Frank
    University of Ulm, Germany
    ffrank@mathematik.uni-ulm.de
15. Gallois David
    France Telecom R&D
    david.gallois@rd.francetelecom.com
16. Gille Wilfried
    University of Halle, Germany
    gille@physik.uni-halle.de
17. Gots Ekaterina
    Voronezh State University, Russia
    kgots@aport2000.ru
18. Hoffmann Lars Michael
    University of Karlsruhe, Germany
    LarsHoffmann@ePost.de
19. Hug Daniel
    University of Freiburg, Germany
    daniel.hug@math.uni-freiburg.de
20. Jónsdóttir Kristjana Ýr
    Aarhus University, Denmark
    kyj@imf.au.dk
21. Karamzin Dmitry
    Computing Centre RAS, Russia
    dmitry_karamzin@mail.ru
22. Kozlova Ekaterina
    Moscow State University Lomonosov,
    Russia
    ekozlova@fors.ru

23. Lautensack Claudia
Inst. Techno- und
Wirtschaftsmathematik Kaiserslautern,
Germany
lautensack@itwm.fraunhofer.de

24. Legland David
INRA, France
david.legland@jouy.inra.fr

25. Lhotsky Jiri
Charles University Prague, Czech
Republic
jiri.lhotsky@email.cz

26. Mannini Claudio
University of Florence, Italy
claudio.mannini@dicea.unifi.it

27. Micheletti Alessandra
University of Milano, Italy
alessandra.micheletti@unimi.it

28. Miori Cinzia
Universidad de Alicante, Spain
cm4@alu.ua.es

29. Morale Daniela
University of Milano, Italy
morale@mat.unimi.it

30. Nazin Sergey
Institute of Control Sciences RAS,
Russia
snazin@ipu.rssi.ru

31. Ortisi Matteo
University of Milano, Italy
ortisi@mat.unimi.it

32. Pantle Ursa
University of Ulm, Germany
pantle@mathematik.uni-ulm.de

33. Salani Paolo
University of Florence, Italy
salani@math.unifi.it

34. Sapozhnikov Artyom
Heriot-Watt University, UK
artyom@ma.hw.ac.uk

35. Scarsini Marco
University of Torino, Italy
marco.scarsini@unito.it

36. Schmaehling Jochen
University of Heidelberg, Germany
jochen.schmaehling@de.bosch.com

37. Schmidt Hendrik
University of Ulm, Germany
hendrik@mathematik.uni-ulm.de

38. Schneider Rolf
University of Freiburg, Germany
rolf.schneider@math.uni-freiburg.de
(**lecturer**)

39. Schuhmacher Dominic
University of Zurich, Switzerland
schumi@amath.unizh.ch

40. Shcherbakov Vadim
CWI, Netherlands
V.Shcherbakov@cwi.nl

41. Sicco Alessandro
University of Torino, Italy
sicco@dm.unito.it

42. Sirovich Roberta
University of Torino, Italy
sirovich@dm.unito.it

43. Solanes Gil
University of Stuttgart, Germany
solanes@mathematik.uni-stuttgart.de

44. Thorarinsdottir Thordis Linda
University of Aarhus, Denmark
disa@imf.au.dk

45. Tontchev Nikolay
University of Berne, Switzerland
nito@stat.unibe.ch

46. Villa Elena
University of Milano, Italy
villa@mat.unimi.it

47. Voss Christian
University of Rostock, Germany
christian.voss@mathematik.uni-rostock.de

48. Weil Wolfgang
University of Karlsruhe, Germany
weil@math.uni-karlsruhe.de (**lecturer, editor**)

49. Winter Steffen
University of Jena, Germany
winter@minet.uni-jena.de

LIST OF C.I.M.E. SEMINARS

Published by C.I.M.E

1954   1. Analisi funzionale
        2. Quadratura delle superficie e questioni connesse
        3. Equazioni differenziali non lineari

1955   4. Teorema di Riemann-Roch e questioni connesse
        5. Teoria dei numeri
        6. Topologia
        7. Teorie non linearizzate in elasticità, idrodinamica, aerodinamic
        8. Geometria proiettivo-differenziale

1956   9. Equazioni alle derivate parziali a caratteristiche reali
       10. Propagazione delle onde elettromagnetiche   automorfe
       11. Teoria della funzioni di più variabili complesse e delle funzioni

1957   12. Geometria aritmetica e algebrica (2 vol.)
       13. Integrali singolari e questioni connesse
       14. Teoria della turbolenza (2 vol.)

1958   15. Vedute e problemi attuali in relatività generale
       16. Problemi di geometria differenziale in grande
       17. Il principio di minimo e le sue applicazioni alle equazioni funzionali

1959   18. Induzione e statistica
       19. Teoria algebrica dei meccanismi automatici (2 vol.)
       20. Gruppi, anelli di Lie e teoria della coomologia

1960   21. Sistemi dinamici e teoremi ergodici
       22. Forme differenziali e loro integrali

1961   23. Geometria del calcolo delle variazioni (2 vol.)
       24. Teoria delle distribuzioni
       25. Onde superficiali

1962   26. Topologia differenziale
       27. Autovalori e autosoluzioni
       28. Magnetofluidodinamica

1963   29. Equazioni differenziali astratte
       30. Funzioni e varietà complesse
       31. Proprietà di media e teoremi di confronto in Fisica Matematica

1964   32. Relatività generale
       33. Dinamica dei gas rarefatti
       34. Alcune questioni di analisi numerica
       35. Equazioni differenziali non lineari

1965   36. Non-linear continuum theories
       37. Some aspects of ring theory
       38. Mathematical optimization in economics

Published by Ed. Cremonese, Firenze

1966   39. Calculus of variations
40. Economia matematica
41. Classi caratteristiche e questioni connesse
42. Some aspects of diffusion theory

1967   43. Modern questions of celestial mechanics
44. Numerical analysis of partial differential equations
45. Geometry of homogeneous bounded domains

1968   46. Controllability and observability
47. Pseudo-differential operators
48. Aspects of mathematical logic

1969   49. Potential theory
50. Non-linear continuum theories in mechanics and physics and their applications
51. Questions of algebraic varieties

1970   52. Relativistic fluid dynamics
53. Theory of group representations and Fourier analysis
54. Functional equations and inequalities
55. Problems in non-linear analysis

1971   56. Stereodynamics
57. Constructive aspects of functional analysis (2 vol.)
58. Categories and commutative algebra

1972   59. Non-linear mechanics
60. Finite geometric structures and their applications
61. Geometric measure theory and minimal surfaces

1973   62. Complex analysis
63. New variational techniques in mathematical physics
64. Spectral analysis

1974   65. Stability problems
66. Singularities of analytic spaces
67. Eigenvalues of non linear problems

1975   68. Theoretical computer sciences
69. Model theory and applications
70. Differential operators and manifolds

Published by Ed. Liguori, Napoli

1976   71. Statistical Mechanics
72. Hyperbolicity
73. Differential topology

1977   74. Materials with memory
75. Pseudodifferential operators with applications
76. Algebraic surfaces

Published by Ed. Liguori, Napoli & Birkhäuser

1978   77. Stochastic differential equations
78. Dynamical systems

1979   79. Recursion theory and computational complexity
80. Mathematics of biology

1980   81. Wave propagation
        82. Harmonic analysis and group representations
        83. Matroid theory and its applications

Published by Springer-Verlag

1981   84. Kinetic Theories and the Boltzmann Equation   (LNM 1048)
        85. Algebraic Threefolds   (LNM 947)
        86. Nonlinear Filtering and Stochastic Control   (LNM 972)

1982   87. Invariant Theory   (LNM 996)
        88. Thermodynamics and Constitutive Equations   (LNP 228)
        89. Fluid Dynamics   (LNM 1047)

1983   90. Complete Intersections   (LNM 1092)
        91. Bifurcation Theory and Applications   (LNM 1057)
        92. Numerical Methods in Fluid Dynamics   (LNM 1127)

1984   93. Harmonic Mappings and Minimal Immersions   (LNM 1161)
        94. Schrödinger Operators   (LNM 1159)
        95. Buildings and the Geometry of Diagrams   (LNM 1181)

1985   96. Probability and Analysis   (LNM 1206)
        97. Some Problems in Nonlinear Diffusion   (LNM 1224)
        98. Theory of Moduli   (LNM 1337)

1986   99. Inverse Problems   (LNM 1225)
       100. Mathematical Economics   (LNM 1330)
       101. Combinatorial Optimization   (LNM 1403)

1987   102. Relativistic Fluid Dynamics   (LNM 1385)
       103. Topics in Calculus of Variations   (LNM 1365)

1988   104. Logic and Computer Science   (LNM 1429)
       105. Global Geometry and Mathematical Physics   (LNM 1451)

1989   106. Methods of nonconvex analysis   (LNM 1446)
       107. Microlocal Analysis and Applications   (LNM 1495)

1990   108. Geometric Topology: Recent Developments   (LNM 1504)
       109. H_∞ Control Theory   (LNM 1496)
       110. Mathematical Modelling of Industrial Processes   (LNM 1521)

1991   111. Topological Methods for Ordinary Differential Equations (LNM 1537)
       112. Arithmetic Algebraic Geometry   (LNM 1553)
       113. Transition to Chaos in Classical and Quantum Mechanics   (LNM 1589)

1992   114. Dirichlet Forms   (LNM 1563)
       115. D-Modules, Representation Theory, and Quantum Groups (LNM 1565)
       116. Nonequilibrium Problems in Many-Particle Systems   (LNM 1551)

1993   117. Integrable Systems and Quantum Groups   (LNM 1620)
       118. Algebraic Cycles and Hodge Theory   (LNM 1594)
       119. Phase Transitions and Hysteresis   (LNM 1584)

1994   120. Recent Mathematical Methods in Nonlinear Wave Propagation (LNM 1640)
       121. Dynamical Systems   (LNM 1609)
       122. Transcendental Methods in Algebraic Geometry   (LNM 1646)

1995   123. Probabilistic Models for Nonlinear PDE's   (LNM 1627)
       124. Viscosity Solutions and Applications   (LNM 1660)
       125. Vector Bundles on Curves. New Directions   (LNM 1649)

1996	126. Integral Geometry, Radon Transforms and Complex Analysis	(LNM 1684)
	127. Calculus of Variations and Geometric Evolution Problems	(LNM 1713)
	128. Financial Mathematics	(LNM 1656)
1997	129. Mathematics Inspired by Biology	(LNM 1714)
	130. Advanced Numerical Approximation of Nonlinear Hyperbolic Equations	(LNM 1697)
	131. Arithmetic Theory of Elliptic Curves	(LNM 1716)
	132. Quantum Cohomology	(LNM 1776)
1998	133. Optimal Shape Design	(LNM 1740)
	134. Dynamical Systems and Small Divisors	(LNM 1784)
	135. Mathematical Problems in Semiconductor Physics	(LNM 1823)
	136. Stochastic PDE's and Kolmogorov Equations in Infinite Dimension	(LNM 1715)
	137. Filtration in Porous Media and Industrial Applications	(LNM 1734)
1999	138. Computational Mathematics driven by Industrial Applications	(LNM 1739)
	139. Iwahori-Hecke Algebras and Representation Theory	(LNM 1804)
	140. Hamiltonian Dynamics - Theory and Applications	(LNM 1861)
	141. Global Theory of Minimal Surfaces in Flat Spaces	(LNM 1775)
	142. Direct and Inverse Methods in Solving Nonlinear Evolution Equations	(LNP 632)
2000	143. Dynamical Systems	(LNM 1822)
	144. Diophantine Approximation	(LNM 1819)
	145. Mathematical Aspects of Evolving Interfaces	(LNM 1812)
	146. Mathematical Methods for Protein Structure	(LNCS 2666)
	147. Noncommutative Geometry	(LNM 1831)
2001	148. Topological Fluid Mechanics	to appear
	149. Spatial Stochastic Processes	(LNM 1802)
	150. Optimal Transportation and Applications	(LNM 1813)
	151. Multiscale Problems and Methods in Numerical Simulations	(LNM 1825)
2002	152. Real Methods in Complex and CR Geometry	(LNM 1848)
	153. Analytic Number Theory	(LNM 1891)
	154. Imaging	to appear
2003	155. Stochastic Methods in Finance	(LNM 1856)
	156. Hyperbolic Systems of Balance Laws	to appear
	157. Symplectic 4-Manifolds and Algebraic Surfaces	to appear
	158. Mathematical Foundation of Turbulent Viscous Flows	(LNM 1871)
2004	159. Representation Theory and Complex Analysis	to appear
	160. Nonlinear and Optimal Control Theory	to appear
	161. Stochastic Geometry	(LNM 1892)
2005	162. Enumerative Invariants in Algebraic Geometry and String Theory	to appear
	163. Calculus of Variations and Non-linear Partial Differential Equations	to appear
	164. SPDE in Hydrodynamics: Recent Progress and Prospects	to appear
2006	165. Pseudo-Differential Operators, Quantization and Signals	announced
	166. Mixed Finite Elements, Compatibility Conditions, and Applications	announced
	167. From a Microscopic to a Macroscopic Description of Complex Systems	announced
	168. Quantum Transport: Modelling, Analysis and Asymptotics	announced

Lecture Notes in Mathematics

For information about earlier volumes
please contact your bookseller or Springer
LNM Online archive: springerlink.com

Vol. 1701: Ti-Jun Xiao, J. Liang, The Cauchy Problem of Higher Order Abstract Differential Equations (1998)

Vol. 1702: J. Ma, J. Yong, Forward-Backward Stochastic Differential Equations and Their Applications (1999)

Vol. 1703: R. M. Dudley, R. Norvaiša, Differentiability of Six Operators on Nonsmooth Functions and p-Variation (1999)

Vol. 1704: H. Tamanoi, Elliptic Genera and Vertex Operator Super-Algebras (1999)

Vol. 1705: I. Nikolaev, E. Zhuzhoma, Flows in 2-dimensional Manifolds (1999)

Vol. 1706: S. Yu. Pilyugin, Shadowing in Dynamical Systems (1999)

Vol. 1707: R. Pytlak, Numerical Methods for Optimal Control Problems with State Constraints (1999)

Vol. 1708: K. Zuo, Representations of Fundamental Groups of Algebraic Varieties (1999)

Vol. 1709: J. Azéma, M. Émery, M. Ledoux, M. Yor (Eds.), Séminaire de Probabilités XXXIII (1999)

Vol. 1710: M. Koecher, The Minnesota Notes on Jordan Algebras and Their Applications (1999)

Vol. 1711: W. Ricker, Operator Algebras Generated by Commuting Projećtions: A Vector Measure Approach (1999)

Vol. 1712: N. Schwartz, J. J. Madden, Semi-algebraic Function Rings and Reflectors of Partially Ordered Rings (1999)

Vol. 1713: F. Bethuel, G. Huisken, S. Müller, K. Steffen, Calculus of Variations and Geometric Evolution Problems. Cetraro, 1996. Editors: S. Hildebrandt, M. Struwe (1999)

Vol. 1714: O. Diekmann, R. Durrett, K. P. Hadeler, P. K. Maini, H. L. Smith, Mathematics Inspired by Biology. Martina Franca, 1997. Editors: V. Capasso, O. Diekmann (1999)

Vol. 1715: N. V. Krylov, M. Röckner, J. Zabczyk, Stochastic PDE's and Kolmogorov Equations in Infinite Dimensions. Cetraro, 1998. Editor: G. Da Prato (1999)

Vol. 1716: J. Coates, R. Greenberg, K. A. Ribet, K. Rubin, Arithmetic Theory of Elliptic Curves. Cetraro, 1997. Editor: C. Viola (1999)

Vol. 1717: J. Bertoin, F. Martinelli, Y. Peres, Lectures on Probability Theory and Statistics. Saint-Flour, 1997. Editor: P. Bernard (1999)

Vol. 1718: A. Eberle, Uniqueness and Non-Uniqueness of Semigroups Generated by Singular Diffusion Operators (1999)

Vol. 1719: K. R. Meyer, Periodic Solutions of the N-Body Problem (1999)

Vol. 1720: D. Elworthy, Y. Le Jan, X-M. Li, On the Geometry of Diffusion Operators and Stochastic Flows (1999)

Vol. 1721: A. Iarrobino, V. Kanev, Power Sums, Gorenstein Algebras, and Determinantal Loci (1999)

Vol. 1722: R. McCutcheon, Elemental Methods in Ergodic Ramsey Theory (1999)

Vol. 1723: J. P. Croisille, C. Lebeau, Diffraction by an Immersed Elastic Wedge (1999)

Vol. 1724: V. N. Kolokoltsov, Semiclassical Analysis for Diffusions and Stochastic Processes (2000)

Vol. 1725: D. A. Wolf-Gladrow, Lattice-Gas Cellular Automata and Lattice Boltzmann Models (2000)

Vol. 1726: V. Marić, Regular Variation and Differential Equations (2000)

Vol. 1727: P. Kravanja M. Van Barel, Computing the Zeros of Analytic Functions (2000)

Vol. 1728: K. Gatermann Computer Algebra Methods for Equivariant Dynamical Systems (2000)

Vol. 1729: J. Azéma, M. Émery, M. Ledoux, M. Yor (Eds.) Séminaire de Probabilités XXXIV (2000)

Vol. 1730: S. Graf, H. Luschgy, Foundations of Quantization for Probability Distributions (2000)

Vol. 1731: T. Hsu, Quilts: Central Extensions, Braid Actions, and Finite Groups (2000)

Vol. 1732: K. Keller, Invariant Factors, Julia Equivalences and the (Abstract) Mandelbrot Set (2000)

Vol. 1733: K. Ritter, Average-Case Analysis of Numerical Problems (2000)

Vol. 1734: M. Espedal, A. Fasano, A. Mikelić, Filtration in Porous Media and Industrial Applications. Cetraro 1998. Editor: A. Fasano. 2000.

Vol. 1735: D. Yafaev, Scattering Theory: Some Old and New Problems (2000)

Vol. 1736: B. O. Turesson, Nonlinear Potential Theory and Weighted Sobolev Spaces (2000)

Vol. 1737: S. Wakabayashi, Classical Microlocal Analysis in the Space of Hyperfunctions (2000)

Vol. 1738: M. Émery, A. Nemirovski, D. Voiculescu, Lectures on Probability Theory and Statistics (2000)

Vol. 1739: R. Burkard, P. Deuflhard, A. Jameson, J.-L. Lions, G. Strang, Computational Mathematics Driven by Industrial Problems. Martina Franca, 1999. Editors: V. Capasso, H. Engl, J. Periaux (2000)

Vol. 1740: B. Kawohl, O. Pironneau, L. Tartar, J.-P. Zolesio, Optimal Shape Design. Tróia, Portugal 1999. Editors: A. Cellina, A. Ornelas (2000)

Vol. 1741: E. Lombardi, Oscillatory Integrals and Phenomena Beyond all Algebraic Orders (2000)

Vol. 1742: A. Unterberger, Quantization and Non-holomorphic Modular Forms (2000)

Vol. 1743: L. Habermann, Riemannian Metrics of Constant Mass and Moduli Spaces of Conformal Structures (2000)

Vol. 1744: M. Kunze, Non-Smooth Dynamical Systems (2000)

Vol. 1745: V. D. Milman, G. Schechtman (Eds.), Geometric Aspects of Functional Analysis. Israel Seminar 1999-2000 (2000)

Vol. 1746: A. Degtyarev, I. Itenberg, V. Kharlamov, Real Enriques Surfaces (2000)

Vol. 1747: L. W. Christensen, Gorenstein Dimensions (2000)

Vol. 1748: M. Ruzicka, Electrorheological Fluids: Modeling and Mathematical Theory (2001)

Vol. 1749: M. Fuchs, G. Seregin, Variational Methods for Problems from Plasticity Theory and for Generalized Newtonian Fluids (2001)

Vol. 1750: B. Conrad, Grothendieck Duality and Base Change (2001)

Vol. 1751: N. J. Cutland, Loeb Measures in Practice: Recent Advances (2001)

Vol. 1752: Y. V. Nesterenko, P. Philippon, Introduction to Algebraic Independence Theory (2001)

Vol. 1753: A. I. Bobenko, U. Eitner, Painlevé Equations in the Differential Geometry of Surfaces (2001)

Vol. 1754: W. Bertram, The Geometry of Jordan and Lie Structures (2001)

Vol. 1755: J. Azéma, M. Émery, M. Ledoux, M. Yor (Eds.), Séminaire de Probabilités XXXV (2001)

Vol. 1756: P. E. Zhidkov, Korteweg de Vries and Nonlinear Schrödinger Equations: Qualitative Theory (2001)

Vol. 1757: R. R. Phelps, Lectures on Choquet's Theorem (2001)

Vol. 1758: N. Monod, Continuous Bounded Cohomology of Locally Compact Groups (2001)

Vol. 1759: Y. Abe, K. Kopfermann, Toroidal Groups (2001)

Vol. 1760: D. Filipović, Consistency Problems for Heath-Jarrow-Morton Interest Rate Models (2001)

Vol. 1761: C. Adelmann, The Decomposition of Primes in Torsion Point Fields (2001)

Vol. 1762: S. Cerrai, Second Order PDE's in Finite and Infinite Dimension (2001)

Vol. 1763: J.-L. Loday, A. Frabetti, F. Chapoton, F. Goichot, Dialgebras and Related Operads (2001)

Vol. 1764: A. Cannas da Silva, Lectures on Symplectic Geometry (2001)

Vol. 1765: T. Kerler, V. V. Lyubashenko, Non-Semisimple Topological Quantum Field Theories for 3-Manifolds with Corners (2001)

Vol. 1766: H. Hennion, L. Hervé, Limit Theorems for Markov Chains and Stochastic Properties of Dynamical Systems by Quasi-Compactness (2001)

Vol. 1767: J. Xiao, Holomorphic Q Classes (2001)

Vol. 1768: M.J. Pflaum, Analytic and Geometric Study of Stratified Spaces (2001)

Vol. 1769: M. Alberich-Carramiñana, Geometry of the Plane Cremona Maps (2002)

Vol. 1770: H. Gluesing-Luerssen, Linear Delay-Differential Systems with Commensurate Delays: An Algebraic Approach (2002)

Vol. 1771: M. Émery, M. Yor (Eds.), Séminaire de Probabilités 1967-1980. A Selection in Martingale Theory (2002)

Vol. 1772: F. Burstall, D. Ferus, K. Leschke, F. Pedit, U. Pinkall, Conformal Geometry of Surfaces in S^4 (2002)

Vol. 1773: Z. Arad, M. Muzychuk, Standard Integral Table Algebras Generated by a Non-real Element of Small Degree (2002)

Vol. 1774: V. Runde, Lectures on Amenability (2002)

Vol. 1775: W. H. Meeks, A. Ros, H. Rosenberg, The Global Theory of Minimal Surfaces in Flat Spaces. Martina Franca 1999. Editor: G. P. Pirola (2002)

Vol. 1776: K. Behrend, C. Gomez, V. Tarasov, G. Tian, Quantum Comohology. Cetraro 1997. Editors: P. de Bartolomeis, B. Dubrovin, C. Reina (2002)

Vol. 1777: E. García-Río, D. N. Kupeli, R. Vázquez-Lorenzo, Osserman Manifolds in Semi-Riemannian Geometry (2002)

Vol. 1778: H. Kiechle, Theory of K-Loops (2002)

Vol. 1779: I. Chueshov, Monotone Random Systems (2002)

Vol. 1780: J. H. Bruinier, Borcherds Products on O(2,1) and Chern Classes of Heegner Divisors (2002)

Vol. 1781: E. Bolthausen, E. Perkins, A. van der Vaart, Lectures on Probability Theory and Statistics. Ecole d' Eté de Probabilités de Saint-Flour XXIX-1999. Editor: P. Bernard (2002)

Vol. 1782: C.-H. Chu, A. T.-M. Lau, Harmonic Functions on Groups and Fourier Algebras (2002)

Vol. 1783: L. Grüne, Asymptotic Behavior of Dynamical and Control Systems under Perturbation and Discretization (2002)

Vol. 1784: L.H. Eliasson, S. B. Kuksin, S. Marmi, J.-C. Yoccoz, Dynamical Systems and Small Divisors. Cetraro, Italy 1998. Editors: S. Marmi, J.-C. Yoccoz (2002)

Vol. 1785: J. Arias de Reyna, Pointwise Convergence of Fourier Series (2002)

Vol. 1786: S. D. Cutkosky, Monomialization of Morphisms from 3-Folds to Surfaces (2002)

Vol. 1787: S. Caenepeel, G. Militaru, S. Zhu, Frobenius and Separable Functors for Generalized Module Categories and Nonlinear Equations (2002)

Vol. 1788: A. Vasil'ev, Moduli of Families of Curves for Conformal and Quasiconformal Mappings (2002)

Vol. 1789: Y. Sommerhäuser, Yetter-Drinfel'd Hopf algebras over groups of prime order (2002)

Vol. 1790: X. Zhan, Matrix Inequalities (2002)

Vol. 1791: M. Knebusch, D. Zhang, Manis Valuations and Prüfer Extensions I: A new Chapter in Commutative Algebra (2002)

Vol. 1792: D. D. Ang, R. Gorenflo, V. K. Le, D. D. Trong, Moment Theory and Some Inverse Problems in Potential Theory and Heat Conduction (2002)

Vol. 1793: J. Cortés Monforte, Geometric, Control and Numerical Aspects of Nonholonomic Systems (2002)

Vol. 1794: N. Pytheas Fogg, Substitution in Dynamics, Arithmetics and Combinatorics. Editors: V. Berthé, S. Ferenczi, C. Mauduit, A. Siegel (2002)

Vol. 1795: H. Li, Filtered-Graded Transfer in Using Noncommutative Gröbner Bases (2002)

Vol. 1796: J.M. Melenk, hp-Finite Element Methods for Singular Perturbations (2002)

Vol. 1797: B. Schmidt, Characters and Cyclotomic Fields in Finite Geometry (2002)

Vol. 1798: W.M. Oliva, Geometric Mechanics (2002)

Vol. 1799: H. Pajot, Analytic Capacity, Rectifiability, Menger Curvature and the Cauchy Integral (2002)

Vol. 1800: O. Gabber, L. Ramero, Almost Ring Theory (2003)

Vol. 1801: J. Azéma, M. Émery, M. Ledoux, M. Yor (Eds.), Séminaire de Probabilités XXXVI (2003)

Vol. 1802: V. Capasso, E. Merzbach, B.G. Ivanoff, M. Dozzi, R. Dalang, T. Mountford, Topics in Spatial Stochastic Processes. Martina Franca, Italy 2001. Editor: E. Merzbach (2003)

Vol. 1803: G. Dolzmann, Variational Methods for Crystalline Microstructure – Analysis and Computation (2003)

Vol. 1804: I. Cherednik, Ya. Markov, R. Howe, G. Lusztig, Iwahori-Hecke Algebras and their Representation Theory. Martina Franca, Italy 1999. Editors: V. Baldoni, D. Barbasch (2003)

Vol. 1805: F. Cao, Geometric Curve Evolution and Image Processing (2003)

Vol. 1806: H. Broer, I. Hoveijn. G. Lunther, G. Vegter, Bifurcations in Hamiltonian Systems. Computing Singularities by Gröbner Bases (2003)

Vol. 1807: V. D. Milman, G. Schechtman (Eds.), Geometric Aspects of Functional Analysis. Israel Seminar 2000-2002 (2003)

Vol. 1808: W. Schindler, Measures with Symmetry Properties (2003)

Vol. 1809: O. Steinbach, Stability Estimates for Hybrid Coupled Domain Decomposition Methods (2003)

Vol. 1810: J. Wengenroth, Derived Functors in Functional Analysis (2003)

Vol. 1811: J. Stevens, Deformations of Singularities (2003)

Vol. 1812: L. Ambrosio, K. Deckelnick, G. Dziuk, M. Mimura, V. A. Solonnikov, H. M. Soner, Mathematical Aspects of Evolving Interfaces. Madeira, Funchal, Portugal 2000. Editors: P. Colli, J. F. Rodrigues (2003)

Vol. 1813: L. Ambrosio, L. A. Caffarelli, Y. Brenier, G. Buttazzo, C. Villani, Optimal Transportation and its Applications. Martina Franca, Italy 2001. Editors: L. A. Caffarelli, S. Salsa (2003)

Vol. 1814: P. Bank, F. Baudoin, H. Föllmer, L.C.G. Rogers, M. Soner, N. Touzi, Paris-Princeton Lectures on Mathematical Finance 2002 (2003)

Vol. 1815: A. M. Vershik (Ed.), Asymptotic Combinatorics with Applications to Mathematical Physics. St. Petersburg, Russia 2001 (2003)

Vol. 1816: S. Albeverio, W. Schachermayer, M. Talagrand, Lectures on Probability Theory and Statistics. Ecole d'Eté de Probabilités de Saint-Flour XXX-2000. Editor: P. Bernard (2003)

Vol. 1817: E. Koelink, W. Van Assche(Eds.), Orthogonal Polynomials and Special Functions. Leuven 2002 (2003)

Vol. 1818: M. Bildhauer, Convex Variational Problems with Linear, nearly Linear and/or Anisotropic Growth Conditions (2003)

Vol. 1819: D. Masser, Yu. V. Nesterenko, H. P. Schlickewei, W. M. Schmidt, M. Waldschmidt, Diophantine Approximation. Cetraro, Italy 2000. Editors: F. Amoroso, U. Zannier (2003)

Vol. 1820: F. Hiai, H. Kosaki, Means of Hilbert Space Operators (2003)

Vol. 1821: S. Teufel, Adiabatic Perturbation Theory in Quantum Dynamics (2003)

Vol. 1822: S.-N. Chow, R. Conti, R. Johnson, J. Mallet-Paret, R. Nussbaum, Dynamical Systems. Cetraro, Italy 2000. Editors: J. W. Macki, P. Zecca (2003)

Vol. 1823: A. M. Anile, W. Allegretto, C. Ringhofer, Mathematical Problems in Semiconductor Physics. Cetraro, Italy 1998. Editor: A. M. Anile (2003)

Vol. 1824: J. A. Navarro González, J. B. Sancho de Salas, $\mathscr{C}^\infty$ – Differentiable Spaces (2003)

Vol. 1825: J. H. Bramble, A. Cohen, W. Dahmen, Multiscale Problems and Methods in Numerical Simulations, Martina Franca, Italy 2001. Editor: C. Canuto (2003)

Vol. 1826: K. Dohmen, Improved Bonferroni Inequalities via Abstract Tubes. Inequalities and Identities of Inclusion-Exclusion Type. VIII, 113 p, 2003.

Vol. 1827: K. M. Pilgrim, Combinations of Complex Dynamical Systems. IX, 118 p, 2003.

Vol. 1828: D. J. Green, Gröbner Bases and the Computation of Group Cohomology. XII, 138 p, 2003.

Vol. 1829: E. Altman, B. Gaujal, A. Hordijk, Discrete-Event Control of Stochastic Networks: Multimodularity and Regularity. XIV, 313 p, 2003.

Vol. 1830: M. I. Gil', Operator Functions and Localization of Spectra. XIV, 256 p, 2003.

Vol. 1831: A. Connes, J. Cuntz, E. Guentner, N. Higson, J. E. Kaminker, Noncommutative Geometry, Martina Franca, Italy 2002. Editors: S. Doplicher, L. Longo (2004)

Vol. 1832: J. Azéma, M. Émery, M. Ledoux, M. Yor (Eds.), Séminaire de Probabilités XXXVII (2003)

Vol. 1833: D.-Q. Jiang, M. Qian, M.-P. Qian, Mathematical Theory of Nonequilibrium Steady States. On the Frontier of Probability and Dynamical Systems. IX, 280 p, 2004.

Vol. 1834: Yo. Yomdin, G. Comte, Tame Geometry with Application in Smooth Analysis. VIII, 186 p, 2004.

Vol. 1835: O.T. Izhboldin, B. Kahn, N.A. Karpenko, A. Vishik, Geometric Methods in the Algebraic Theory of Quadratic Forms. Summer School, Lens, 2000. Editor: J.-P. Tignol (2004)

Vol. 1836: C. Năstăsescu, F. Van Oystaeyen, Methods of Graded Rings. XIII, 304 p, 2004.

Vol. 1837: S. Tavaré, O. Zeitouni, Lectures on Probability Theory and Statistics. Ecole d'Eté de Probabilités de Saint-Flour XXXI-2001. Editor: J. Picard (2004)

Vol. 1838: A.J. Ganesh, N.W. O'Connell, D.J. Wischik, Big Queues. XII, 254 p, 2004.

Vol. 1839: R. Gohm, Noncommutative Stationary Processes. VIII, 170 p, 2004.

Vol. 1840: B. Tsirelson, W. Werner, Lectures on Probability Theory and Statistics. Ecole d'Eté de Probabilités de Saint-Flour XXXII-2002. Editor: J. Picard (2004)

Vol. 1841: W. Reichel, Uniqueness Theorems for Variational Problems by the Method of Transformation Groups (2004)

Vol. 1842: T. Johnsen, A.L. Knutsen, K3 Projective Models in Scrolls (2004)

Vol. 1843: B. Jefferies, Spectral Properties of Noncommuting Operators (2004)

Vol. 1844: K.F. Siburg, The Principle of Least Action in Geometry and Dynamics (2004)

Vol. 1845: Min Ho Lee, Mixed Automorphic Forms, Torus Bundles, and Jacobi Forms (2004)

Vol. 1846: H. Ammari, H. Kang, Reconstruction of Small Inhomogeneities from Boundary Measurements (2004)

Vol. 1847: T.R. Bielecki, T. Björk, M. Jeanblanc, M. Rutkowski, J.A. Scheinkman, W. Xiong, Paris-Princeton Lectures on Mathematical Finance 2003 (2004)

Vol. 1848: M. Abate, J. E. Fornaess, X. Huang, J. P. Rosay, A. Tumanov, Real Methods in Complex and CR Geometry, Martina Franca, Italy 2002. Editors: D. Zaitsev, G. Zampieri (2004)

Vol. 1849: Martin L. Brown, Heegner Modules and Elliptic Curves (2004)

Vol. 1850: V. D. Milman, G. Schechtman (Eds.), Geometric Aspects of Functional Analysis. Israel Seminar 2002-2003 (2004)

Vol. 1851: O. Catoni, Statistical Learning Theory and Stochastic Optimization (2004)

Vol. 1852: A.S. Kechris, B.D. Miller, Topics in Orbit Equivalence (2004)

Vol. 1853: Ch. Favre, M. Jonsson, The Valuative Tree (2004)

Vol. 1854: O. Saeki, Topology of Singular Fibers of Differential Maps (2004)

Vol. 1855: G. Da Prato, P.C. Kunstmann, I. Lasiecka, A. Lunardi, R. Schnaubelt, L. Weis, Functional Analytic

Methods for Evolution Equations. Editors: M. Iannelli, R. Nagel, S. Piazzera (2004)

Vol. 1856: K. Back, T.R. Bielecki, C. Hipp, S. Peng, W. Schachermayer, Stochastic Methods in Finance, Bressanone/Brixen, Italy, 2003. Editors: M. Fritelli, W. Runggaldier (2004)

Vol. 1857: M. Émery, M. Ledoux, M. Yor (Eds.), Séminaire de Probabilités XXXVIII (2005)

Vol. 1858: A.S. Cherny, H.-J. Engelbert, Singular Stochastic Differential Equations (2005)

Vol. 1859: E. Letellier, Fourier Transforms of Invariant Functions on Finite Reductive Lie Algebras (2005)

Vol. 1860: A. Borisyuk, G.B. Ermentrout, A. Friedman, D. Terman, Tutorials in Mathematical Biosciences I. Mathematical Neurosciences (2005)

Vol. 1861: G. Benettin, J. Henrard, S. Kuksin, Hamiltonian Dynamics – Theory and Applications, Cetraro, Italy, 1999. Editor: A. Giorgilli (2005)

Vol. 1862: B. Helffer, F. Nier, Hypoelliptic Estimates and Spectral Theory for Fokker-Planck Operators and Witten Laplacians (2005)

Vol. 1863: H. Fürh, Abstract Harmonic Analysis of Continuous Wavelet Transforms (2005)

Vol. 1864: K. Efstathiou, Metamorphoses of Hamiltonian Systems with Symmetries (2005)

Vol. 1865: D. Applebaum, B.V. R. Bhat, J. Kustermans, J. M. Lindsay, Quantum Independent Increment Processes I. From Classical Probability to Quantum Stochastic Calculus. Editors: M. Schürmann, U. Franz (2005)

Vol. 1866: O.E. Barndorff-Nielsen, U. Franz, R. Gohm, B. Kümmerer, S. Thorbjønsen, Quantum Independent Increment Processes II. Structure of Quantum Levy Processes, Classical Probability, and Physics. Editors: M. Schürmann, U. Franz, (2005)

Vol. 1867: J. Sneyd (Ed.), Tutorials in Mathematical Biosciences II. Mathematical Modeling of Calcium Dynamics and Signal Transduction. (2005)

Vol. 1868: J. Jorgenson, S. Lang, $Pos_n(R)$ and Eisenstein Sereies. (2005)

Vol. 1869: A. Dembo, T. Funaki, Lectures on Probability Theory and Statistics. Ecole d'Eté de Probabilités de Saint-Flour XXXIII-2003. Editor: J. Picard (2005)

Vol. 1870: V.I. Gurariy, W. Lusky, Geometry of Müntz Spaces and Related Questions. (2005)

Vol. 1871: P. Constantin, G. Gallavotti, A.V. Kazhikhov, Y. Meyer, S. Ukai, Mathematical Foundation of Turbulent Viscous Flows, Martina Franca, Italy, 2003. Editors: M. Cannone, T. Miyakawa (2006)

Vol. 1872: A. Friedman (Ed.), Tutorials in Mathematical Biosciences III. Cell Cycle, Proliferation, and Cancer (2006)

Vol. 1873: R. Mansuy, M. Yor, Random Times and Enlargements of Filtrations in a Brownian Setting (2006)

Vol. 1874: M. Yor, M. Andr Meyer - S minaire de Probabilités XXXIX (2006)

Vol. 1875: J. Pitman, Combinatorial Stochastic Processes. Ecole dŠEté de Probabilités de Saint-Flour XXXII-2002. Editor: J. Picard (2006)

Vol. 1876: H. Herrlich, Axiom of Choice (2006)

Vol. 1877: J. Steuding, Value Distributions of L-Functions(2006)

Vol. 1878: R. Cerf, The Wulff Crystal in Ising and Percolation Models, Ecole d'Et de Probabilits de Saint-Flour XXXIV-2004. Editor: Jean Picard (2006)

Vol. 1879: G. Slade, The Lace Expansion and its Appli- cations, Ecole d'Eté de Probabilités de Saint-Flour XXXIV-2004. Editor: Jean Picard (2006)

Vol. 1880: S. Attal, A. Joye, C.-A. Pillet, Open Quantum Systems I, The Hamiltonian Approach (2006)

Vol. 1881: S. Attal, A. Joye, C.-A. Pillet, Open Quantum Systems II, The Markovian Approach (2006)

Vol. 1882: S. Attal, A. Joye, C.-A. Pillet, Open Quantum Systems III, Recent Developments (2006)

Vol. 1883: W. Van Assche, F. Marcell n (Eds.), Orthogonal Polynomials and Special Functions, Computation and Application (2006)

Vol. 1884: N. Hayashi, E.I. Kaikina, P.I. Naumkin, I.A. Shishmarev, Asymptotics for Dissipative Nonlinear Equations (2006)

Vol. 1885: A. Telcs, The Art of Random Walks (2006)

Vol. 1886: S. Takamura, Splitting Deformations of Degenerations of Complex Curves (2006)

Vol. 1887: K. Habermann, L. Habermann, Introduction to Symplectic Dirac Operators (2006)

Vol. 1888: J. van der Hoeven, Transseries and Real Differential Algebra (2006)

Vol. 1889: G. Osipenko, Dynamical Systems, Graphs, and Algorithms (2006)

Vol. 1890: M. Bunge, J. Frunk, Singular Coverings of Toposes (2006)

Vol. 1891: J. B. Friedlander, D. R. Heath-Brown, H. Iwaniec, J. Kaczorowski, Analytic Number Theory, Cetraro, Italy, 2002. Editors: A. Perelli, C. Viola (2006)

Vol. 1892: A. Baddeley, I. Bárány, R. Schneider, W. Weil, Stochastic Geometry, Martina Franca, Italy, 2004. Editor: W. Weil (2007)

Recent Reprints and New Editions

Vol. 1618: G. Pisier, Similarity Problems and Completely Bounded Maps. 1995 – Second, Expanded Edition (2001)

Vol. 1629: J.D. Moore, Lectures on Seiberg-Witten Invariants. 1997 – Second Edition (2001)

Vol. 1638: P. Vanhaecke, Integrable Systems in the realm of Algebraic Geometry. 1996 – Second Edition (2001)

Vol. 1702: J. Ma, J. Yong, Forward-Backward Stochastic Differential Equations and their Applications. 1999. – Corrected 3rd printing (2005)

4. Manuscripts should in general be submitted in English. Final manuscripts should contain at least 100 pages of mathematical text and should always include

  – a general table of contents;

  – an informative introduction, with adequate motivation and perhaps some historical remarks: it should be accessible to a reader not intimately familiar with the topic treated;

  – a global subject index: as a rule this is genuinely helpful for the reader.

  Lecture Notes volumes are, as a rule, printed digitally from the authors' files. We strongly recommend that all contributions in a volume be written in the same LaTeX version, preferably LaTeX2e. To ensure best results, authors are asked to use the LaTeX2e style files available from Springer's web-server at

  ftp://ftp.springer.de/pub/tex/latex/mathegl/mono.zip (for monographs) and
  ftp://ftp.springer.de/pub/tex/latex/mathegl/mult.zip (for summer schools/tutorials).

  Additional technical instructions, if necessary, are available on request from:

  lnm@springer-sbm.com.

5. Careful preparation of the manuscripts will help keep production time short besides ensuring satisfactory appearance of the finished book in print and online. After acceptance of the manuscript authors will be asked to prepare the final LaTeX source files (and also the corresponding dvi-, pdf- or zipped ps-file) together with the final printout made from these files. The LaTeX source files are essential for producing the full-text online version of the book. For the existing online volumes of LNM see:

  http://www.springerlink.com/openurl.asp?genre=journal&issn=0075-8434.

  The actual production of a Lecture Notes volume takes approximately 8 weeks.

6. Volume editors receive a total of 50 free copies of their volume to be shared with the authors, but no royalties. They and the authors are entitled to a discount of 33.3 % on the price of Springer books purchased for their personal use, if ordering directly from Springer.

7. Commitment to publish is made by letter of intent rather than by signing a formal contract. Springer-Verlag secures the copyright for each volume. Authors are free to reuse material contained in their LNM volumes in later publications: A brief written (or e-mail) request for formal permission is sufficient.

Addresses:

Professor J.-M. Morel, CMLA,
École Normale Supérieure de Cachan,
61 Avenue du Président Wilson, 94235 Cachan Cedex, France
E-mail: Jean-Michel.Morel@cmla.ens-cachan.fr

Professor F. Takens, Mathematisch Instituut,
Rijksuniversiteit Groningen, Postbus 800,
9700 AV Groningen, The Netherlands
E-mail: F.Takens@math.rug.nl

Professor B. Teissier, Institut Mathématique de Jussieu,
UMR 7586 du CNRS, Équipe "Géométrie et Dynamique",
175 rue du Chevaleret, 75013 Paris, France
E-mail: teissier@math.jussieu.fr

For the "Mathematical Biosciences Subseries" of LNM :
Professor P. K. Maini, Center for Mathematical Biology,
Mathematical Institute, 24-29 St Giles,
Oxford OX1 3LP, UK
E-mail : maini@maths.ox.ac.uk

Springer, Mathematics Editorial I, Tiergartenstr. 17,
69121 Heidelberg, Germany,
Tel.: +49 (6221) 487-8410
Fax: +49 (6221) 487-8355
E-mail: lnm@springer-sbm.com